RADIO ASTRONOMY AND COSMOLOGY

INTERNATIONAL ASTRONOMICAL UNION
UNION ASTRONOMIQUE INTERNATIONALE

SYMPOSIUM No. 74

HELD AT CAVENDISH LABORATORY, CAMBRIDGE, ENGLAND
AUGUST 16–20, 1976

RADIO ASTRONOMY
AND COSMOLOGY

EDITED BY

DAVID L. JAUNCEY

CSIRO Division of Radiophysics,
P.O. Box 76, Epping, NSW, Australia

D. REIDEL PUBLISHING COMPANY

DORDRECHT-HOLLAND / BOSTON - U.S.A.

1977

Library of Congress Cataloging in Publication Data

Main entry under title:

Radio astronomy and cosmology.

      (Symposium – International Astronomical Union; no. 74)
      'Held at Cavendish Laboratory, Cambridge, England,
August 16–20, 1976.'
      Bibliography: p.
      1.   Radio astronomy—Congresses. 2. Cosmology—Congresses.
I. Jauncey, David L. II. Series: International Astronomical Union. Symposium; no. 74.
QB475.A1R29   522.682   77-24152
ISBN 90-277-0838-X
ISBN 90-277-0839-8 pbk.

Published on behalf of
the International Astronomical Union
by
D. Reidel Publishing Company, P.O. Box 17, Dordrecht, Holland

Sold and distributed in the U.S.A., Canada, and Mexico
by D. Reidel Publishing Company, Inc.
Lincoln Building, 160 Old Derby Street, Hingham,
Mass. 02043, U.S.A.

Printed in The Netherlands

TABLE OF CONTENTS

EDITORIAL NOTES

Much of the initiative in bringing Radio Astronomy into the study of Cosmology belongs to the Cavendish Laboratory and it is indeed appropriate that this Symposium was held there. This volume contains all of the invited papers presented, with the exception of that by D. Walsh. The discussion and contributed papers are also included but sections have been re-ordered so as to relate more closely to the invited papers to which they refer.

It is to be hoped that this volume captures some of the flavour and spirit of the Symposium. The fundamental problems of understanding the origin and evolution of our universe are far from being solved but it is clear that Radio Astronomy is having a significant impact on Cosmology.

SCIENTIFIC ORGANISING COMMITTEE

J.G. BOLTON: CSIRO Division of Radiophysics, Epping, Australia.

G.R. BURBIDGE: Physics Dept., University of California, San Diego, USA.

K.I. KELLERMANN: National Radio Astronomy Observatory, Green Bank, USA.

H. van der LAAN: Huygens Laboratorium, Sterrewacht Leiden, Netherlands.

M.S. LONGAIR (Chairman): Cavendish Laboratory, University of Cambridge,UK.

I.D. NOVIKOV: Space Research Inst., Academy of Sciences of USSR, Moscow.

Yu. N. PARIJSKIJ: Pulkovo Observatory, 196140 Leningrad M-140, USSR.

LOCAL ORGANISING COMMITTEE

M.S. LONGAIR: Cavendish Laboratory, University of Cambridge, UK.

J.M. RILEY: Cavendish Laboratory, University of Cambridge, UK.

J.R. SHAKESHAFT: Cavendish Laboratory, University of Cambridge, UK.

A.S. WEBSTER: Institute of Astronomy, Cambridge, UK.

SECRETARY of the SYMPOSIUM

Miss S. FIELDHOUSE: Cavendish Laboratory, Cambridge, UK.

ACKNOWLEDGMENTS

Many people contributed towards making this a successful Symposium. In particular we would like to thank Professor Sir Brian Pippard for his opening remarks; Miss Shirley Fieldhouse for her untiring organisational efforts; the Scientific Organising Committee for bringing us all together; the Local Organising Committee and the Cavendish Laboratory for hosting the Symposium; Tim Pearson, Mike McEllin, Ian McHardy and Mike Scott for their editorial assistance; and finally Malcolm Longair for staging the practical matters of this Symposium.

Thanks are also due to Mrs. Sue Tuffs, Mrs. Elaine Grimmitt and the staff at CSIRO Division of Radiophysics for their efforts in translating a successful Symposium into the present volume.

LIST OF PARTICIPANTS

AMES, Susan, Los Alamos Scientific Laboratory, New Mexico, 87545, USA.

ANDREW, B.H., Herzberg Institute of Astrophysics, NRC, Ottawa, Canada.

ANGULO, C.M., Universidad Politécnica de Barcelona, Barcelona 3, Spain.

ARP, H.C., The Hale Observatories, 813 Santa Barbara St., Pasadena, USA.

BAHCALL, J.N., Institute for Advanced Study, Princeton, 08540, USA.

BALDWIN, J., Lick Observatory, University of California, Santa Cruz, USA.

BALDWIN, J.E., Cavendish Laboratory, University of Cambridge, UK.

BASH, F.N., Astronomy Dept., University of Texas, Austin, 78712, USA.

BERTOLA, F., Instituto di Astronomia, 35100 Padova, Italy.

BOKSENBERG, A., Physics & Astronomy Dept., University College, London.

BOLTON, J.G., CSIRO Division of Radiophysics, Epping, 2121, Australia.

BONOMETTO, S., Instituto di Fisica Galileo Galilei, 35100, Padova, Italy.

BRANDIE, G.W., Physics Dept., Queen's University, Kingston, Canada.

BRIDLE, A.H., Astronomy & Physics, Queen's University, Kingston, Canada.

BROWNE, I.W.A., Nuffield Radio Astronomy Laboratories, Macclesfield, UK.

BURBIDGE, G.R., Physics, University of California, La Jolla, USA.

BURBIDGE, Margaret, Physics, University of California, La Jolla, USA.

BURKE, B.F., Physics Dept., M.I.T., Cambridge, 02139, USA.

CALLAHAN, P.S., Theoretical Physics Dept., Oxford University, UK.

CANNON, R.D., Royal Observatory, Blackford Hill, Edinburgh, Scotland.

CARSWELL, R.F., Physics & Astronomy Dept., University College, London.

CAVALIERE, A.G., Laboratorio Astrofisica Spaziale CNR, Frascati, Italy.

CHEVALIER, R.A., Kitt Peak National Observatory, Tucson, 85726, USA.

CODE, A.D., Washburn Observatory, University of Wisconsin, Madison, USA.

COHEN, M.H., Astrophysics Dept., California Institute of Technology,
    1201 East California Boulevard, Pasadena, 91127, USA.

CONDON, J.J., Physics Dept., Virginia Polytechnic Inst., Blacksburg, USA.

CONWAY, R.G., Nuffield Radio Astronomy Laboratories, Macclesfield, UK.

COTTON, W.D., Earth & Planetary Sciences Dept., M.I.T., Cambridge, USA.

DAGKESAMANSKI, R.D., Radio Astronomy Observatory of FIAN, P.N. Lebedev
    Institute of Physics, 142292 Pushchino, Moscow Region, USSR.

DAVIS, M.M., Arecibo Observatory, Puerto Rico, 00612, USA.

de FELICE, F., Instituto di Fisica Galileo Galilei, 35100 Padova, Italy.

de RUITER, H.R., Huygens Laboratorium, Sterrewacht Leiden, Netherlands.

De YOUNG, D.S., N.R.A.O., Edgemont Rd., Charlottesville, 22901, USA.

DOUGLAS, J.N., Astronomy Dept., University of Texas, Austin, 78712, USA.

EDWARDS, P.J., Physics Dept., University of Otago, Dunedin, New Zealand.

EKERS, R.D., Sterrenkundig Laboratorium "Kapteyn", Groningen, Netherlands.

FANTI, Carla, Laboratorio di Radioastronomia, CNR, 40126 Bologna, Italy.

FANTI, R., Laboratorio di Radioastronomia CNR, 40126 Bologna, Italy.

FOGARTY, W.G., Electrical Engineering Dept., Radio Laboratory, Helsinki
    University of Technology, 02150 Otaniemi, Finland.

GALT, J.A., Herzberg Institute of Astrophysics, NRC, Penticton, Canada.

GILLESPIE, A.R., Physics Dept., Queen Mary College, Mile End Rd., London.

GOLDSTEIN Jr., S.J., Astronomy, University of Virginia, Charlottesville,
                                                                    USA.
GRUEFF, G., Laboratorio di Radioastronomia CNR, 40126 Bologna, Italy.

GULA, R., Obserwatorium Astronomiczne, Uniwersytetu Jagiellonskiego,
    ul. Kopernika 27, 31-501 Krakow, Poland.

GULKIS, S., Jet Propulsion Laboratory, Pasadena, 91103, USA.

GULL, S.F., Cavendish Laboratory, University of Cambridge, UK.

HARROWER, G.A., Physics Dept., Queen's University, Kingston, Canada.

HAZARD, C., Institute of Astronomy, Cambridge, UK.

HEESCHEN, D.S., N.R.A.O., Edgemont Rd., Charlottesville, 22901, USA.

HEIDMANN, J., Observatoire de Paris, 92190, Meudon, France.

HENRIKSEN, R.N., Physics Dept., Queen's University, Kingston, Canada.

HEWISH, A., Cavendish Laboratory, University of Cambridge, UK.

HÖGBOM, J.A., Stockholm Observatory, S-13300 Saltsjöbaden, Sweden.

ILLARIONEV, A., Space Research Inst., Profsoyuznaya 88, Moscow 117485.

IRWIN, J.B., Earth & Planetary Sciences, Kean College, Union, 07083, USA.

JACOBS, K.C., Leander McCormick Observatory, Charlottesville, 22903, USA.

JAFFE, W., School of Natural Sciences, Institute for Advanced Study, Princeton, 08540, USA.

JAUNCEY, D.L., CSIRO Division of Radiophysics, Epping, 2121, Australia.

JOHANSEN, Karen, Copenhagen University Observatory, Tølløse, Denmark.

KAPAHI, V.K., Tata Inst. of Fundamental Research, Bombay, India.

KATGERT, P., Laboratorio di Radioastronomia CNR, Bologna, Italy.

KATGERT-MERKELIJN, J.K., Lab. di Radioastronomia CNR, Bologna, Italy.

KAUFMANN, P., Universidade Mackenzie, CRAAM, 01000 São Paulo, Brazil.

KELLERMANN, K.I., National Radio Astronomy Observatory, Green Bank, USA.

KENDERDINE, S., Cavendish Laboratory, University of Cambridge, UK.

KNAPP, Gillian R., Owens Valley Radio Observatory, California Institute of Technology, Pasadena, 91125, USA.

KREMPEC, Janina, Polska Akademia Nauk, Zaklad Astronomii, Pracownia Astrofizyki I, 87-100 Torun, ul. Chopina 12/18, Poland.

KRISTIAN, J., Hale Observatories, 813 Santa Barbara St., Pasadena, USA.

KRONBERG, P.P., David Dunlap Observatory, Richmond Hill, Canada, and Max-Planck-Institut für Radioastronomie, Bonn, West Germany.

KRYGIER, B., Uniwersytet M. Kopernika, Instytut Astronomii, Torun, Poland

KUHR, H., Max-Planck-Institut für Radioastronomie, Bonn, West Germany.

KUS, A.J., Inst. of Astronomy, Copernicus University, Torun, Poland.

LARI, C., Laboratorio di Radioastronomia CNR, 40126 Bologna, Italy.

LELIEVRE, G., Observatoire de Paris, 92190 Meudon, France.

LEPOOLE, R.S., Huygens Laboratorium, Sterrewacht Leiden, Netherlands.

LEWIS, B.M., Carter Observatory, P.O. Box 2909, Wellington, New Zealand.

LOCKE, J.L., Herzberg Institute of Astrophysics, NRC, Ottawa, Canada.

LONGAIR, M.S., Cavendish Laboratory, University of Cambridge, UK.

LYNDEN-BELL, D., Institute of Astronomy, Cambridge, UK.

McCREA, W.H., 87 Houndean Rise, Lewes, Sussex BN7 1EJ, UK.

McCULLOCH, P.M., Physics, University of Tasmania, Hobart, Australia.

MACHALSKI, J., Observatorium Astronomiczne, Uniwersytetu
    Jagiellonskiego, ul. Kopernika 27, 31-501 Krakow, Poland.

McLEOD, J.M., Herzberg Institute of Astrophysics, NRC, Ottawa, Canada.

MASLOWSKI, J., Observatorium Astronomiczne, Uniwersytetu Jagiellonskiego,
    Krakow, Poland, and M-P-I für Radioastronomie, Bonn, West Germany.

MATSUDA, T., Aeronautical Engineering Dept., Kyoto University, Japan, and
    Applied Mathematics & Astronomy, University College, Cardiff, UK.

MENON, T.K., Tata Institute of Fundamental Research, Bombay, India.

MENZEL, D.H., Center for Astrophysics, 60 Garden St., Cambridge, USA.

MILEY, G.K., Huygens Laboratorium, Sterrewacht Leiden, Netherlands.

MILLER, J.S., Lick Observatory, University of California, Santa Cruz, USA.

MILLS, B.Y., Astrophysics Dept., University of Sydney, 2006, Australia.

MOFFET, A.T., Owens Valley Radio Observatory, California Institute of
    Technology, Pasadena, 91125, USA.

MURDOCH, H.S., Astrophysics Dept., University of Sydney, Australia.

NARLIKAR, J.V., Tata Institute of Fundamental Research, Bombay, India.

NE'EMAN, Y., Physics & Astronomy Dept., Tel-Aviv University, Israel.

NICOLSON, G.D., Hartebeesthoek Radio Astronomy Observatory, National Inst. for Telecommunications Research, Johannesburg, South Africa.

NIETO, J.L., Institut d'Astrophysique, Boulevard Arago, Paris, France.

OORT, J.H., Huygens Laboratorium, Sterrewacht Leiden, Netherlands.

OSTERBROCK, D.E., Lick Observatory, Santa Cruz, 95064, USA.

PADRIELLI, Lucia, Laboratorio di Radioastronomia CNR, Bologna, Italy.

PALMER, H.P., Nuffield Radio Astronomy Laboratories, Macclesfield, UK.

PARTRIDGE, R.B., Astronomy Dept., Haverford College, Haverford, 19041, USA

PAULINY-TOTH, I., Max-Planck-Inst. für Radioastronomie, Bonn, F.R.G.

PENZIAS, A.A., Bell Laboratories, P.O. Box 400, Holmdel, 07733, USA.

PEROLA, G.C., Universita Degli Studi di Milano, Instituto di Scienze Fisiche, Via Celoria 16, 20133, Milano, Italy.

PETERSON, B.A., Anglo-Australian Observatory, Epping, 2121, Australia.

PETROSIAN, V., Inst. for Plasma Research, Stanford University, USA.

PHILLIPPS, S., Physics Dept., University of Durham, DH1 3LE, UK.

POOLEY, G.G., Cavendish Laboratory, University of Cambridge, UK.

PREUSS, E., Max-Planck-Inst. für Radioastronomie, Bonn, West Germany.

QUIGLEY, M.J.S., Appleton Laboratory, S.R.C., Ditton Park, Slough, UK.

RADHAKRISHNAN, V., Raman Research Institute, Hebal P.O., Bangalore, India.

READHEAD, A.C.S., Cavendish Laboratory, University of Cambridge, UK.

REES, M.J., Institute of Astronomy, Cambridge, UK.

REINHARDT, M., Astronomisches Institut der Universitat, Bochum, F.R.G.

RILEY, Julia M., Cavendish Laboratory, University of Cambridge, UK.

ROBSON, I., Physics Dept., Queen Mary College, Mile End Rd., London, UK.

ROEDER, R.C., Scarborough College, University of Toronto, M1C 1A4, Canada.

ROOD, H.J., Astronomy & Astrophysics Dept., Michigan State University, USA, and 52 Elizabeth St., 5 Bound Brook, N.J., 08880, USA.

ROWAN-ROBINSON, M., Applied Mathematics, Queen Mary College, London, UK.

RYLE, Sir Martin, Cavendish Laboratory, University of Cambridge, UK.

SASLAW, W., National Radio Astronomy Observatory, Charlottesville, USA.

SCALISE Jr., E., Universidade Mackenzie, CRAAM, Sao Paulo, Brazil.

SCARGLE, J.D., Ames Research Center, N.A.S.A., Moffett Field, 94035, USA.

SCHEUER, P.A.G., Cavendish Laboratory, University of Cambridge, UK.

SCHILIZZI, R., Owens Valley Radio Observatory, California Institute of
    Technology, Pasadena, 91125, USA.

SCHMIDT, M., Physics, Mathematics & Astronomy, California Institute of
    Technology, Pasadena, 91109, USA.

SELDNER, M., Physics Dept., Princeton University, Princeton, 08540, USA.

SETTI, G., Laboratorio di Radioastronomia CNR, Bologna, Italy.

SHAFFER, D.B., National Radio Astronomy Observatory, Green Bank, USA.

SHAKESHAFT, J.R., Cavendish Laboratory, University of Cambridge, UK.

SMITH, H.E., Physics, University of California, San Diego, La Jolla, USA.

SMITH, M.G., Observatorio Interamericano de Cerro Tololo, La Serena, Chile.

SOFUE, Y., Physics Dept., Nagoya University, Nagoya 464, Japan.

STANNARD, D., Nuffield Radio Astronomy Laboratories, Macclesfield, UK.

STROM, R.G., Radiosterrenwacht, Dwingeloo, The Netherlands.

SULLIVAN III, W.T., Astronomy, University of Washington, Seattle, USA.

SWARUP, G., Tata Institute of Fundamental Research, Tamil Nadu, India.

TALLQVIST, S., Radio Laboratory, Helsinki University of Technology,
    Otakaari 5A, 02150 Otaniemi, Finland.

THUAN, T.X., Astronomy, California Institute of Technology, Pasadena, USA.

TINSLEY, Beatrice M., Astronomy Dept., Yale University, New Haven, USA.

TIURI, M., Radio Laboratory, Helsinki University of Technology,
    Otakaari 5A, 02150 Otaniemi, Finland.

TURNER, E.L., School of Natural Sciences, Institute for Advanced Study,
    Princeton, N.J., 08540, USA.

TURNER, K.C., Dept. of Terrestrial Magnetism, 5241 Broad Branch Rd.,
  Washington, D.C., USA; and Instituto Argentina de Radioastronomia.

TURTLE, A.J., School of Physics, University of Sydney, 2006, Australia.

ULFBECK, O., Niels Bohr Institute, 17 Blegdamsvej, Copenhagen, Denmark.

ULRICH, Marie-Helene, Astronomy Dept., University of Texas, Austin, USA.

URBANIK, M., Observatorium Astronomiczne, Uniwersytetu Jagiellonskiego,
  ul. Kopernika 27, 21-501 Krakow, Poland.

VALENTIJN, E.A., Huygens Laboratorium, Sterrewacht Leiden, Netherlands.

van der KRUIT, P.C., Kapteyn Astronomical Inst., Groningen, Netherlands.

van der LAAN, H., Huygens Laboratorium, Sterrewacht Leiden, Netherlands.

VANDERMOLEN, J.C., Institut d'Astronomie et de Géophysique George
  Lemaître, 2 Chemin du Cyclotron, B-1348 Louvain-la-Neuve, Belgium.

VERON, M.-P., Observatoire de Paris, 92190, Meudon, France.

VERON, P., Observatoire de Paris, 92190, Meudon, France.

VIGOTTI, M., Laboratorio de Radioastronomia CNR, 40126, Bologna, Italy.

WALL, J.V., Cavendish Laboratory, University of Cambridge, UK.

WALSH, D., Nuffield Radio Astronomy Laboratories, Macclesfield, UK.

WAMPLER, E.J., Lick Observatory, University of California, Santa Cruz, USA.

WARDLE, J.F.C., Physics Dept., Brandeis University, Waltham, Mass., USA.

WEBSTER, A.S., Institute of Astronomy, Cambridge, UK.

WIELEBINSKI, R., Max-Planck-Inst. für Radioastronomie, Bonn, West Germany.

WILKINSON, Althea, Hale Observatories, California Institute of Technology,
  813 Santa Barbara St., Pasadena, 91125, USA.

WILLIS, A.G., Huygens Laboratorium, Sterrewacht Leiden, Netherlands.

WILLS, Beverley J., Astronomy Dept., University of Texas, Austin, USA.

WILLS, D., Astronomy Dept., University of Texas, Austin, USA.

WILSON, A.S., Astronomy Centre, University of Sussex, Brighton, UK.

WITTELS, Jill J., Earth & Planetary Sciences, M.I.T., Cambridge, USA.

WITZEL, A., Max-Planck-Inst. für Radioastronomie, Bonn, West Germany.

WOLFE, A.M., Physics & Astronomy Dept., University of Pittsburg, USA.

OBSERVERS

ARONS, J., University of California, Berkeley, USA.

BENFORD, G.A., University of California, Irvine, USA.

CAVALLO, G., Institute of Astronomy, Cambridge, UK.

CASWELL, J.L., CSIRO Radiophysics Division, Epping, 2121, Australia.

FEIGELSON, E., Harvard University, Cambridge, USA.

FLIN, P., Uniwersytet Jagiellonski, Krakow, Poland.

JAROSZYNSKI, M., Warsaw, Poland.

KAFKA, P., Max-Planck-Institut, Munich, West Germany.

MESZAROS, P., Max-Planck-Institut, Bonn, West Germany.

OKOYE, S., University of Nigeria, Nsukka, Nigeria, West Africa.

REAVES, G., University of Southern California, Los Angeles, USA.

SANITT, N., Institute of Astronomy, Cambridge, UK.

SPYROU, N., Institute of Astronomy, Cambridge, UK.

STARK, R., University of Oxford, UK.

STRIMPEL, O., University of Oxford, UK.

TRIMBLE, Virginia L., University of California, Irvine, USA.

I

SURVEYS OF RADIO SOURCES, SOURCE COUNTS AND ANISOTROPIES

THE 6C SURVEY

J.E. Baldwin
Mullard Radio Astronomy Observatory, Cavendish Laboratory,
Cambridge, U.K.

A new survey of radio sources at 151 MHz, which has not been described previously, is in progress at Cambridge. There are several of us working on it including Warner, Kenderdine, Waggett, Masson and Mayer. The results of the first observations are at present in a preliminary state but we hope that in time they will form the first part of the 6C survey. The purpose of the survey is not to reach the faintest sources detected so far in aperture synthesis observations but to study moderately faint sources at a low observing frequency and to cover a large part of the northern sky rapidly. The deepest survey made so far at a low frequency is that of Ryle and Neville (1962) at 178 MHz over a region of 50 square degrees near the north celestial pole. The faintest sources detected had flux densities of 0.25 Jy, corresponding to a source density of 10^4 sr^{-1}. It is already 15 years since that survey, which was the first trial of aperture synthesis using the earth's rotation, and much more is now technically possible. One of the most interesting features of a low frequency survey is its ability to detect preferentially sources with steep radio spectra and to be sensitive to sources of very low surface brightness. We know that in many cases these two properties go together and are associated with old radio sources, or at least with those parts of sources which are old. Many of the weak radio galaxies in nearby clusters are obvious examples of this type of source while the final, and so far unidentified, stages of the development of the most powerful double sources may be exciting candidates for discovery.

The telescope on which the survey is based is an east-west earth rotation synthesis instrument. It differs from other telescopes of this type in the low observing frequency, the large number of interferometer spacings and the simultaneous coverage of all baselines. The main characteristics are summarised in Table 1.

D. L. Jauncey (ed.), Radio Astronomy and Cosmology, 3-7. All Rights Reserved.
Copyright © 1977 by the IAU.

Table 1

Observing frequency	151.5 MHz
Bandwidth	800 kHz
Maximum baseline	1.37 km
Number of interferometer spacings	446
Smallest spacing	6λ
Angular resolution	$3'.7 \times 3'.7 \text{ cosec } \delta$
Primary beam	$17^{\circ} \times 17^{\circ}$

Desirable characteristics of telescopes are cheapness, speed and sensitivity.   In this telescope the fifty elements of the interferometer are simple arrays each comprising four Yagi aerials, resulting in a rather low total cost of £3 x 10^4 in 1973-5.   The speed of observing was an important feature of the design, the aim being to cover the sky north of δ = +20° in about two years.   Fig. 1 shows profiles of a map of an area of sky about 15° x 15° centred on the north celestial pole, obtained by averaging 10 separate 12-hour observations.   This averaging procedure was a necessary feature of the design for the following reason.   Fig. 1 illustrates the large dynamic range, and hence good sidelobe level, which is needed in the maps.

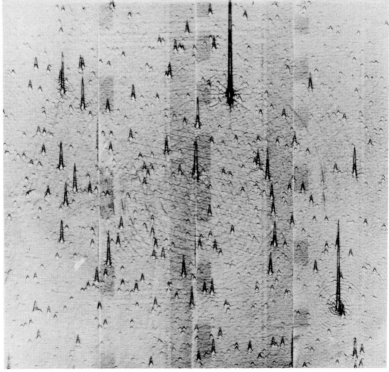

Fig. 1.   Profiles of the first 151 MHz map centred on the north celestial pole.

The brightest source has a flux density of 30 Jy, the faintest visible
about 0.3 Jy and even fainter ones, not seen in this mode of
presentation, about 0.05 Jy.   At 151 MHz the effects of ionospheric
fluctuations on the interferometer phase are important in determining
the sidelobe level.   So the size of the elements of the interferometer
was chosen to be quite small so that, in a single 12-hour observation,
the rms noise and the rms sidelobe levels on the map were roughly equal.
The field of view is then large and enables averaging of many days
observations to be carried out without seriously affecting the speed of
the survey.

     Near the lower left hand corner is an extended source which is
shown again as a contour map in Fig. 2 together with a map with closely
similar angular resolution at 1419 MHz made by Waggett with the Half
Mile telescope.   This source is of interest intrinsically for its very

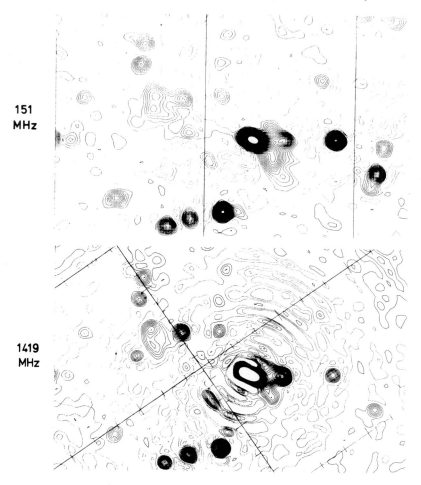

Fig. 2.   Comparison of maps at 151 MHz and 1419 MHz of a small region.

large size and low surface brightness but the map is presented here only to illustrate the faintest sources on the 151 MHz map which are believed to be real.   The contour interval at 151 MHz corresponds to 30 mJy for a point source at the centre of the primary beam of the telescope and the contour interval at 1419 MHz has been chosen to give a similar appearance for a source of average spectrum.   In places, sources of about 60 mJy are seen to be confirmed by the 1419 MHz map. The latter map has not been corrected for the primary beam attenuation so that an apparent absence of sources in the outer parts is to be expected.   50 mJy was the design limit of the 151 MHz survey.   At present it appears that such sources can be detected on the maps but satisfactory measurements of them cannot yet be made.   Fig. 1 shows the presence of intruding features, such as imperfectly removed grating rings and interference, whose effects can be, and in some cases have been, improved by further work.   For instance, the parallel ridges across the field centre were due to undersampling of part of the data at an intermediate stage of the computation.   At the expected sensitivity limit of 50 mJy there should be roughly 35,000 sources sr^{-1}. The very flat slope of the source counts at this level suggests that confusion, even with only 25 beam areas per source, is unlikely to be a very serious problem.   The survey limit corresponds to a surface brightness of 3.5 mJy $(arc min)^{-2}$ at 151 MHz, somewhat fainter than is easily reached by other synthesis telescopes for sources with normal spectra.   We expect that this sensitivity will be particularly useful both for studies of old radio galaxies and also for the disks and haloes of normal spirals.

    We have made a preliminary source count which is shown in Fig. 3 as differential counts normalized in the usual way to a uniform static

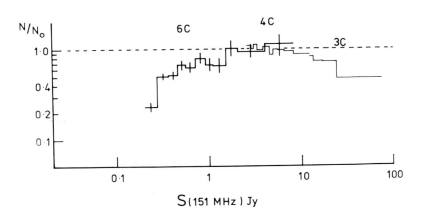

Fig. 3. Preliminary differential source counts from the present survey and from the 3C and 4C surveys adjusted to 151 MHz. The counts are normalised to 2400 sr^{-1} for S > 1 Jy.

Euclidean model. The differential counts from the 3C and 4C surveys at 178 MHz, as presented by Longair (1974), have been incorporated in the diagram after multiplying the 178 MHz flux densities by 1.13 to allow for a mean spectral index of 0.77. The normalisation corresponds to a source density of 2400 sr^{-1} for S > 1 Jy at 151 MHz. The 6C values are based on a very conservative search within 7° of the celestial pole for sources having S ⩾ 200 mJy, omitting small areas near the two bright 3C sources in the field. The convergence in the source counts looks rather dramatic at the last value plotted but at this level there is no question about the completeness of the survey. Problems arise connected with the angular size of sources and especially over the question, common to many surveys, of whether two particular sources are individuals or components of a double source. With that proviso, the counts shown in Fig. 3, although preliminary, are unlikely to be amended very much. The changes we can anticipate are an improvement by a factor of four in the limiting flux density which we hope to obtain quite soon and on eventual improvement by a factor of 80 in the statistics when the survey is complete. We do not dare to predict a date for that.

REFERENCES

Longair, M.S., 1974, IAU Symp. 63. Confrontation of Cosmological Theories with Observational Data. p.93.
Ryle, M. & Neville, A.C., 1962, Monthly Notices R. astr. Soc., 125, 39.

DISCUSSION

van der Laan: Do you use, or intend to use any "clean" techniques for dynamic range enhancement, or just to avoid the area around strong sources?

Baldwin: At present we exclude areas around the two strongest sources. We don't intend, at the moment, to use any "clean" techniques.

SURVEY OF DATA FOR DETERMINING SCALES OF THE ABSOLUTE FLUX DENSITIES IN 10-180 MHz RANGE AND SOURCE SPECTRA IN THE DECLINATION STRIP 10° - 20°.

S. Ya. Braude
Academy of Sciences of the Ukrainian SSR,
Institute of Radio Physics and Electronics
12 Acad. Proskura Street, Kharkov 85, USSR 310085

To determine the frequency spectra of radio sources the flux densities of these sources should be measured at the widest possible frequency range. The use of different telescope types, methods of measurements and calibrations led in a number of cases to the considerable difference in the data of various observatories. At the present time for frequencies above 400 MHz the scales of flux densities - S for - $S \geqslant 1$ Jy $= 10^{-26}$ w m^{-2}Hz^{-1} presented by different authors coincide accurate to 5%. Up to now for frequencies below 200 MHz a single scale of fluxes recognized by all radioastronomers is absent. We attempted to determine such a scale for frequencies 180-10 MHz. We used both analysis of the published data and the results of the new measurements obtained with the UTR-2 Radiotelescope in the declination strip 10° - 20° at frequencies 10.0; 12.6; 14.7; 16.7; 20, and 25 MHz. The method of these measurements and the obtained results are described in detail in [1]. The UTR-2 Radiotelescope has 5 beams in declination with half power beam width in a zenith direction 20' x 20' at 25 MHz [2]. The absolute values of the flux densities of all observed discrete radio sources are defined during the experiments. Minimum flux density measured at frequency 25 MHz - 15-20 Jy, with a signal-to-noise ratio equal to 3-4. During measurements about 300 radio sources were found. Comparing these data with our earlier measurements obtained with the UTR-1 Telescope [3] as well as with corrected data of the both Pentincton [4] at 10.02 and 22.25 MHz and the Clark-Lake Observatories at 26.3 MHz [5] and the results obtained in Cambridge at 38 MHz [6] and 178 MHz [7, 8] we have derived some correcting factors allowing to bring the data of different catalogues to a single scale. The method for determining such factors is described in [9]. It was shown that below 200 MHz where the influence of ionosphere is great, it is necessary to take into account two corrections : regular scale shift of the one catalogue to another $\tau_{12} = 1/\tau_{21}$ (τ_{12} is the average value of the relation between flux densities of the first catalogue and the second one) and a scale shift caused by the multiplicative scatter. A value is greater than unity and it can be determined comparing with each other three different catalogues. The comparison results of a number of catalogues with data obtained with UTR-2 are

D. L. Jauncey (ed.), Radio Astronomy and Cosmology, 9-13. All Rights Reserved.

shown in table I.

TABLE 1

1	2	3	4	5	6
Surveys	f_{MHz}	$z_{12} = <\frac{F}{S_{UTR2}}>$	x_o	N	3C461 3C405 3C274
UTR-1 [3]	12.6	2.52±0.33	1.25	13	1.13±0.10
	14.7	2.06±0.18	1.09	12	1.20±0.06
	16.7	1.62±0.09	1.05	17	1.05±0.12
	20.0	1.56±0.09	1.05	19	1.00±0.10
	25.0	1.28±0.13	1.13	14	1.07±0.10
[10]	10.02	1.39±0.17	-	9	0.77±0.17
[4]	22.25	1.14±0.06	1.06	21	1.05±0.05
[5]	26.3	1.07±0.04	1.04	50	0.99±0.03
[11]	38	0.97±0.06	1.04	49	0.97±0.05
UTR-2	-	-	1.01	-	-

In Table 1 : 1 - catalogues, 2 - frequency (MHz), 3 and 4 are correcting factors, z_{12} and x_o, here F - is the flux density of the data survey S_{UTR2} is average flux density of UTR-2 obtained from spectra at frequency of this survey for comparable radio sources determined in the 10-1400 MHz range [12]; 5 - is a number of common radio sources of both catalogues. The sixth column shows the average relation of the flux densities between three the most intense sources. As it follows from the Table 1 the data obtained with UTR-2 are in a good agreement with surveys [4, 5, 11] (down to the fluxes of 50 Jy), but differ from those of the catalogue [10] and UTR-1. The reasons of these disagreements are considered in [9]. Now we believe that the absolute scales of the low frequency surveys correlate well. Data of measurements obtained at 10-1400 MHz and correcting factors table 1 allowed to form spectra of 266 sources placed in the declination strip 10° - 20° [12]. From these spectra 47 standard linear (in logarithmic scale) spectra were chosen which may be used as the reference standards [9]. All obtained spectra we separated in three groups: linear spectra with zero-order curvature (type S) account for 86%, spectra with positive (type C+) and those with negative curvature (type C−) amount to 10% and 4% respectively. Thus, as it follows from these data, linear spectra are predominant down to 10 MHz frequency. This confirms once more the fact mentioned earlier [13] that in the discrete radio sources (a part of which has a set of details with small angular sizes) in a flux of the whole source neither reabsorption nor

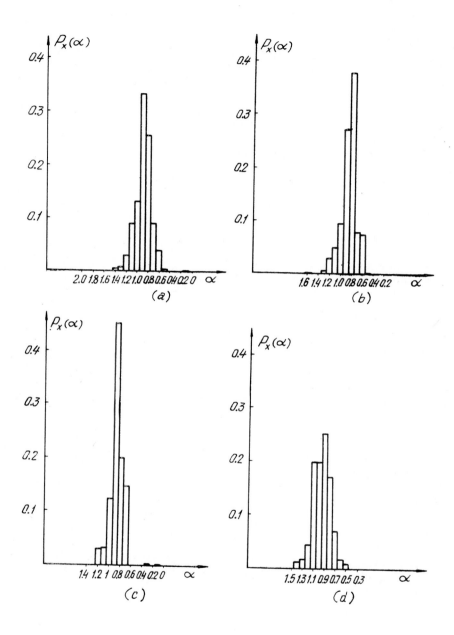

Fig. 1. Histograms for distribution of spectral indices for:
1a - all spectra; 1b - radio galaxies; 1c - quasars; 1d - unidentified objects.

absorption in HII, nor other physical processes, which could lead to
the spectrum bending at decametric waves, are observed.   The absence
of such bending is possibly connected with the fact that the high and
low radio frequencies radiated from different regions of the radio
sources.   In this case, the brightness distribution over the source
and its effective size should be dependent on the frequency.   So it
seems to be rather important to measure the brightness distribution
over the source with necessary resolution at the lowest possible
frequency.   The second particularity of the obtained data is that the
spectra of type C^+ prevail over that of type C^- even though the type
C^+ spectra percentage is considerably less than it was indicated in [3].

A distribution of the spectral indices α for all spectra is given
in Fig. 1a.   Figures 1b, 1c, 1d show the distributions of α for radio
galaxies, quasars and unidentified objects.   We determined the average
value of the spectral index $<\alpha>$ its error $\nabla\alpha$ and dispersionσ.   In the
range of 10-1400 MHz for type S spectra (they are 227) we have $<\alpha>$ =
0.91 ± 0.01, and σ = 0.16.   At low frequencies 10-25 MHz for type C^+
spectra (they are 27) we receive that $<\alpha>$ = 2.22 ± 0.16 and σ = 0.31;
and for the range of high frequencies 25 - 1400 MHz $<\alpha>$ = 0.78 ± 0.05,
and σ = 0.25.   When separating radio sources with type S spectrum
into Galaxies, quasars, and unidentified objects the average spectral
indices are as follows.   For Galaxies (they are 71) $<\alpha>$ = 0.86 ± 0.02
at σ = 0.18;   for quasars (they are 48) $<\alpha>$ = 0.89 ± 0.02 at σ = 0.14,
and for unidentified objects (they are 108) $<\alpha>$ = 0.96 ± 0.02 at σ =
0.17.   Thus the unidentified sources have more steep spectral index
than galaxies and quasars whose spectral indices are close one another.
For low frequency spectral indices of the type C^+ spectra are rather
steep whereas the high frequency spectra of these sources are more
flat than those of spectra of the type S.

Only 4% from 304 sources, measured in the declination strip 10° -
20° are new sources, which are missing for the presented catalogues
of the North Sky.

REFERENCES

1.   S. Ya. Braude, I.N. Zhouck, A.V. Megn, B.P. Ryabov, N.K. Sharykin.
        Preprint IRE AN USSR (1976) N 61.

2.   S. Ya. Braude, Yu. M. Bruck, P.A. Melyanovsky, A.V. Megn, L.G.
        Sodin, N.K. Sharykin. Preprint IRE AN USSR (1971) N 7.

3.   S. Ya. Braude, O.M. Lebedeva, A.V. Megn, B.P. Ryabov, I.N. Zhouck.
        Mont. Not. Roy. Astr. Soc. (1969) 143 289, 301.

4.   R.S. Roger, A.H. Bridle, C.H. Costain. Astron. Journ. (1973)
        78, 1030.

5.   M.R. Viner, W.C. Erickson. Astron. Journ. (1975) 80, 931.

6.   P.S. Williams, S. Kenderdine, J.E. Baldwin.   Mem. Roy. Astron. Soc.
        (1966) 70, 53.

7. J.D.H. Pilkington, R.F. Scott. Mem. Roy. Astron. Soc. (1965) <u>69</u>, 183.

8. J.F.R. Gower, R.F. Scott, D. Wills. Mem. Roy. Astr. Soc. (1967) <u>71</u>, 49.

9. S.Ya. Braude, I.N. Zhouck, B.P. Ryabov, N.K. Sharykin. Preprint IRE AN USSR (1976) N 62.

10. A.H. Bridle, C.R. Purton. Astron. Journ. (1968) <u>73</u>, 717.

11. K.I. Kellermann, I.I.K. Pauling-Toth, P.J.S. Williams. Astroph. Journ. (1969) <u>157</u>, 1.

12. S.Ya. Braude, I.N. Zhouck, A.V. Megn, B.P. Ryabov, K.P. Sokolov, N.K. Sharykin. Preprint IRE AN USSR (1976) N 64.

13. S.Ya. Braude, I.N. Zhouck, O.M.Lebedeva, A.V. Megn, B.P. Ryabov. Preprint IRE (1970) N 3.

INTERIM REPORT ON THE TEXAS SURVEY

James N. Douglas and Frank N. Bash
University of Texas Radio Astronomy Observatory
Department of Astronomy, The University of Texas at Austin

The University of Texas Radio Astronomy Observatory (UTRAO) is engaged in a survey of the entire sky north of -35° declination at various frequencies in the range 335-380 MHz. Primary goals are (i) determination of accurate (~1") positions for about 50,000 sources, followed by (ii) optical identification of the sources on the basis of exact radio-optical position coincidence; (iii) provision of rough structure models for all listed sources; and (iv) monitor the sky for variable sources on the time scale of 1 to 2 years. The survey is not expected to be a reliable source of absolute flux density information except for those sources known to be unresolved from other work.

1. THE TEXAS INTERFEROMETER

The Texas Interferometer consists of 5 fan-beam meridian-transit antennas arranged in a 2 by 2 mile diamond as indicated in Figure 1. The antennas are interconnected to form 8 interferometer baselines: 2 are North-south and 6 are oblique. A large (30 MHz or 8%) bandwidth causes the UV plane response of each baseline to be spread along the baseline vector; the spatial frequency response of the eight baselines falls into three patches in the UV plane, as shown in Figure 2. Each patch is a <u>sub beam</u> in angle space; sub-beams A and C have oblique interferometer fringes, and envelope beamwidths of 4.82 x 7.44 arc-minutes; sub-beam B has a N-S baseline and an envelope beamwidth of 9.65 x 5.42 arcminutes. When the three sub-beams are added together to produce the synthesized beam, the envelope antenna solid angle is 20.18 square arcminutes, or 180 beams/square degree; this corresponds to about 144 beam areas per source at the surface density of the survey.

Instantaneous coverage of a declination strip about 10° wide (actually 9° at zenith to 17° at -26°) is achieved by operating 120 complex correlators at different time delays; this multi-beam back-end can handle either one north-south or two oblique baselines. Thus five days' observing are required to complete the instrument response to the total solid angle of the declination strip (e.g. 9° x 360° or 3081

D. L. Jauncey (ed.), Radio Astronomy and Cosmology, 15-24. All Rights Reserved.
Copyright © 1977 by the IAU.

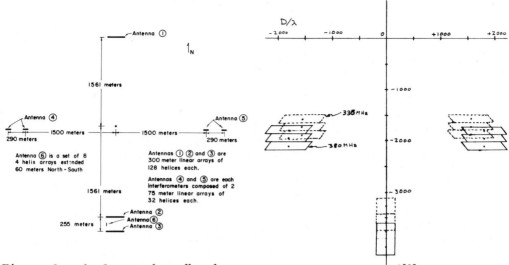

Figure 1.   6-element broadband
synthesis interferometer,
UTRAO.

Figure 2.   U-V plane response at
zenith, UTRAO 5-element synthesis
telescope.

square degrees for the 18° declination strip).

This five-day sequence produces data which is dominated by noise;
to attain the sensitivity necessitated by our survey goals 8-12 such
sequences must be averaged (40-60 days' observing); this averaged data
is called one cycle for the declination strip in question. The sensitivity of a cycle is such that all sources ultimately included in the
catalogue will appear at reasonable signal-to-noise ratio in one cycle.
Quite apart from improved sensitivity, the averaged data has also
smoothed out the effects of irregular ionospheric refraction which
would otherwise limit performance, and has made negligible the effect
of RFI and equipment malfunction in creating gaps in the data.

In order to check for systematic diurnal refraction effects,
investigate source variability, and improve sensitivity, three observing cycles are scheduled for each declination strip, spaced by 6-8
months. All the sources catalogued in a declination strip (∼5,000) can
be used to reduce the systematic errors in position and flux to a standard systematic error system; say, that of the first observed cycle.
The errors of this system are then removed by reference to external
position and flux density calibrators.

One of the three observing cycles is scheduled to be at 335 MHz,
while two are at 380 MHz. This procedure is necessary to reduce lobe-
shift incidence for faint sources to an acceptable level; it additionally

Strip	Steradians	Epoch of 380 MHz Obs.			Epoch of 335 MHz Obs.
		1	2	3	
-26	1.606	1975.30			
-12	1.293	1975.44			
-01	1.139	1975.94	(1976.66)		
+09	1.043	1975.78			
+18	0.949	1974.21	1974.99	1975.60	(1976.83)
+27	0.872	1973.20	1973.49	1973.82	
+36	0.794	1974.32	1974.79	1975.15	
+45	0.715	1976.10			
+55	0.622	1976.24			
+65	0.500	1976.36			
+77	0.320	1976.47			

Table 1.   Progress of Texas Survey through 1976.

gives further information on source structure as one can see from Figure 2; overall UV plane coverage is increased by about 60%.

Final catalogued positions will be a mean of the three observing cycles; source models and spectral index will be based on sub-beam fringe visibilities at two frequencies, and sources that vary significantly over a 1-2 year period will be noted.

A more detailed description of the instrument may be found in Douglas et al. (1973).

2. PROGRESS OF OBSERVATIONS

This is plainly a time-consuming data-acquisition and a tedious data-analysis project. The status of the data acquisition phase (as projected through the end of 1976) is shown in Table 1. Data acquisition is not yet complete for any declination strip, although +18 will be complete by late fall, and +27 and +36 should be completed in early 1977. The entire sky available to us (9.89 ster) has been observed on one cycle, and 26% of that (or 2.62 ster) has been observed on three cycles. We are just at the half-way mark in observations; completion is estimated to be in 1978.

3. PRELIMINARY RESULTS OF DATA ANALYSIS

Data analysis (apart from quick-look procedures) has not yet been completed for any declination strip, or indeed for any cycle of any dec strip. However some indicative interim results from two cycles of the 18 degree strip are now available, which allow us to characterize instrument performance, particularly for the brighter (> 0.4 Jy) sources in the ultimate catalogue.

Flux Density	# Compared	Observed Noise Error for 1 Cycle			Expected Noise Error for 3 Cycles		
		ΔS(%)	Δα(")	Δδ(")	ΔS(%)	Δα(")	Δδ(")
S<0.435 Jy	50	14.5%	2".75	1".95	8.4%	1".59	1".12
0.435≤S<0.761 Jy	225	8.5%	1".88	1".12	4.9%	1".09	0".65
S≥0.761 Jy	325	4.0%	0".92	0".52	2.3%	0".53	0".30

Table 2. Texas Survey, +18° strip noise error.

3.1 Internal position and flux accuracy

Let us start with the primary goal of the survey: accurate positions. An assessment of internal noise and systematic errors can be obtained by differencing positions obtained in those parts of the two 18 degree cycles already reduced. The data in each are totally independent, although calibration and data reduction procedures are identical. The results are summarized in Table 2. The estimates for a single cycle assumed the two cycles compared had the same variance; those for three cycles simply are single-cycle results reduced by $\sqrt{3}$. Not shown are the systematic differences. These are small (∼1") and slowly varying, and are both understandable and easily removable. The smaller error in δ than α is a direct consequence of the greater extent of our system in the NS direction, and a greater redundancy in δ-determination. All the observed noise errors in the table are consistent with the formal noise estimates produced by our reduction program. The position accuracy goal of the survey is being met.

3.2 External position accuracy and optical identifications

To assess external position errors, and check the rate and content of optical identifications, the region from 1417 to 1600 hours in right ascension (14° to 22°.3 in declination) was chosen for a pilot identification study. The preliminary survey listed 199 sources in this region; 5 were subsequently found to be lobe-shifted, leaving a sample of 194 sources to be studied.

Optical identifications were sought in each radio source field using the UTRAO computer-interactive laser measuring engine and the glass copies of the NGS-Palomar Sky Survey. Reference stars were taken from the SAO catalogue; following a 6-parameter adjustment, a finding photograph was taken at the radio source position. Such a photograph is shown in Figure 3, which illustrates the utility of accurate radio positions in optical identifications. The source name from the UT catalogue appears in the lower right hand corner (1559+173 = 4C 17.65); this object is a QSO with z = 1.944.

Of the 194 fields examined, optical candidates within 5 arc-seconds of the radio position were noted in 84 cases; a more detailed breakdown is shown in Table 3.

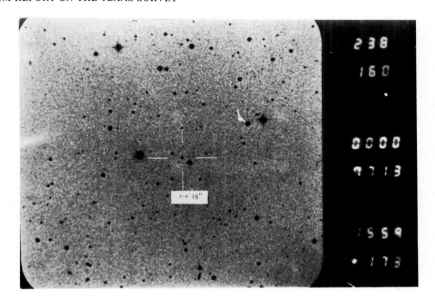

Figure 3. Finding chart for 1559+173.

Of these 84 objects within 5", 50 were within 3". An estimate of reliability is based on the work of Frank Ghigo (1976) who using the same instrument investigated the incidence of objects to the PSS plate limit in 600 random fields near radio sources. He found $4.76(10)^{-4}$ objects per square arcsecond in random fields; thus in 194 5" radius fields one would expect 7.25 objects (84 observed) and within 3" one would expect 2.61 objects (50 observed). The 84 objects are thus about 90% reliable; the 50 about 95% reliable.

As can be seen from Table 3, the mean optical-radio position difference is negligible, and the position scatter is approximately 2 arcseconds in each coordinate. This scatter is significantly higher than the sum of radio and optical position errors, and may reflect an intrinsic variance in radio and optical centroid position. The radio

	S≥0.761 Jy	0.435≤S<0.761 Jy	S<0.435 Jy	Total
# Sources	83	85	26	194
# Objects	45	29	10	84
Mean $\Delta\alpha$	0".56±0".28	0".02±0".47	-1".35±0".83	0".14±0".24
Mean $\Delta\delta$	0".03±0".27	-0".42±0".45	0".76±0".56	-0".08±0".22
$\sigma_{\Delta\alpha}$	1".83	2".43	2".34	2".18
$\sigma_{\Delta\delta}$	1".78	2".35	1".60	1".98

Table 3. Texas Survey optical sample region
1417<α<1600   14°0<δ<22°3

positions seem to be absolved in any event, since the error distribution is symmetric in α and δ, rather than being significantly smaller in δ as would have been expected from the radio position errors.

The identification rate of 84/194 = 43% is close to that found by Frank Ghigo in his dissertation (48%). Identification content of this sample is not yet fully determined, but preliminary indications are that about half are galaxies, and half QSO's.

3.3 Source variability

The 600 sources which were intercompared to estimate systematic and noise errors were observed in cycles 9 months apart; no sources were found to vary by more than 25% over this time scale. A few sources may vary at the 20% level; the best case is 0119+191, whose flux density increased by 22±6% in 9 months.

3.4 Comparisons with MC3

A comparison has been made of the available segments of the survey with the Molonglo MC3 catalogue (Sutton et al., 1974) to assess the relative completeness and lobe-shift incidence in our preliminary data. The results are summarized in Table 4. Of the 445 MC3 sources in our survey region, 321 were found; the number we missed is consistent with the angular structure selection effect in the Texas Interferometer. Of the 321 found, 196 were sufficiently unresolved to produce good positions without structure modeling; 4% of these were lobe shifted. Most of the lobe shifts were of course at low flux density, and in all but one case, the second most likely position based on our data was correct. It should be emphasized that the powerful additional help of the 335 MHz observations has not yet been brought into play; this should bring a further order-of-magnitude reduction in lobe shift incidence. Complete elimination of lobe-shifts in a dilute transit interferometer is of course impossible; one can only demand (1) that the probability a position is affected by lobe-shift is both small and calculable and (2) that the second most likely position is correct in a majority of the lobe-shift cases.

A number of sources ($\sim$40) were found in the MC3 region which were not in the MC3 catalogue; in most cases these sources fell below their

| | MC3 Flux Density | | | |
	S<0.4 Jy	0.4 Jy$\leq$S<0.7 Jy	0.7 Jy$\leq$S	Total
# MC3 Sources	91	178	176	445
# found	33	122	166	321
# positions	7	71	118	196
% lobe shifts	$\sim$30%	15%	2%	4%

Table 4. Texas Survey 18C1, comparison with MC3.

0.4 Jy completeness limit. Position differences of the common sources shows the Molonglo positions to be free of systematic error, and to have a scatter approximately correctly described by their estimated error.

3.5 Remaining data analysis problems and prospectus

The preliminary results thus far available refer mostly to the brightest 2/3 of sources in the ultimate catalogue (surface density of 3500/steradian); our automatic reduction programs need a further stage of fine adjustment to achieve reliable operation at the 5000 source/ steradian level ultimately desired.

Furthermore, structure modeling has not yet been incorporated into the automatic program; when available it will increase the number and reliability of the positions listed, as well as giving the most likely asymmetric double model for moderately resolved sources.

Both of these problems should be solved during this year, and the first completed section of the survey will be available in January of 1977.

REFERENCES

Douglas, J.N., Bash, F.N., Ghigo, F.D., Moseley, G.F., and Torrence, G.W.: 1973, Astron.J., 78, 1-17.

Ghigo, F.D.: 1976, 'Identification of Radio Sources Using Accurate Radio and Optical Positions', University of Texas at Austin (Ph.D. Thesis).

Sutton, J.M., Davies, I.M., Little, A.G., and Murdoch, H.S.: 1974, 'The Molonglo Radio Source Catalogues 2 and 3', publication of the School of Physics, University of Sydney.

DISCUSSION

Webster: I am interested in analysing surveys for evidence of clustering of sources, and wish to enquire how uniform this survey is likely to be, and how many sources per beamwidth you expect at the catalogue limit.

Bash: This survey which is made with interferometers is sensitive to the angular structure of radio sources. Although the brighter, resolved sources can be recovered, we will lose the resolved sources near the bottom of the catalogue.

G. Burbidge: Can you say how many sources you have looked at for variation. You have apparently found only one variable if I understand you correctly.

Bash: We examined 600 sources and found only one which varies by more than 20% (by at least 3 sigma). This is a smaller number of variables than that suggested by Cotton (1976, Ap. J., <u>204</u>, L63); however, our two measurements are separated by only 9 months which is a shorter time than the typical time scale for variations found by him.

Cannon: Would not part of the scatter between the radio and optical positions be due to the inclusion of spurious optical identifications?

Bash: Yes, but only a very small part. Ghigo has measured the background density in random fields as a function of galactic latitude to the Palomar Sky Survey plate limit. From the background counts we would expect 7 of the 84 identifications within 5" of the radio source and 3 of the 50 identifications within 3" of the radio source to the background objects.

Baldwin: For sources which are seen to be extended, what information do you plan to publish?

Bash: We are considering publishing the best-fitting double source model for the resolved sources and the centroid position for the double.

LOW FREQUENCY VARIABLES

W.D. Cotton

A sample of about 1500 radio sources selected from surveys with both long and short baseline interferometers was monitored for about two years (1973-1975) at 365 or 380 MHz at the University of Texas Radio Astronomy Observatory. About 800 were small enough, strong enough and well enough observed for variability on the order of 20% to 50% to be detected. Twenty two sources appear likely variables: 0051+317, 0113+154, 0127+233 (3C43), 0251+200, 0348+175, 0402+160, 0422+178, 0618+145, 0631+191, 0735+178, 0958+256, 1019+222, 1123+303, 1422+202, 1611+343, 1633+382, 1729+211, 2015+131, 2033+187, 2056+445, 2147+145 (3C437.1), 2338+132 (Cotton, 1976, Ap. J., <u>204</u>, 163). Analysis of a sample relatively unbiased towards small angular size indicates that at least 2% of all sources selected at low frequency are variable.

Most of the variable sources were observed in June 1976 at 7.8 and 15.5 GHz to determine the high frequency spectrum. A few spectra were flat but most were steep at low frequency becoming flat at high frequency. A couple of sources appear to have steep spectra out to 15.5 GHz, which according to the usual interpretation indicates the absence of a strong high frequency compact component.

G. Burbidge: Is any one of the variable sources identified with a genuine Galaxy of stars whose redshift may not be in question?

Cotton: One of the sources is identified with a galaxy but it is one of the more questionable variables.

Conway: What are the names of the 2 variables with steep, straight
spectra?

Cotton: 3C 43 and 437.1

Stannard: Two comments on low frequency variability from Jodrell Bank
studies at 408 and 962 MHz:
1) 3C454.3, which Hunstead reports to vary by about 30% at 408 MHz,
   shows double structure at low radio frequencies, with an extended
   steep spectrum component 5.3 arc sec from the nuclear component in
   position angle 131°. Studies with LB interferometers show, however,
   that it is the flat spectrum nuclear component which is responsible
   for the variations at 408 MHz.
2) One of the most active sources at low frequencies is BL Lac, which
   has varied between ~ 2 and 5.5 Jy at both 408 and 962 MHz.

Turtle: Bruce McAdam and myself have made regular transit observations
with the Molonglo Cross (408 MHz) of a fixed sequence of over 250 sources
for 3 or 4 days every month for a year. We included about 30 sources
which were suspected to vary. A partial analysis (W.B. McAdam (1976),
Proc. Astron. Soc. Austral. _3_,(in press)) has revealed at least 13 sources
which show marked (> 8%) changes in flux density compared to their
neighbours in the sequence. These changes occur over a period of a few
months - all sources remaining constant during a 3-4 day session.

0056 - 001	8%	1148 - 001	16%
0101 - 128	11%	1504 - 167	25%
0736 + 017	39%	1510 - 089	9%
0753 + 023	10%	1854 - 663	9%
0833 - 450	54%	2052 - 474	10%
(Vela Pulsar)		2251 + 158	15%
1036 - 697	9%		
1055 + 018			

Swarup: What is your estimate of the diameter of these variable sources
at meter wavelengths? If it's less than a milli-arc second, there are
possibilities of observing interstellar scintillations.

Cotton: The estimated diameters from the timescale of the variations
are on the order of a milli-arc second or less.

Condon: Bob Brown, Dave de Young, and I have made a survey of several
dozen compact sources to search for evidence of interstellar scintillation.
Flux densities in 100 5 MHz-wide channels from 500-1000 MHz were measured;
intensity fluctuations would appear as departures (on the scale of the
scintillation decorrelation bandwidth) from a power-law spectrum. Upper
limits of $\mu \leqslant 0.05$ were set, which limit source brightness temperatures
to less than 10^{18} K. New observations are planned which should go as low
as 10^{15} K.

Murdoch: 1504-16.7 which varied by a ratio 2 to 1 continues to vary at the same rate. It was originally identified in the PKS catalogue as a galaxy but it is almost certainly a quasar.

Readhead: Wilkinson, Purcell and I have made VLBI observations, at 610 MHz and on three baselines, of a number of sources - some of which are suspected of showing low frequency flux density variation. The observations were made at two epochs: December 1973 and March 1975. We found no evidence of variation in either the total flux density or the visibility of these sources with the possible exception of 3C 273 in which there is a hint of variation on the longest baseline. The ratio of the mean visibilities $\frac{1}{n} \sum_{IHA} \frac{\gamma(1973)}{\gamma(1975)}$ for each source on each baseline are as follows:

Source	NRAO-OVRO	NRAO-Ft. Davis	OVRO-Ft. Davis
CTA 21	1.03 ± 0.02	1.03 + 0.01	1.03 ± 0.01
3C 84	1.06 ± 0.04	1.03 ± 0.05	1.03 ± 0.03
3C 147	0.94 ± 0.04	0.97 ± 0.04	0.96 ± 0.05
3C 273	0.93 ± 0.02	0.96 ± 0.02	1.00 ± 0.01
3C 286	0.98 ± 0.02	0.98 ± 0.02	0.98 ± 0.01
3C 345	1.02 ± 0.02	1.04 ± 0.02	1.03 ± 0.04
CTA 102	1.05 ± 0.04	0.99 ± 0.01	0.95 ± 0.02
3C 454.3	0.99 ± 0.02	0.99 ± 0.01	0.99 ± 0.02

Thus no variation was seen in the three sources CTA 21, CTA 102 and 3C 454.3 which is curious in view of the very large (~ 50%) variations reported for these sources at earlier epochs.

Goldstein: Are there no stars among the Texas variable sources?

Cotton: None of the variable sources have been identified as stars.

THE BOLOGNA SURVEY OF RADIO SOURCES AT 408 MHz

Carla Fanti and Carlo Lari
Laboratorio di Radioastronomia, Bologna

The Bologna survey (Colla et al. 1970, 1972, 1973) does not represent a particularly new achievement since the first records refer to 1968. We will summarize here the main properties for a statistical use of the survey, the differential counts and the isotropy characteristic of this flux level, the latter being completely new.

The three major sections of the Bologna survey (conventionally named B2.1 ($S_{408} \geq .2$ Jy) and B2.2, B2.3 ($S_{408} \geq .25$ Jy)) cover roughly 1.5 steradians in the northern sky. The survey was projected to furnish a wide sample (9475 sources) with good positions and fluxes and well-defined completeness down to faint levels for statistical studies. In recent years the optical identification work and spectral indices analysis has shown the correspondence of the results to the initial objectives.

Well-defined completeness does not mean complete straightaway: there is not such a complete catalogue. Apart from statistical confusion and systematic noise effects, nearby strong and relatively strong sources there is always a defect of faint ones. For the B2 survey we have forced this effect, excluding areas around sources in a predictable and homogeneous way. For the strongest sources, the avoidance of faint ones can reach 1 degree. In this manner the 9475 found sources would correspond to 11,600 if this (forced) bias was not present. One must keep this in mind in order to use the catalogue properly; for particular studies this is somewhat complicated without a computer and, of course, the original formula of exclusion. Among those particular studies there is certainly clustering or spectral indices distribution for samples defined at different frequencies.

THE COUNTS

To get proper counts one must allow for: a) completeness, b) resolution and c) confusion and systematic noise effects. The resolution effect is not very significant for the bulk of all radio sources, apart from the nearby bright radio galaxies. In order to avoid an uncertainty in the

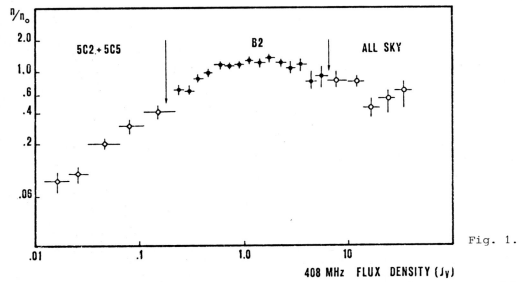

Fig. 1.

correction of the order of the effect, we have preferred not to consider it.

Because the completeness is well controlled a priori, if one assumes no clustering at small scales, there is no better way to determine confusion and noise effect (using the actual noise) than a Montecarlo estimate. We performed three different Montecarlo estimates which furnished not only the correction but also the uncertainty in the correction itself.

The correction is important, of course, near the limit of the catalogue: for example, in the interval .20 - .25 Jy, confusion and noise cause an over-estimate of the actual differential counts of a factor of 1.35 ± .076. As one can see, the error in the correction is quite large and represents the major sources of uncertainty for faint sources. After applying all the corrections, the B2 catalogues would have to contain 9153 sources.

Fig. 1 shows the differential count, normalized to an "euclidean" 5/2 slope that could give 4725 sources/ster. of .31 Jy. We have adopted the Robertson flux scale (1973), whose counts are reported at high fluxes together with 5C2 + 5C5 as reported by Pearson (1975).

ISOTROPY

Webster (1976) has given a good discussion for the distribution of radio sources in the 4C, Molonglo and GB surveys, using a powerful tool like his power spectrum analysis. He has shown that no believable clustering is found. On the contrary, most of the surveys, aside from the Molonglo one, which is far from the confusion limit, show anti-clustering, i.e. a lack of faint sources around the stronger.

We have performed the power spectrum analysis (P.S.A.) on the three sections of the B2 separately, not to make small possible calibration differences simulate clustering at long wavelengths. We have also made an extensive bin analysis from 0.5 square degrees up to the size of the catalogues, not only to check the P.S.A. results but also because it is much easier to take into account the controlled "anti-clustering" present in our survey.

Referring to Webster's paper for the explanation of the P.S.A. method, we recall here only that he introduces the quantity $Q \equiv \Sigma_I/\Sigma_\nu$, Σ_I being the sum of the normalized square Fourier amplitudes of source positions (up to a given wave number $1/\lambda$), and Σ_ν twice the summed number of Fourier terms. Q has as the expectation value the average number of sources per constellation and is 1 for uniform random distribution. We state consequently that Q would correspond to the ratio between the number of sources actually found and the predicted number, taking account of exclusion areas.

The following table gives the results for four different limiting fluxes and the combinations of the three sections.

S_o	Σ_I	Q	Q_{exp}	$1/\lambda$ deg^{-1}
≥ 1.005	1278	1.037	.988	.3
$\geq$ .505	3484	1.002	.986	.5
$\geq$ .355	5831	.985	.964	.7
$\geq$ .255	8510	.970	.922	.8

The fourth column gives the expected values from our estimates of the excluded areas around the sources. The excess found for each column is not significant, being of the order of 1 standard error except for the last flux which would be at 3 standard errors. However, one must consider that the excess if of the order of 5% only, corresponding to 400 sources involved, and that the error in the correction given by statistical fluctuation of the overall number may well be of that order. Also, for well-known clustering of the Abell type, the physical doubles are expected to contribute to short scales of that amount. Finally, a confidence for such small effects must stand up to Montecarlo checks of the effect from fainter sources than the survey limit, because a confusion effect from faint sources is expected to depend upon the actual local density of sources, giving rise to non-linear behaviour.

Fig. 2 shows Q' and Q trend superimposed against $1/\lambda$ for $S_o \geq$ .255. The excess found at short spatial wavelengths never reaches 3 standard errors. The behaviour of the other fluxes is pretty similar.

The Bin Analysis (B.A.) confirms rather well that which was found from P.S.A., and the test for different bin sizes never gives a

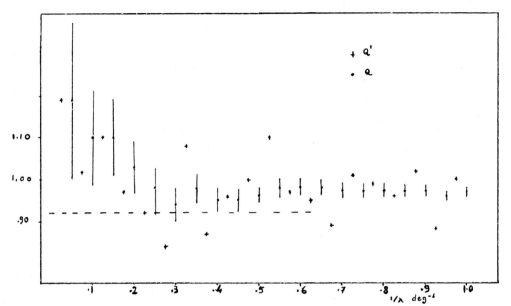

Fig.

probability less than 1% for uniform distribution, and it is normally
higher than 5% for the small scales as well. Using the B.A. we were also
able to test differentially for the different fluxes that showed statistically uncorrelated results. The latter is not only a property of the
spatial distribution of the radio sources but it is also a property of
the luminosity function that appears to have all the properties of a
statistical distribution function for luminosities for this purpose.

     As a last remark on the very large scales, we must state that the
densities of sources for the three sections are scattered with a dispersion of 2.8% while the statistically expected dispersion would be 2%.
This set a limit for a different scale between the three surveys of 1.5%.

CONCLUSIONS

No large deviation from the uniform distribution has been found in the B2
survey, setting a limit of a few percent for the clustered population.
In reality, when we look around at the positions from which clustering is
expected, i.e. Abell cluster centers, we pick up the effect very clearly.
This happens to scales that are a fraction of the Abell radius or a few
minutes of arc. The only chance to explore this effect for an eventual
proto-cluster of that size would be to look in surveys which are deep
enough to detect this population and on such a small scale as to have
few clusters in the column of the universe visible at that flux level.
This is well below the performance of the present radio telescopes.

     Our limits on the clustering refer to large scale effects in the
universe owing to the deepness of the radio sky. These scales are
generally larger than known superclusters and give stringent limits to

the homogeneity of the universe, both in content of galaxies and in evolutionary characteristics. No theoretical analysis of the cosmological meaning of such isotropy is known to the authors.

REFERENCES

Colla, A. et al., 1970, Astron. Astrophys. Suppl. 1, 281.
Colla, A. et al., 1972, Astron. Astrophys. Suppl. 7, 1.
Colla, A. et al., 1973, Astron. Astrophys. Suppl. 11, 291.
Pearson, T.F., 1975, Mon. Not. Roy. Astron. Soc. 171, 475.
Robertson, J.G., 1973, Aust. J. Phys. 26, 403.
Webster, A.S., 1976, Mon. Not. Roy. Astron. Soc. 175, 61.
Webster, A.S., 1976, Mon. Not. Roy. Astron. Soc. 175, 71.

DISCUSSION

Jauncey: Have you really detected Abell clustering in the Bologna survey?

Lari: Yes. We have covered enough Abell clusters to overcome the background fluctuations, and we find a very peaked distribution near the centres of the clusters. More exactly, we have added together all distance 5 Abell clusters and we find a total number of 96 B2 sources within one Abell radius. The expected number from the overall background is 42, leaving an excess of 54, although the average excess density is less than one source per cluster. The distribution falls off very rapidly, with a peak only 0.1 Abell radius wide and a central density about 20 times larger than the background. Preliminary results from Westerbork show many of the B2 sources in the clusters breaking up into two sources both associated with galaxies in the cluster.

THE MOLONGLO RADIO SOURCE SURVEYS AT 408 MHz

B.Y. Mills,
School of Physics, University of Sydney, Australia.

One of the basic programs undertaken with the 1 mile Cross-type radio telescope at the Molonglo Radio Observatory has been a systematic survey of the sky south of +18° declination. Commencing in late 1967, the survey has been continuing at a low priority until it is now complete to a declination of -85°, except for some small areas affected by interference or instrumental problems which are being re-observed. A program of 'deep surveys' is also being undertaken; this involves multiple observations within narrow strips at selected declinations. The signal-to-noise ratio is some three to five times better than in normal surveying and the confusion limit is approached.

The radio telescope output is stored on magnetic tapes in digital form and on paper faximile charts in analogue form. These are analysed and catalogues prepared as time permits. The present status is shown in the table below.

Molonglo Catalogues

Catalogue	Declination	Area sr	Number of Sources	Reference
MC1	-20°	0.21	1545	Davies *et al.*, 1973
MC2	+11°	0.11	609	Sutton *et al.*, 1974
MC3	+17°	0.15	657	Sutton *et al.*, 1974
MC4	-70°	0.23	1340	Clarke *et al.*, 1976
MC5	+11°	0.14		In Preparation
1 Jy Catalogue	South of +18°	8		In Preparation
Deep Survey 1	-20°	0.020	373	Robertson 1976(a)
Deep Survey 2	-62°	0.0055	95	Robertson 1976(b)

D. L. Jauncey (ed.), Radio Astronomy and Cosmology, 31-37. All Rights Reserved.
Copyright © 1977 by the IAU.

Four catalogues of small areas totalling about 0.69 steradians have
been published (MC1, 2, 3 and 4). These contain sources down to a level
of about 0.2 Jy, depending on the declination and the instrumental sensitivity at the time of the observations. The catalogues range in source
density from about 5000 to 7400 sources per steradian. Several more
catalogues of this kind are planned in astronomically interesting areas
and areas in which the instrumental sensitivity is high. In addition,
a catalogue is currently in preparation of the whole area accessible to
the instrument, excluding a 6° wide strip along the Galactic plane. This
catalogue is complete to 1 Jy and extends somewhat lower in most areas;
it will contain approximately 1000 sources per steradian. The very small
'deep survey' areas are restricted to sources less than 1 Jy and are
complete to between 80 and 90 mJy; they contain about 20000 sources per
steradian. All flux densities in the Molonglo catalogues are given in
the Wyllie (1969) scale.

An opportunity to sample the reliability and completeness of the
MC series of catalogues is afforded by the first 'deep survey' which
overlaps MC1. There are 140 MC1 sources in the area of overlap and the
catalogues are compared in the table below on the assumption that any
discrepancies reflect errors in MC1.

A Test on a Sample of MC1 Sources

Flux Density (Deep Survey) Jy	Total Source Numbers	MC1 Completeness $\frac{\text{Sources in MC1}}{\text{Total Numbers}} \times 100$ %	MC1 Reliability $\frac{\text{MC1 Sources Confirmed}}{\text{Sources in MC1}} \times 100$ %
> 0.5	54	98	100
> 0.25	129	89	99
> 0.15	238	57	95

One may conclude that the MC1 catalogue is reliable and reasonably
complete above 0.25 Jy and the completeness falls off rapidly at lower
flux densities. The reliability also decreases, but is still 95% for
the whole sample. Later catalogues have been truncated at a higher
signal-to-noise ratio and should be correspondingly more reliable.

The catalogues have been used to make source identifications, to
determine spectra in conjunction with higher frequency data and to
examine the distribution of sources in depth (source counts) and their
distribution on the celestial sphere.

SOURCE COUNTS

The differential source counts for MC1 have been presented by Mills *et
al.* (1973). Robertson (1976a, b, c) has now included MC2, MC3 and the
deep surveys, improving the statistics at medium flux densities and

extending the counts to much lower levels. His results are shown in
Figure 1; the so-called convergence of the counts at low levels is
confirmed and the region of turn-over well defined.

It is particularly significant that all the measurements have been
made with a single radio telescope, except for about seventy of the
strong sources from the 'all sky' catalogue of Robertson (1973). The
rather large statistical uncertainties just below 10 Jy should be much
reduced soon by using the Molonglo 1 Jy catalogue.

The counts expected from two simplified cosmological models are
also shown in Figure 1 (Robertson, paper in preparation). There is
little difficulty in fitting the counts at low flux densities with a
uniform density non-evolving model but, as is well known, the number
of sources with high flux densities is then far too small to be explained plausibly as a chance fluctuation in a Universe containing a
completely random distribution of sources.

Numerous models can be proposed in which the number of radio sources was greater when the Universe was several times smaller than it is
now, possibly as the result of a period of rapid source formation.
Thereafter the rate of source production and/or the luminosity of individual sources decreased rapidly to the present level. The second curve
is representative of this class of model. There is sufficient freedom
in the choice of undefined parameters to fit the counts well. However,
before one can accept this kind of result as representing the real Universe it is necessary to examine the possibility that physical source
groupings may permit the high flux density results with reasonable
plausibility in a non-evolving or slowly evolving Universe. Information
about source groupings is also, of course, intrinsically important.

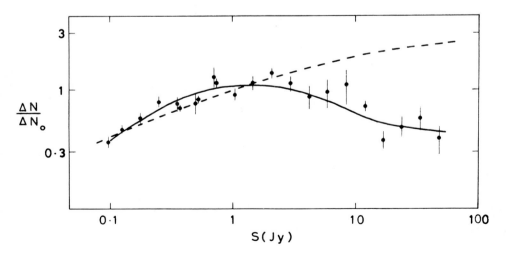

Figure 1. Combined differential source counts. Theoretical curves for
a non-evolving model (dashed) and an evolving model are shown.

SOURCE GROUPINGS

As they are produced, all Molonglo catalogues are examined for small
scale clustering using simple 'binning' analysis. No significant non-
randomness has been found for the MC series at the most common flux
density of 0.2 Jy to 0.5 Jy, and this result has been confirmed for
MC1 by a more sophisticated power spectrum analysis (Webster, 1976).
The deep survey catalogues do show signs of small scale clustering but
at a low significance level (5%-10%) which, in view of the large number
of tests performed, cannot be given much weight.

Visual examination of the distribution of strong sources in the 'all
sky' catalogue of Robertson (1973), indicates a certain amount of pairing
and statistical analysis confirms that this is probably marginally significant for source separations less than two or three degrees. The implication is that some widely separated doubles are catalogued as individual
sources at high flux densities whereas similar pairs, when more distant,
weaker and closer, might be catalogued as single sources. The selection
effect is of the right kind, but totally inadequate in magnitude, to
account for the form of the high flux density counts.

In contrast to the above very weak evidence, there is strong evidence for large scale anisotropy in the Molonglo catalogues, but it is
not yet firmly established whether this is caused by a real anisotropy
in the source distribution or by systematic calibration errors. Inter-
comparing the MC series at 1 Jy and above, where all the catalogues are
believed to be accurate, complete and reliable, it is found that there
are no significant differences between MC1 and MC4 or MC2 and MC3, but
both MC2 and MC3 have a substantially higher source density than the
others. A comparison of MC1 with MC2 and MC3 combined gives, for
sources $\geq$ 1 Jy:

$$\begin{array}{ll} \text{MC1} & 895 \pm 66 \text{ sr}^{-1} \\ \text{MC2 + MC3} & 1217 \pm 69 \text{ sr}^{-1} \end{array}$$

The ratio is 1.36 and the formal significance 0.1%. At lower flux
densities the ratio and the significance decrease, which may reflect a
greater homogeneity for the weaker sources or, perhaps, some incompleteness in MC2 and MC3, for which the radio telescope sensitivity was much
reduced.

Murdoch (1976a, b) reached essentially the same conclusion and, in
addition, compared the two portions of MC2 and MC3 which are in the
North and South Galactic Hemispheres. Above 0.5 Jy the catalogue source
density in the northern hemisphere exceeds that in the southern by a
factor of 1.23 at a significance level of 1%. Although the Virgo cluster
is included in the area of high source counts it is not in a particularly
dense region.

Both these indications of large scale anisotropies are subject to
calibration uncertainties. The antenna gain of the Molonglo instrument

is a function of declination and the internal calibration is slightly dependent on temperature, so that small diurnal variations do occur. However, it seems impossible for calibration differences to be large enough to account for the observed results. Nevertheless the uncertainties reduce the statistical significance of the indicated anisotropies; for example, if the standard error in the relative calibration of the MC1 and MC2 + MC3 catalogues was as high as 5%, the significance of the anisotropy would be reduced from the 0.1% level to the 1% level. The importance of very accurate relative calibrations over the whole sky is clear and further calibration programs are planned to improve the present situation.

Despite these uncertainties, the present results do suggest that radio sources are not randomly distributed throughout the Universe. If this is so, the source counts at high flux densities may provide little or no information about source evolution unless the relationship of the local source density to the mean can be established in other ways.

Finally, another possible large scale anisotropy should be mentioned. Robertson (1976c), in comparing his deep surveys with the Cambridge 5C surveys, found that the Molonglo catalogues contain more sources than the Cambridge, above equivalent levels, by a factor of 1.29. The formal significance is about 1% but, in view of the increased probability of calibration discordances between two quite different catalogues, this indication of anisotropy is very weak.

REFERENCES

Clarke, J.N., Little, A.G. and Mills, B.Y.: 1976, 'The Molonglo Radio Source Catalogue 4; The Magellanic Cloud Region', *Australian J. Phys. Astrophys. Suppl.* (in press).
Davies, I.M., Little, A.G. and Mills, B.Y.: 1973, *Australian J. Phys. Astrophys. Suppl.* No. 28, 1.
Mills, B.Y., Davies, I.M. and Robertson, J.G.: 1973, *Australian J. Phys.* 27, 417.
Murdoch, H.S.: 1976a, 'Radio Source Spectra for Sources Selected at 408 MHz', *Monthly Notices Roy. Astron. Soc.* (in press).
Murdoch, H.S.: 1976b, 'Radio Spectra and Optical Identification of Sources from the MC2 and MC3 Catalogues', *Proc. Astron. Soc. Australia* (in press).
Robertson, J.G.: 1973, *Australian J. Phys.* 26, 417.
Robertson, J.G.: 1976a, 'The Molonglo Deep Sky Survey of Radio Sources I. Declination Zone - 20°', *Australian J. Phys.* (in press).
Robertson, J.G.: 1976b, 'The Molonglo Deep Sky Survey of Radio Sources II. Declination Zone -62°', *Australian J. Phys.* (in press).
Robertson, J.G.: 1976c, 'The Molonglo Deep Sky Survey of Radio Sources III. Source Counts', *Australian J. Phys.* (in press).
Sutton, J.M., Davies, I.M., Little, A.G. and Murdoch, H.S.: 1974, *Australian J. Phys. Astrophys. Suppl.* No. 33, 1.
Webster, A.: 1976, *Monthly Notices Roy. Astron. Soc.* 175, 71.
Wyllie, D.: 1969, *Monthly Notices Roy. Astron. Soc.* 142, 229

DISCUSSION

Grueff: Supposing that the clustering found at about 1 Jy is real, have you any idea about the angular size involved?

Mills: There appears no clustering below a few degrees and the significant differences in source density occur between catalogues about 35° apart in declination and between portions of the same catalogue about 90° apart in R.A.

Wall: You commented that MC2 and MC3 were "statistically undistinguishable", but that MC1 differed in source density from MC2 + MC3 at a significance of 0.1%. Is this not an invalid statistical procedure in that it is "a posteriori" statistics?

Mills: This question illustrates the subjective nature of significance tests. The actual procedure was to compare first MC1 with MC2 + MC3 to check whether these two well separated sky regions of approximately equal area had the same source density. The internal comparison MC2 with MC3 followed and MC4 was also checked, but the latter catalogue is regarded as atypical because it includes the Magellanic Clouds; it is not used for the source counts. I believe that the conclusions in the paper are valid.

Ekers: The MC2 and MC3 surveys pass through the region of the Virgo cluster. Can you estimate from your identifications how many sources are associated with this cluster?

Mills: There is no particular excess in the Virgo cluster region, either of radio sources or of galaxy identifications, but there may be in the 1 Jy catalogue in the vicinity of the Hercules cluster. Comparison of strong sources between Molonglo at 408 MHz and Arecibo at 430 MHz suggests no evidence for declination dependence of the calibration except possibly for a small error in the periodic term of the calibration curve. This suggests a possible calibration error in MC2 and MC3 of about 2%, that would affect the counts by 3% which is negligible compared to the 40% difference with MC1.

THE 5C6 AND 5C7 SURVEYS

A.J. Kus and T.J. Pearson

We have recently completed two new deep surveys at Cambridge: 5C6 and 5C7. They lie at $\delta=32°$, $b=-27°$ and $\delta=27°$, $b=30°$; and being similar in their observational properties they are ideal for the study of the isotropy of faint sources. The two source counts at 408 MHz are in excellent agreement with each other, but the source density is slightly higher than in 5C2 and 5C5, in agreement with the new results from Molonglo and Westerbork. At 1407 MHz the source counts agree with each other and with 5C5, and the combined count from all three surveys shows that convergence continues down to the survey limit at about 2 mJy.

The 5C counts agree with the Westerbork 1 survey but are lower than the corrected Westerbork 3 counts.

The spectral index distributions derived from 5C5, 5C6 and 5C7 are all similar, with median $\alpha(408,1407) \simeq 0.80 \pm 0.03$.

WESTERBORK SURVEYS OF RADIO SOURCES AT 610 AND 1415 MHZ

A.G. Willis, C.E. Oosterbaan, R.S. Le Poole, H.R. de Ruiter,
R.G. Strom and E.A. Valentijn, Sterrewacht Leiden

P. Katgert, J.K. Katgert-Merkelijn
Laboratorio di Radioastronomia CNR, Bologna

1. INTRODUCTION

The Westerbork Synthesis Radio Telescope (WSRT) has now been used to make source surveys at frequencies of 610 and 1415 MHz. This paper summarizes the results concerning source counts and anisotropies in the distribution of sources from those surveys not concerned with clusters of galaxies.

2. 610 MHZ SURVEYS

The 610 MHz source counts described here are a combined count derived from the work of three separate groups. Although they were originally searching for radio counterparts to unidentified 3-U X-ray sources, Harris et al. (1976) have derived a source count for all the radio sources detected in the five fields they observed. Valentijn et al. (1976) were mainly concerned with observing the Coma cluster of galaxies but have computed a source count for those radio sources detected which they consider to be unassociated with the cluster itself. These sources are either unidentified optically or identified with objects having magnitude fainter than 17.5. Katgert-Merkelijn (1976) has observed several regions. Four fields lie in the area of the first 1415 MHz WSRT survey (Katgert et al,1973), one covers the region of the second 1415 MHz survey (Katgert-Merkelijn and Spinrad,1974) and three are high declination fields mostly observed for testing purposes when the 610 MHz system was first installed on the WSRT in 1973. At 610 MHz the primary beam pattern of the WSRT has a half power width of 83 arcmin, while the synthesized antenna pattern half power width is 55" (RA) x 55"cosecδ (DEC). The completeness limits for the counts of the different groups are: Harris et al., 15 mJy; Valentijn et al., 6.3 mJy; Katgert-Merkelijn, 22 mJy. Figure 1 shows the differential count derived by combining the various observations and, to lower the statistical error, using only those sample bins found to contain more than 10 sources. The total number of sources contributing to the count is 472. The error bars on the measurements represent sampling errors proportional to $\sqrt{n}$ where n is the number of sources found within a sampling bin. An excellent least squares fit to the observed counts

D. L. Jauncey (ed.), Radio Astronomy and Cosmology, 39-45. All Rights Reserved.

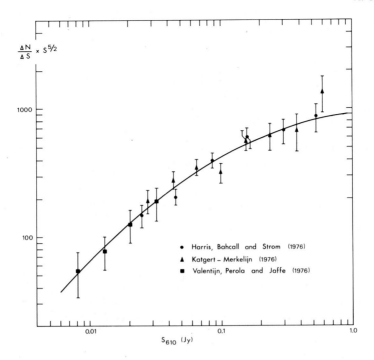

Fig. 1. WSRT 610 MHz differential counts. The solid line describes eq.(1).

is given by

$$dN/dS = 912.06 \ S^{(-2.39 - 0.101 \ \ln S)}$$

(1)

One must note, however, that there is no a priori astrophysical reason
for the use of this particular type of equation. It is apparent that we
are seeing strong convergence of the counts at the lower flux density
levels. The slope of the differential count continually flattens, reaching
a value of -1.46 at 10 mJy. Note that the fit to the counts for equation
(1) did not include the Katgert-Merkelijn data point at 600 mJy because
this point is strongly influenced by the count from the 1415 MHz survey 1
area (see below). An anomalously large number of sources stronger than
100 mJy is found in this region at 1415 MHz. We find a similar excess to
occur at 610 MHz when the Katgert-Merkelijn integral count above 200 mJy
(corresponding roughly to 100 mJy at 1415 MHz) from the four fields in
the 1415 MHz survey 1 area is compared with the count of Harris et al..
Katgert-Merkelijn obtains 8300 ± 1470 sr^{-1} while Harris et al. find
5530 ± 790 sr^{-1}. The excess, 1.7σ, is not statistically significant,
unlike the result at 1415 MHz, but this may be due to the small control
sample presently available. This conclusion is supported by the fact that
the counts from the other areas studied by Katgert-Merkelijn are also low
relative to the counts in the 1415 MHz survey 1 area. Two of the other
fields have some interference and thus may suffer from some incompleteness.
Therefore these fields were not directly added to the control sample.

3. 1415 MHZ SOURCE SURVEYS

Seven surveys have now been completed using the WSRT at this frequency. Table 1 gives a brief description of their characteristics. Properties of the telescope common to all surveys are 1) primary beam half power width is 36 arcmin and 2) synthesized antenna pattern half power width is 23" (RA) x 23"cosecδ (DEC). Detailed descriptions of the reduction procedures employed for the various surveys are given in the references listed in Table 1. Source counts for surveys 1 to 4 are also found in the references.

Figure 2 shows the counts from survey 5 (Oosterbaan, 1977) combined with preliminary counts from the deep surveys 6 and 7 (Le Poole and Willis, 1977). The counts from these surveys have not yet been corrected for biases due to noise and resolution. Corrections for these effects, particularly the latter, will probably cause the counts to migrate upwards

Table 1 : 1415 MHz WSRT Source Surveys

Survey	Reference	Area (deg^2)	No. Sources	S_{lim} (mJy)
1	Katgert et al., 1973	25	224	7
2	Katgert-Merkelijn and Spinrad, 1974	4.5	58	7
3	Katgert, 1975, 1976	18	238	6.25
4	van Vliet et al., 1976	20	130	7.5 to 20
5	Willis et al., 1976	90	1075	3 to 10
6	Le Poole, 1977	2	60	1.5
7	Le Poole and Willis, 1977	1	50	1

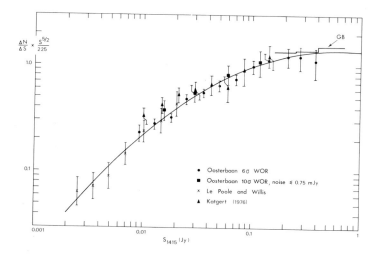

Fig. 2. 1415 MHz differential source count from surveys 3, 5, 6 and 7. The solid line describes eq. (2). For comparison, the counts from Maslowski's (1973) 1400 MHz Green Bank (GB) survey are also plotted.

since the separate count for the survey 5 sources having signal to noise
ratios larger than 10 all move above the fitted line (eq. (2) below) in
the direction of Katgert's (1976) counts for survey 3. The survey 3 counts
have been corrected for noise and resolution bias. A fit to the counts
from surveys 5, 6 and 7, which must therefore be regarded with caution, is

$$dN/dS = 310.72 \, S^{(-2.59 - 0.106 \ln S)} \tag{2}$$

The counts at 1415 MHz, like those at 610 MHz, show strong convergence
at low flux densities, and the slope of the differential count flattens
to -1.50 by 6 mJy.

Because there are now WSRT 1415 MHz surveys composed of fields
scattered over the sky (surveys 4 and 5) as well as of fields concentrated
in a small area (surveys 1, 2 and 3) we can make some tests for anisotropy
in the source counts on both a large and small scale.

Firstly, if radio sources are randomly distributed over the entire
sky, the number of sources stronger than some limiting flux density
detected within a uniform area should be poisson distributed. Thus the
number of sources found within a WSRT field out to a distance of 0.55°
from the field centre should have this distribution. Survey 5 provides an
excellent sample to test this hypothesis since the fields are reasonably
well distributed over the sky. Figure 3 shows distributions of the number
of sources detected per field for 85 fields from survey 5. We have plotted
(a) the number of sources per field having a peak $S \geq 10$ mJy, (b) the
number having an intrinsic sky $S \geq 100$ mJy and (c) sources contained in
group (a) but not in (b). Distribution (a) essentially tests for variations
in the integral count above the 10 mJy level, (b) for fluctuations in
the integral count above 100 mJy and (c) for variations in the differential

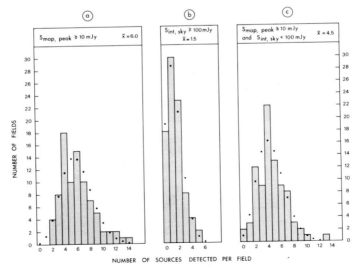

Fig. 3. Distributions of radio sources detected per field at 1415 MHz.

count between 10 and 100 mJy. Figure 3 also lists the mean number of
sources detected per field in each group and the dots show the theoretical
poisson distribution expected on the basis of the observed mean number
of sources. The observed distributions obviously agree very well with
those expected. This is confirmed by a χ^2 test. Thus there does not seem
to be any obvious evidence of large scale anisotropy. The poisson test
is, however, insensitive to effects such as a gradient in the density
distribution over the sky. We have divided the 85 fields into groups and
searched for gradients in the number of sources detected as functions of
galactic latitude, right ascension and declination. A gradient in the
declination distribution would not be unexpected due to the changing
resolution of the WSRT. No significant departures from random fluctuations
were found in any of the groupings.

One can search for small scale anisotropy by comparing the number
density of sources in each of the surveys composed of fields concentrated
in a small area (1, 2 and 3 in Table 1) with the density found in control
samples produced from other surveys listed in Table 1. There is certainly
one significant anomaly. The survey 1 area contains too large a number of
sources having intrinsic flux densities stronger than 100 mJy. 37 sources
stronger than 100 mJy are found in the 14 fields from the survey 1 area
which do not overlap each other. Using a control sample of 112 fields from
surveys 3, 4 and 5 we find that a mean of 19.6 $\pm$ 1.6 sources is expected
in 14 fields. The difference between the two samples is thus 2.8σ, which
is significant at about the 0.6% level. The optical identifications, 9
galaxies and two stellar objects, are also anomalous. Accurate optical
identifications of survey 5 sources stronger than 100 mJy (de Ruiter et
al.,1976) lead us to expect about 4 stellar objects and only two galaxies
to be identified in a 37 source sample. Thus the survey 1 identifications
have about a 2σ excess of galaxies. Although these data suggest the
possibility that strong sources cluster on a scale of $\sim$ 10 to 15 deg^2,
more surveys of a similar size are definitely needed to confirm such a
result. It is certainly not improbable that the survey 1 anomaly
constitutes a "statistical mishap" with an occurrence of about 1 in 100
to 200 cases. Note that the source excess does not extend to flux densities
below 100 mJy. The counts from survey 1 agree with those from the other
surveys below this limit.

4. COMPARISON OF RESULTS AT DIFFERENT FREQUENCIES

Before one can accurately intercompare source counts made at two
different frequencies one needs to know the two point spectral index
distribution between the frequencies. This distribution has only been
determined in a preliminary way for WSRT 610 and 1415 MHz samples (P.
Katgert et al., this volume). We have simply scaled the counts using
an effective (or δ function) spectral index, α_e. To make the WSRT 610 MHz
and 1415 MHz surveys 5,6 and 7 counts presented in sections 2 and 3
compatible with each other one must change α_e from 0.6 to 0.8 as the
1415 MHz flux density increases from 2.6 mJy to $\sim$ 400 mJy. Excellent
agreement between the counts is then obtained. This variation in α_e is
consistent with that found by Fomalont et al. (1974). Note, however,

that the corrected counts of 1415 MHz survey 3 (Katgert,1976) imply that α_e actually becomes considerably flatter than 0.6 in the 1415 MHz flux density range from $\sim$ 10 to 30 mJy.

A slight discrepancy is found when the WSRT 610 MHz counts are compared with the 5C2 and 5C5 counts (Pearson, 1975) made at the nearby frequency of 408 MHz. The 5C differential counts appear to be system-atically lower by $\sim$ 20% and to exhibit less curvature over the 408 MHz flux density range from $\sim$ 20 to 200 mJy than do the 610 MHz counts scaled with α_e equal to 0.6. At this time one cannot say whether the disagreement is due to instrumental effects or is indicative of a real anomaly.

Although they were only derived from the WSRT data, equations (1) and (2) are found to give a reasonable mathematical description of the source counts up to S $\sim$ 10 Jy and $\sim$ 15 Jy at 1415 and 610 MHz respectively. Integration of S·dN gives an estimate of the total sky brightness temperature due to extragalactic radio sources. If we assume equations (1) and (2) to describe the counts to S = 0 Jy integration from S = 0 to ∞ yields brightness temperatures of 0.063 K and 0.651 K at 1415 and 610 MHz respectively. These integrals are not very sensitive to the lower limit. If we begin the integration at 2 mJy at 1415 MHz and 6 mJy at 610 MHz the resulting temperatures are 0.059 K and 0.599 K respectively. Using the assumption that the first two temperatures are reasonable estimates of the total brightness temperatures we compute the temperature spectral index between 1415 and 610 MHz to be 2.78. The brightness temperature at 178 MHz is then expected to be $\sim$ 20 K. This value is in quite reasonable agreement with Bridle's(1967) estimate of 30 ± 7 K at this frequency.

The Westerbork Radio Observatory is operated by the Netherlands Foundation for Radio Astronomy with the financial support of the Netherlands Organization for the Advancement of Pure Research (Z.W.O.).

REFERENCES

Bridle, A.H., 1967, Mon. Not. Roy. Astron. Soc. <u>136</u>, 219.
Fomalont, E.B.,Bridle,A.H.,Davis,M.M., 1974, Astron. Astrophys. <u>36</u>, 273.
Harris,D.E.,Bahcall,N.,Strom,R.G., 1976, Astron. Astrophys. submitted.
Katgert,P.,Katgert-Merkelijn,J.K.,Le Poole,R.S.,Laan,H. van der, 1973, Astron. Astrophys. <u>23</u>, 171.
Katgert,P., 1975, Astron. Astrophys. <u>38</u>, 87.
Katgert,P., 1976, Astron. Astrophys. <u>49</u>, 221.
Katgert-Merkelijn,J.K.,Spinrad,H., 1974, Astron. Astrophys. <u>35</u>, 393.
Katgert-Merkelijn,J.K.,1976, in preparation.
Maslowski,J., 1973, Astron. Astrophys. <u>26</u>, 343.
Pearson,T.J., 1975, Mon. Not. Roy. Astron. Soc. <u>171</u>, 475.
Le Poole,R.S., 1977, in preparation.
Le Poole,R.S.,Willis,A.G., 1977, in preparation.
de Ruiter,H.R.,Willis,A.G.,Arp,H.C., 1976, Astron. Astrophys. Suppl. in press.

Valentijn,E.A.,Perola,G.C.,Jaffe,W.J., 1976, Astron. Astrophys. Suppl. submitted.

Vliet,W. van,Harten,R.,Miley,G.K.,Albers,H., 1976, Astron. Astrophys. 47, 345.

Willis,A.G.,Oosterbaan,C.E.,de Ruiter,H.R., 1976, Astron. Astrophys. Suppl. in press.

DISCUSSION

Longair: If you omit the corrections from Westerbork 3 survey, do the numbers agree with all the other surveys?

Katgert: In view of the fact that a large percentage of the weak sources have angular sizes of the order of the beam, it would seem that a correction for resolution effects must be applied. The magnitude of the correction can be estimated only through model calculations, and it is possible that our use of (unequal) double sources has led to slight overestimation of the correction. Yet it would seem that such effects cannot really explain the higher source densities in the 3rd survey.

Pearson: Concerning the difference between the Westerbork 3 survey and the 5C surveys at 1400 MHz. The 5C source counts have not been corrected for resolution. Although these corrections are necessary, they are not significant above 10 mJy.

Katgert: It is true that corrections are important only near the flux density limit, but their effect on the counts is present also at higher flux densities, due to the effect of the envelope attenuation.

Jauncey: To increase the significance of the Westerbork results from the present 2 to 3 sigma to a more believable 5 or 6 sigma is going to require 4 times as much data from Westerbork.

Petrosian: At what flux level would you expect the normal galaxies to dominate the counts?

Willis: I expect the normal galaxies at about 1 mJy.

Longair: No. You only expect normal galaxies at about 0.1 mJy and then they will be at cosmological distances anyway.

Lynden-Bell: Since Westerbork survey 1 has too many Galaxy identifications, has there been a Galaxy count over this region to see whether there are too many optical Galaxies?

Katgert-Merkelijn: No.

THE GREEN BANK SURVEYS AT 1400 MHZ

J. Maslowski
Max-Planck-Institut für Radioastronomie, Bonn, West Germany
and Astronomical Observatory, Jagellonian University,
Cracow, Poland

The present 1400 MHz complete pencil-beam survey, made with the NRAO 300-foot telescope*, consists of four parts which cover four different regions of the sky, as shown in Fig. 1.

The first survey (GA, unpublished), made by M.M. Davis in 1968, covers 0.45 sr in the zone defined by $23°5 \leq \delta \leq 30°5$, $b^{II} > 20°$. According to Fomalont et al. (1974), the survey is to be complete above 0.55 Jy and lists 185 sources with $0.55 \leq S \leq 2.0$ Jy.

The second survey (GB), made by Maslowski (1971, 1972) in 1970 covers 0.1586 sr in the zone defined by $7^h15^m \leq \alpha < 16^h20^m$, $45°9 \leq \delta \leq 51°8$, $b^{II} > 20°$. The catalogue lists 1086 sources with $S \geq 0.09$ Jy but the survey is to be essentially complete above 0.16 Jy. So there are 602 sources with $0.16 \leq S < 2.0$ Jy which can be used for

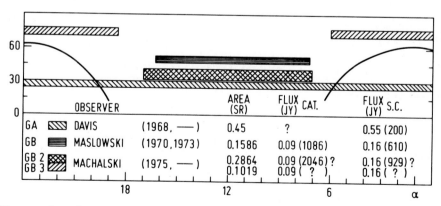

Figure 1. A map of the regions covered by the Green Bank surveys at 1400 MHz

* National Radio Astronomy Observatory is operated by Associated Universities Inc. under contract with the National Science Foundation

D. L. Jauncey (ed.), Radio Astronomy and Cosmology, 47-54. All Rights Reserved.
Copyright © 1977 by the IAU.

source counting purposes or rather 518 sources, if the appropriate corrections resulting from noise and confusion are taken into account (Maslowski, 1973).

The third survey (GB2), made by Machalski in 1975, covers 0.2864 sr in the zone defined by $7^h08^m.1 \leq \alpha \leq 16^h58^m.8$, $31°.8 \leq \delta \leq 39°.6$, $b^{II} > 20°$. The observational data are already reduced but the results have not been published yet. The numbers presented here, prior to publication (Machalski, 1976, personal information), may change slightly during the final stage of data analysis for several reasons. Therefore, the present results should be treated with some caution. The number of sources found in the survey is 2046 with $S \geq 0.09$ Jy, but the survey is to be essentially complete above 0.16 Jy. So there are 910 sources with $0.16 \leq S < 2.0$ Jy available for source counting purposes or about 782 sources, if the corrections similar to that used in case of the GB survey (Maslowski, 1973) are applied.

The fourth survey (GB3) has also been made by Machalski in 1975, but the observational data have not been reduced yet. The survey covers 0.102 sr in the zone defined by $18^h30^m \leq \alpha \leq 5^h58^m$, $70°.0 \leq \delta \leq 76°.8$.

1. SOURCE COUNTS

At present, the data from the first three Green Bank surveys, i.e. GA, GB and GB2, may be used either separately to derive a source count for a given region of the sky (the size ranges from 0.16 sr to 0.45 sr) from which the data were collected in a given flux density range, since each of the surveys contains a sufficiently large number of sources, or to amalgamate them simply in order to obtain an improved source count, since the surveys are statistically independent. The source counts mentioned above or any combination of them can be extended toward a high flux density region, using the improved data which have been assembled by Bridle et al. (1972), Bridle and Fomalont (1974) and Fomalont et al. (1974), and toward a low flux density range, using the data from the Westerbork synthesis surveys (Katgert et al., 1973; Katgert, 1975).

According to Fomalont et al. (1974) all the surveys, except GB2 (Machalski, 1975), were carefully checked and they are already on the same KPW flux density scale (Kellermann et al., 1969) after the correction of the Davis flux density scale. Concerning the Machalski scale, I believe that his survey is also on the KPW scale although, at least at present, some overestimation of the flux density of GB2 sources cannot be excluded.

The present situation for the source counts at 1400 MHz is shown in Fig. 2. All the source counts are presented in the differential form and normalised to the differential form arising from the integral count $N = 200 \cdot S^{-1.5}$, expected in the Euclidean Universe without any

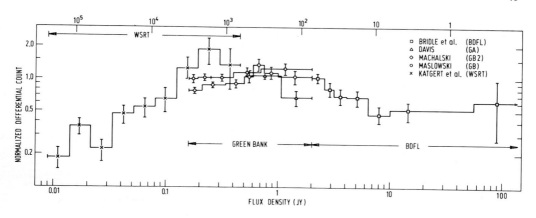

Figure 2.   The combined differential counts at 1400 MHz

kind of evolution.

In spite of the significant increase of the statistics in the intermediate flux density range due to Machalski's survey, the situation is not quite clear and may be summarized as follows.

In the flux density range 0.55 - 2.0 Jy, there are three quite independent surveys of three different regions of the sky (GA, GB and GB2) made with the same instrument, so that a comparison among them can be made, using the results of the differential counts given in Table 1. The χ^2 test applied to this contingency table shows that the three counts (slopes) differ significantly at the level of 1% (χ^2 = 13.47 with 4 degrees of freedom). The source densities (last column in Table 1) in all the surveys in the range 0.55 to 2.0 Jy are almost the same.

In the flux density range 0.50 to 2.0 Jy, we have now the two different surveys, GB and GB2, the count results of which can be compared. The differential counts from these surveys are given in Table 2. The χ^2 test applied to the data in Table 2 shows that the difference between the two counts (in the slopes) is statistically significant at

Table 1.   Comparison of the GA, GB and GB2 source counts

| Survey | Flux density range | | | Σ | Density/sr |
	0.555 - 0.704	0.705 - 1.062	1.063 - 1.994		$(0.55 \leq S < 2.0)$
GA	67	84	34	185	411 ± 30
GB	24	27	26	77	485 ± 55
GB2	63	43	38	144	502 ± 42

Table 2. Comparison of the GB and GB2 source counts

Survey	Flux density range			Σ	Density/sr
	0.50 - 0.62	0.63 - 0.89	0.90 - 2.00		$(0.50 \leq S < 2.0)$
GB	31	21	41	93	586 ± 61
GB2	48	67	47	162	566 ± 44

the level of 0.6% (χ^2 = 10.17 with 2 degrees of freedom). The source density in both the surveys is again the same in the range of 0.50 to 2.0 Jy.

So if the Machalski flux density scale does not differ signifi- cantly from the KPW scale, then the presented data of the three surveys with source density of about 500 sr^{-1} suggest that none of the three differential counts (in the slope) is to be representative. There is, of course, the matter of personal judgement whether the observed dif- ferences can be considered as being statistically significant or not, i.e. whether they are just statistical fluctuations. We would like to emphasize only that the observed differences between the three surveys (GA, GB and GB2) at a source density of $\sim$ 500 sr^{-1} are statistically more significant than the difference between the observed and the predicted count in the high flux density range (BDFL data). It is obvious that the careful check of the GB2 flux density scale is in- dispensable before any further and more detailed investigations of the GB2 source count can be made.

2. SPECTRAL INDEX DISTRIBUTIONS

The question of dependence of the mean spectral slope on flux density (α - S) is very important for both the cosmology and the source evolution itself. Significant evidence of the variation of the spec- tral index distribution with flux density for a complete sample of sources, selected at high frequencies (5 GHz and 2.7 GHz), has been found by Pauliny-Toth and Kellermann (1972a), Wall (1972), Pauliny-Toth et al. (1974), Fanti et al. (1974), Condon and Jauncey (1974) and Balonek et al. (1975). This variation is such that the mean spectral slope is considerably steeper for the weak sources than for the strong sources in a lower frequency range, i.e. between the survey frequency and a low frequency (178 MHz or 408 MHz). In the higher frequency range, i.e. between the survey frequency and a high frequency (10.7 GHz or 15 GHz), the variation of the spectral index distribution with flux density is not significant. The situation is much less clear for samples selected at low frequencies, i.e. below 1.4 GHz. Many inves- tigations have been made to detect the α - S dependence for sources selected either at 178 MHz or 408 MHz (Pauliny-Toth and Kellermann

1972b; Willson 1972) but no significant change in the spectral index distributions with decreasing S has been found. However, some results obtained from a study of the Molonglo catalogues MC2 and MC3 (Sutton et al. 1974) have shown that the increase in the mean spectral slope for weaker sources in the sample complete at 408 MHz is significant (Murdoch 1976) if comparison is made with an all-sky catalogue of strong sources (Robertson 1973).

As far as I am aware, no such investigations have been made for large samples of sources complete at 1400 MHz down to 0.2 Jy in the high frequency range. Some effect, namely the increase in the mean spectral slope for weaker sources, has been reported by Willson (1973) in the low frequency range, but his result must be viewed with some caution, because it is based on sources taken from the 5C surveys which suffer a serious, radially dependent systematic bias in the aperture synthesis flux density (Condon and Jauncey 1973).

At present both the GB and GB2 surveys provide a sufficiently large sample of sources for a study of this effect and very preliminary results are presented here for obvious reasons. All GB2 sources (Machalski, private information) found in the 1400 MHz survey to be stronger than 0.5 Jy were measured at 5 GHz using the NRAO 300-foot telescope to provide the two-point spectral indices, $\alpha(1400, 5000)$, (the spectral index, α, is defined by $S \propto \nu^{-\alpha}$ and the notation $\alpha(\nu_1, \nu_2)$ used here means α between ν_1 and ν_2 for sources selected at ν_1) in the high frequency range. Concerning the low frequency range, the Bologna 408 MHz flux densities from the B2 survey (Colla et al. 1970; Colla et al. 1972) are available for the GB2 sources, so that the two-point spectral indices $\alpha(1400, 408)$ could be determined.

The $\alpha - S$ dependences in the complete sample of GB2 sources with $0.5 \leq S < 2.0$ Jy were investigated by breaking the sample into two population subsamples in the two flux density intervals, separately in the low and high frequency ranges. The results of the investigations are summarized in Table 3 in which the following parameters, determined for each subsample, are given: the median and width of the spectral index distribution of the normal component (upper number), using a

Table 3. Spectral parameters of the sources ($S \geq 0.5$) in the GB2 survey

Flux at 1400 MHz	α_L (1400,408)			α_H (1400,5000)			Number of sources
	Median	Width	F	Median	Width	F	
0.70 - 2.00	$0.75 \pm .017$	.130		$0.89 \pm .019$	.145		
			$.20 \pm .044$			$.17 \pm .042$	81
	$0.74 \pm .028$	.250		$0.82 \pm .031$	.275		
0.50 - 0.69	$0.85 \pm .014$	.110		$0.86 \pm .019$	.150		
			$.21 \pm .045$			$.12 \pm .035$	85
	$0.80 \pm .028$	.260		$0.84 \pm .025$	.225		

method of passing mathematical "windows" of various widths $\Delta\alpha$ (Condon and Jauncey 1974), the median and width of the whole spectral index distribution (lower number), the fraction (F) of flat spectrum sources ($\alpha < 0.5$) and the number of sources in the subsample. Several features are readily apparent from Table 3. Firstly, there is a strong mean spectral curvature for the sources stronger than 0.7 Jy in the normal component. The sources with S < 0.7 Jy have, on the average, straight spectra over the range of 408 - 1400 - 5000 MHz. Secondly, there is a rapid change in the median $\alpha(1400, 408)$ of the normal component in the lower frequency range. The spectral steepening of the weaker sub-sample (S < 0.7 Jy) seems to be statistically significant, while in the higher frequency range there is no spectral difference between the strong (S > 0.7 Jy) and the weak (S < 0.7 Jy) sources in the normal components. Thirdly, it is to be noted that no significant differences exist among the subsamples if they are compared in the χ^2 test. More details of the α - S dependence will be given by Machalski in his forthcoming paper.

In the case of GB sources, the analysis of the α - S dependence has to be restricted to the higher frequency range, since the low frequency flux densities are not available for a majority of them. All sources found in the GB survey to be stronger than 0.2 Jy at 1400 MHz were observed at 2695 MHz using the NRAO 300-foot telescope (Maslowski, unpublished). These observations provided the two-point spectral indices $\alpha(1400, 2695)$ for the sample of GB sources. Several investigations of the distribution of $\alpha(1400, 2695)$ have been made to find, if any, the variation of the spectral index distribution with flux density. The search for the intensity dependence of the $\alpha(1400, 2695)$ spectral index distribution was performed by splitting the sample into four almost equal subsamples in four flux density intervals and determining, as previously, median and width for both the normal com-

Table 4. Spectral parameters of the GB sources.

Flux at 1400 MHz	α (1400, 2695)			Number of sources
	Median	Width	F	
0.50 - ∞	0.92 ± .022	.180		
			.21 ± .039	99
	0.89 ± .036	.360		
0.33 - 0.49	0.81 ± .017	.135		
			.20 ± .039	102
	0.79 ± .035	.350		
0.24 - 0.32	0.94 ± .016	.170		
			.11 ± .030	114
	0.89 ± .025	.270		
0.20 - 0.23	0.92 ± .020	.210		
			.20 ± .038	107
	0.85 ± .036	.375		

ponent and the whole spectral index distribution in each of the sub-samples. The results of the analysis are summarized in Table 4. It can be seen from the table that there is a rapid change in the median of the normal component which occurs over a very narrow range in flux density, namely from 0.33 Jy to 0.50 Jy. The median spectral slope of the normal component is considerably steeper for the sources stronger than 0.49 Jy and weaker than 0.33 Jy. When the subsamples are compared in the χ^2 test, they do not show significant differences.

I thank Dr. J. Machalski for permission to quote results prior to publication.

REFERENCES

Balonek, T.J., Broderick, J.J., Condon, J.J., Drawford, D.F. and Jauncey, D.L. 1975, Astrophys. J. 201, 20
Bridle, A.H., Davis, M.M., Fomalont, E.B. and Lequeux, J. 1972, Astron. J. 77, 405
Bridle, A.H., Davis, M.M., Fomalont, E.B. and Lequeux, J. 1972, Nature Phys. Sci. 235, 123
Bridle, A.H. and Fomalont, E.B. 1974, Astron. J. 79, 1000
Colla, G. et al. 1970, Astron. Astrophys. Suppl. 1, 281
Colla, G. et al. 1972, Astron. Astrophys. Suppl. 7, 1
Condon, J.J. and Jauncey, D.L. 1973, Astrophys. J. Letters 184, L33
Condon, J.J. and Jauncey, D.L. 1974, Astron. J. 79, 437
Davis, M.M. 1969, unpublished
Fanti, R., Ficarra, L., Formiggini, I., Gioia, I. and Pardrielli, L. 1974, Astron. Astrophys. 32. 155
Fomalont, E.B., Bridle, A.H. and Davis, M.M. 1974, Astron. Astrophys. 36, 273
Katgert, P. 1976, Astron. Astrophys. 38, 87
Katgert, P., Katgert-Merkelijn, J.K., LePoole, R.S. and van der Laan, H. 1973, Astron. Astrophys. 23, 171
Kellermann, K.I. 1972, Astron. J. 77, 531
Kellermann, K.I., Pauliny-Toth, I.I.K. and Williams, P.J.S. 1969, Astrophys. J. 157, 1
Machalski, J. 1976, unpublished
Maslowski, J. 1971, Astron. Astrophys. 14, 215
Maslowski, J. 1972, Acta Astron. 22, 197
Maslowski, J. 1973, Astron. Astrophys. 26, 343
Murdoch, H.S. 1976, preprint
Pauliny-Toth, I.I.K. and Kellermann, K.I. 1972a, Astron. J. 77, 560
Pauliny-Toth, I.I.K. and Kellermann, K.I. 1972b, Astron. J. 77, 797
Pauliny-Toth, I.I.K., Witzel, A. and Preuss, E. 1974, Astron. Astrophys. 35, 421
Robertson, J.G. 1973, Australian J. Phys. 26, 403
Sutton, J.M., Davies, I.M., Little, A.G. and Murdoch, H.S. 1974, Australian J. Phys. Astrophys. Suppl. No. 33
Wall, J.V. 1972, Australian J. Phys. Astrophys. Suppl. No. 24
Willson, M.A.G. 1972, Monthly Notices Roy. Astron. Soc. 156, 7

DISCUSSION

Davis: Do you still find a difference between the GB A and B surveys?
I don't feel there is a significant difference any longer?

Maslowski: It is significant at the 2.5% level after the GA flux
density corrections have been made.

THE AGGREGATE FLUX OF WEAK POINT SOURCES AT 1404 MHz
S.J. Goldstein, Jr., A.P. Marscher, R.T. Rood.
Astrophys. J. 210, 321 (1976).

THE PARKES 2700 MHz SURVEY: COUNTS OF THE SOURCES AND THEIR DISTRIBUTION ON THE SKY

J.V. Wall
Mullard Radio Astronomy Observatory,
Cavendish Laboratory,
Cambridge, England.

The Parkes 2700-MHz survey for extragalactic sources now extends over virtually the whole sky available to the 64-m telescope at galactic latitudes $|b| > 10^{\circ}$. Eleven parts of the catalogue have been published as Astrophysical Supplements of the Aust. J. Phys. (Numbers 19, 21, 26, 30, 32, 34, and 39); part 11 and subsequent parts consist of re-surveys of some areas to fainter flux density limits. The current version of the composite catalogue for the 7.36 sr covered (Fig. 1) contains about 7500 sources, together with the results of identification programmes and flux density measurements at other frequencies.

The survey was undertaken to examine the extent and nature of the population of "flat-spectrum" or "cm-excess" sources. The increase in representation of this type of source as the survey frequency is raised is clear from studies of sources detected both in this survey and in the NRAO 5000-MHz survey, and is well demonstrated in 'two spectral-index' diagrams (Wall 1975). The extent of the flat-spectrum population, which constitutes $\sim 40\%$ of sources found at 2700 MHz and $\sim 60\%$ at 5000 MHz, makes it evident that the trials of surveyors at low frequencies — the survey, spectral observations, positions/identifications/redshifts, anisotropy/association searches, source-count

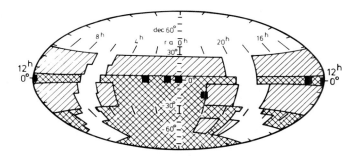

Fig. 1. The Parkes 2700 MHz survey. Hatched areas are complete to 160 sources/sr, cross-hatched areas to 500 – 900 sources/sr, and filled areas to 2500 sources/sr.

D. L. Jauncey (ed.), Radio Astronomy and Cosmology, 55-61. All Rights Reserved.
Copyright © 1977 by the IAU.

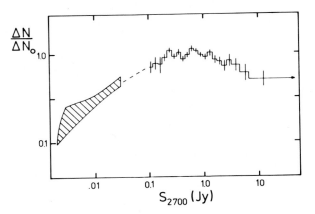

Fig. 2.   The 2700 MHz N(S) relation.

compilation and analysis – must be endured by surveyors at high
frequencies also. Now, however, the interpretation is harder because
at least two source-populations are involved, so that there is even
greater incentive to be certain of the observations and initial
premises before proceeding. This contribution is therefore confined
to the basic data, and considers the morphology of the source counts
and the uniformity of source distribution over the sky.

1. THE 2700-MHz N(S) RELATION

Fig. 2 shows the 2700-MHz source count in relative differential form,
extending down to $S_{2700} = 0.10$ Jy (2500 sources sr^{-1}). (Details of the
compilation will be presented elsewhere). The survey parameters are
such that the confusion bias at 2500 sources sr^{-1} is far smaller than
the statistical errors, so no such correction has been made.
The faint end of the N(S) relation is delineated by a P(D) (background

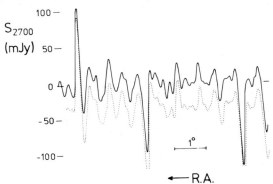

Fig. 3.   Two independent observations of the confusion limit along the
same r.a. scan, shown displaced.

deflection) analysis. The peculiarly-shaped area in Fig. 2 is from preliminary consideration of new data obtained at Parkes by D.J. Cooke and myself, and supersedes a previous determination (Wall and Cooke 1975) with which it is consistent. Some of the new data are shown in Fig. 3. These were obtained with the twin-beam system used for the survey, for which the receiver output is the difference between the signals received by the two beams. The independent observations of the same area show that the deflections are genuine and much larger than noise errors, and therefore the sampling of such scans yields a distribution of deflections due primarily to faint extragalactic sources; analysis of this sets limits on the N(S) relation to a source density as high as one per beam area, or 2×10^5 sr^{-1}, corresponding to $S_{2700} \sim 2$ mJy.

Qualitatively the N(S) relation of Fig. 2 is similar to the low-frequency relations, showing an initial rise indicative of an integral slope steeper than the $-3/2$ (Euclidean) value, followed by a turnover and a fall at the lowest flux densities. The 2700-MHz count is smooth, in that fluctuations do not exceed those expected from the statistics of limited numbers. In particular there are no regions of abrupt changes in slope, which have been observed at both lower and higher frequencies (Jauncey 1975), and which have suggested to some authors that the counts do not have cosmological relevance. Qualitatively there is an obvious difference between the 2700-MHz relation and those at lower frequencies in that the region of turnover is much broader at 2700 MHz. This effect, noted independently by Fomalont et al. (1974) and by Wall and Cooke (1975), is very apparent in Fig. 4, a collection

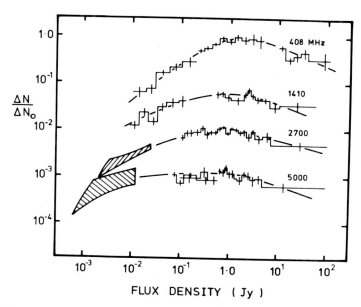

Fig. 4. N(S) relations at four frequencies; references to original data in Wall and Cooke (1975).

of N(S) relations at four frequencies. Included in this diagram is the
preliminary result of a recent P(D) analysis at 5000 MHz which provides
striking confirmation of the trend, and shows that the relation at this
frequency is close to Euclidean in slope over a range of three decades
in flux density.

These results imply that the flat-spectrum sources favoured by
high-frequency surveys have an N(S) relation which differs from that
for steep-spectrum sources. The effect is most marked at the faint
end of the relations. Nevertheless it can be demonstrated directly at
the bright end, using the 5000-MHz flux densities available for most
sources of the 2700-MHz survey with $S_{2700} > 0.35$ Jy. Division of the
sample at $\alpha_{2700}^{5000} = 0.5$ ($S \propto \nu^{-\alpha}$) yields the counts shown in Fig. 5.
The flat-spectrum sources give a flatter count in accordance with
Fig. 4, with the spectral-index – flux density dependence found by
Wall (1972) and by Condon and Jauncey (1974), and with the earliest
results obtained from the 2700 MHz survey (Shimmins et al. 1968).
The division at $\alpha_{2700}^{5000} = 0.5$ is of physical significance in that
virtually all flat-spectrum sources thus defined have structures
dominated by compact components, and in regions of low galactic
obscuration, J.G. Bolton and B.A. Peterson have shown that most of
these can be identified with QSOs.

Potentially the 2700-MHz count offers a powerful means of exploring
the evolution of the flat-spectrum population, but its validity for
cosmological exploration is clearly suspect if sources show serious
anisotropy in their distribution over the sky.

2. THE DISTRIBUTION OF THE SOURCES ON THE SKY

In the course of the survey we obtained subjective impressions of areas
disproportionately rich in flat-spectrum sources, and furthermore, we
found the curious result of Fig. 6, wherein the first 2 sr of the survey
produced both a significantly higher source density and a flatter count
than the remaining 5 sr. It is impossible to attribute this result to
instrumental effects, and it is important to note that the first 2 sr

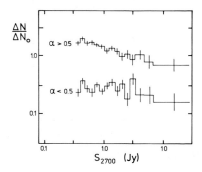

Fig. 5. Counts for steep-spectrum and flat-spectrum sources, displaced
for clarity.

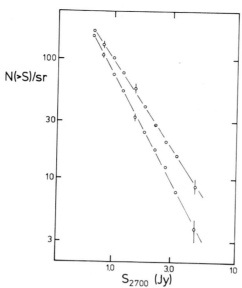

Fig. 6. Integral source counts for first 2 sr of survey (upper; maximum
-likelihood slope -1.48±.08) and remaining 5 sr (slope -1.91±.06).

and the remaining 5 sr are not contiguous areas, but are made up of
regions typically 0.3 sr in size and strewn all over the sky. Several
analyses of source distribution were undertaken to investigate these
effects.

On the largest possible scales, Yahil (1972) suggested that
differences in the source counts between galactic hemispheres were
present in several surveys, and results from the NRAO 5000-MHz survey
(Pauliny-Toth and Kellermann 1972) added weight to the suggestion. The
maximum-likelihood slopes of counts in the hemispheres from the full 7
sr of the 2700 MHz survey ($S_{2700} > 0.7$ Jy) are as follows, with the
numbers of sources in brackets:

		Whole sky	b^+	b^-
All sources	PKS	1.77±.07 (1111)	1.85±.10 (432)	1.70±.06 (679)
	NRAO	1.76±.11 (271)	2.18±.23 (107)	1.58±.14 (155)
$\alpha < 0.5$	PKS	1.65±.10 (346)	1.64±.15 (130)	1.66±.12 (216)
	NRAO	1.56±.14 (138)	1.78±.29 (49)	1.56±.19 (84)
$\alpha > 0.5$	PKS	1.80±.06 (765)	1.96±.12 (302)	1.72±.08 (463)
	NRAO	1.90±.06 (133)	2.41±.38 (58)	1.54±.24 (71)

The effects found in the NRAO survey are still present in the 2700-MHz
data, but the weight of numbers has reduced the statistical
significance to the level below which is is of interest.

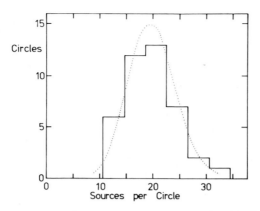

Fig. 7.   Sources with $S_{2700} > 0.7$ Jy in 42 independent small circles 10° in radius, ~ 0.1 sr in area.

Fig. 6 suggested anisotropies on the smaller scales of 0.1 to 1 sr. A simple binning analysis produced the result of Fig. 7; the numbers of sources in 42 independent small circles on the celestial sphere show no deviation from the Poissonian distribution expected for random distribution of the sources.

To examine clustering on all scales but the largest, the most powerful technique is that of power-spectrum analysis (PSA) as applied to the problem by Webster (1976). Several zones of the 2700 MHz survey were subjected to this test (Webster, in press), and no deviation from isotropy is apparent. Since this analysis, more survey data have become available, completing to $S_{2700} = 0.35$ Jy a region bounded by R.A. 22^{h} through to 05^{h}, dec. $+4^{\circ}$ south to -65°. Such a "square" is highly suitable for PSA. The results (in Webster's notation) appear in Fig. 8,

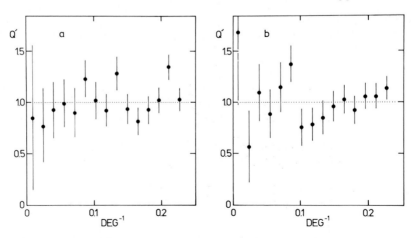

Fig. 8.   PSA for R.A. 22^{h} to 05^{h}, dec. -65° to $+4^{\circ}$, $S_{2700} > 0.35$ Jy. (a) all 836 sources, (b) the 225 flat-spectrum sources.

and provide no indication that the sources are distributed in any way other than uniformly, randomly, and independently on the sky.

3. CONCLUSIONS

The absence of positive results in the clustering/anisotropy searches suggest (but do not prove) that the simplest hypothesis is correct: the sources are indeed randomly distributed on the sky, and the result of Fig. 6 and the differences between galactic hemispheres represent statistical mishaps. This is a working hypothesis in that it admits conventional analysis of the 2700 MHz N(S) relation. This relation is now defined down to 2×10^5 sr^{-1}, is continuous, and shows the same general features as low-frequency N(S) relations. There is, however, the qualitative difference − the greatly extended region of turnover − which reflects the increasing influence of flat-spectrum sources with increasing survey frequency, and implies that luminosity functions and/ or evolution for these sources must differ from those of the steep-spectrum population.

*****

I thank John Bolton, Ann Savage, Alan Wright and Jenny Trett for data in advance of publication, and Adrian Webster for help with the PSA.

REFERENCES

Condon, J.J. and Jauncey, D.L.: 1974, Astron. J. 79, 437.
Fomalont, E.B., Bridle, A.H. and Davis, M.M.: 1974, Astron. Astrophys. 36, 273.
Jauncey, D.L.: 1975, Ann. Rev. Astron. Astrophys. 13, 23.
Pauliny-Toth, I.I.K. and Kellermann, K.I.: 1972, Astron. J. 77, 797.
Shimmins, A.J., Bolton, J.G. and Wall, J.V.: 1968, Nature 217, 818.
Wall, J.V.: 1972, Australian J. Phys. Astrophys. Suppl. No. 24.
Wall, J.V.: 1975, Observatory 95, 196.
Wall, J.V. and Cooke, D.J.: 1975, Monthly Notices Roy. Astron. Soc. 171, 9.
Webster, A.S.: 1976, Monthly Notices Roy. Astron. Soc. 175, 61.
Yahil, A.: 1972, Astrophys. J. 178, 45.

DISCUSSION

van der Laan: Have you done any nearest neighbour analysis or triple and other compact group tests?

Wall: No, but the power spectrum analysis encompasses such tests and is considerably more powerful, as Webster has shown.

Kellermann: Is the galactic hemisphere difference less significant in the 2700 MHz survey than in the 5000 MHz survey?

Wall: Yes, and we have tried various flux density cut-off levels.

SURVEYS OF RADIO SOURCES AT 5 GHZ

I.I.K. Pauliny-Toth

Max-Planck-Institut für Radioastronomie

Bonn, FRG

1. THE SURVEYS

A number of surveys have been carried out at a frequency of 5 GHz at the National Radio Astronomy Observatory (NRAO) and at the Max-Planck-Institut für Radioastronomie (MPIfR), with the aim of determining the number-flux density relation for the sources detected and also of obtaining their radio spectra and optical identifications. The surveys fall into two categories: first, the strong source (S) surveys, which are intended in due course to cover the whole northern sky and to be complete above a flux density of about 0.6 Jy; second, surveys of limited areas of sky down to lower levels of the flux density.

The regions of sky covered by the strong source surveys are shown in Figure 1. Three of these, denoted by S1, S2 and S3, have already

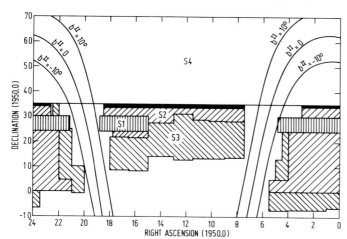

Figure 1. Areas of sky covered by the S surveys. The narrow strip shaded in black is the area of the "I" survey.

D. L. Jauncey (ed.), Radio Astronomy and Cosmology, 63-74. *All Rights Reserved.*
Copyright © 1977 by the IAU.

been reported (Kellermann et al. 1968, Pauliny-Toth et al. 1972, Pauliny-Toth and Kellermann 1972). All three were made with the NRAO 140-ft. telescope and consisted of a fast finding survey, followed by reobservation of the sources detected. In the case of the more recent S4 survey (Pauliny-Toth et al. 1977), the finding survey was made with the NRAO 300-ft. telescope and the sources were reobserved with the MPIfR 100-m antenna. This survey covers the sky between declinations of 35^0 and 70^0 and is somewhat more sensitive than the other S surveys.

Further data of comparable sensitivity for the southern sky are provided by the Parkes 2.7 GHz surveys, a number of which are complete above a flux density of 0.35 Jy at 2.7 GHz (Wall et al. 1971, Shimmins 1971, Shimmins and Bolton 1972b and 1974, Bolton and Shimmins 1973, Bolton et al. 1975, Wall et al. 1976). Flux densities at 5 GHz for the sources in these surveys have been reported either in the above references or by Shimmins and Bolton (1972a) and Shimmins et al. (1969). The distribution of the spectral indices of these sources, derived from the data at 2.7 and 5 GHz shows that a sample of sources having a flux density S(2.7 GHz)>0.35 Jy would be virtually complete at 5 GHz above 0.6 Jy. This sample, taken from the Parkes survey, covers a solid angle of 2.6 sr when regions of overlap with the S surveys, areas within the Magellanic clouds, or areas within 10^0 of the galactic plane are excluded.

Table I gives the solid angle of sky covered by each strong source survey, as well as the number of sources found above the given limit of completeness, both for the whole survey, and for the areas in the south and north galactic hemispheres having galactic latitudes greater than 10^0. In the Table, the S3 survey, which overlapped part of the more sensitive S4 survey, has been correspondingly reduced in area, while sources from the "I" survey (see below) having flux densities greater than 0.5 Jy have been used to fill the gap between the S4 and the other S surveys.

Table I
5 GHz strong source surveys.

Survey	Solid angle covered / Number of sources			Completeness Limit (Jy)
	Total	$b^{II}{<}{-}10^0$	$b^{II}{>}{+}10^0$	
S1	0.273/ 23	0.185/ 14	0.088/ 9	0.8
S2	0.973/136	0.760/100	0.213/.36	0.6
S3	1.059/123	0.399/ 51	0.660/ 72	0.6
S4	1.713 208	0.296 42	1.416 166	0.5[+]
I	0.074	0.015	0.059	
PKS	2.604/302	1.922/217	0.682/ 85	0.6

[+]Numbers of sources are given for S > 0.6 Jy. The total number above 0.5 Jy is 269.

The complete sample of sources stronger than 0.6 Jy thus contains 769 sources and covers 6.48 sr, that is, more than half of the sky at galactic latitude greater than 10^0.

The surveys of limited areas of sky, complete down to lower flux densities, consist of:

a) an "intermediate source" (I) survey, complete down to 0.25 Jy and covering 0.079 sr (Pauliny-Toth et al. 1972),

b) a deep (D) survey, made with the 300-ft. telescope, and covering 3.77×10^{-2} sr down to a limiting flux density of 0.067 Jy (Davis 1971), and

c) two deeper surveys made with the MPIfR 100-m telescope. The first of these extends over an area of 4.53×10^{-3} sr near the North Celestial Pole (Pauliny-Toth, Witzel and Baldwin 1977) and is believed to be complete to 0.020 Jy; the second covers ten small regions near declination 30^{0} and is complete to 0.015 Jy over an area of 2.91×10^{-3} sr (Pauliny-Toth, Witzel and Preuß 1977). These surveys contain 59 and 58 sources above their respective limiting flux densities.

For all these surveys, measurements of the flux densities of the sources at other frequencies have been, or are being made. Virtually all the sources in the S surveys have been observed at 2.7 GHz, either with the 300-ft. telescope (Davis, unpublished data) or with the 100-m telescope (Pauliny-Toth et al. 1977), and measurements of most of the sources at 10.7 GHz are available (Pauliny-Toth et al. 1972, and 1977). Sources from the D survey have been measured at both these frequencies (Pauliny-Toth et al. 1974), and observations of sources from the MPIfR deep surveys are in progress (Davis 1977).

Optical identifications for sources in the first three S surveys have been reported with the survey results, and also by Johnson (1974) for the S3 survey. For sources in the S4 survey, accurate optical positions for objects in the fields of those sources which had not previously been identified have been measured (Kühr 1977) and identifications have been suggested. These new identifications are based on accurate radio positions measured by Fomalont with the NRAO three-element interferometer for about one-third of the sources, and on positions measured with the 100-m telescope with an accuracy of about 3.5 arc sec for a further third, the rest having been previously identified. Suggested optical identifications are thus available for almost all sources in the complete S survey sample.

2. THE SOURCE COUNTS

The composite source counts at 5 GHz are shown in Figure 2 in the form of a differential plot, normalised to a uniform Euclidean model having an integral count $N_0 = 60 \, S^{-1.5}$. The flux density intervals within each survey have been chosen so that this model should give approximately the same number of sources in each interval.

For the stronger sources, having S (5 GHz) > 0.6 Jy, a number-flux relation of the form $N = K \, S^x$ has been fitted to the ungrouped data, using the method of maximum likelihood (Crawford et al. 1970),

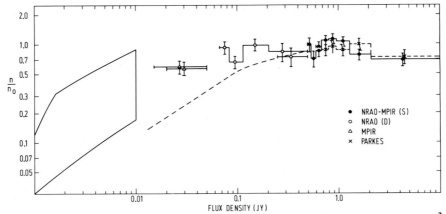

Figure 2. Differential counts at 5 GHz, normalised to $N_o=60\ S^{-1.5}$.

with the results shown in Table II.

Table II
Number-flux relation for S(5GHz)>0.6 Jy

Sample	K	x	Number of sources
All data	50.9±1.6	-1.66±0.06	769
S surveys	51.0±2.4	-1.71±0.08	467
Parkes	52.0±2.5	-1.57±0.09	302

The difference between the slope of the counts for the stronger sources and the Euclidean value is formally significant at a level of about 2.5 σ. It must be noted, however, that this difference is about the same as that between the slopes for the S surveys and the Parkes sample, and that an even larger difference exists between the slopes of the source counts in the two galactic hemispheres (Section 5). The real significance of the steep slope found for the stronger sources at 5 GHz is thus far from clear. If it is due to evolution of the sources, (e.g. Longair 1974), then the evolution differs for different directions in the sky. If, on the other hand, it is caused by a local "hole" in their distribution (e.g. Pauliny-Toth and Kellermann 1974) then the cosmological interpretation for the redshifts of a large fraction of the sources must be abandoned.

In the intermediate range of flux densities (0.1 to 1.5 Jy), the slope of the number counts is close to -1.5. Below this range, the deep survey data indicate that it flattens to about -1.3. The further convergence of the counts is supported by the P(D) analysis of Wall and Cooke (1975) whose limits on the number-flux relation between 0.01 and 0.001 Jy are indicated by the area outlined in Figure 2.

The observed convergence of the source counts at 5 GHz is, however, not consistent with the prediction of the evolutionary model of Fanaroff and Longair (1973), shown as a dashed curve in Figure 2. The model, which is based on the luminosity function and sepctral index distribution at 178 MHz, predicts counts at 5 GHz for the weak sources which are consistently lower than those observed, the discrepancy reaching a factor of about 3 at the limit of the deep surveys. This discrepancy has also been noted by Davis and Taubes (1974), on the basis of their P(D) analysis of data from the D survey. The model of Fanaroff and Longair predicts that the fraction of sources having flat or inverted spectra between 5 GHz and 178 MHz should decrease with S (5GHz), except for the strongest sources. Such a decrease is in fact observed (Pauliny-Toth and Kellermann 1972, Condon and Jauncey 1974), down to S(5GHz) in the range 0.25 to 0.067 Jy. As Davis and Taubes have suggested, a possible explanation for the discrepancy in the source counts is that the predicted decrease in the flat-spectrum population does not continue with a further decrease in the flux density. A direct test of this suggestion will be provided by the measurements of the deep survey sources at other frequencies.

3. IDENTIFICATION CONTENT

The statistics of the identifications for the complete S survey sample, excluding a few optically crowded fields and two sources identified with planetary nebulae, are given in Table III for sources identified with quasistellar objects (QSS), radio galaxies (GAL) or lying in empty fields (EF).

Table III
Identification content of the S surveys

Flux density range (Jy	GAL		QSS		EF	
	N	%	N	%	N	%
0.6 to 0.8	57	34	60	36	51	30
0.8 to 1.3+	62	35	77	44	38	21
> 1.3+	50	44	48	43	15	13

+Includes sources from S1 survey.

The identification content of the strong source sample is very different from that found for metre-wavelength surveys: a much larger fraction of the sources are identified with QSS. This difference is related to the spectral-index distribution (Section 4) which shows that a larger fraction of the sources in high-frequency surveys have flat or inverted spectra and that most of these sources are QSS.

The number-flux relation for the sources identified with quasi-stellar objects or radio galaxies has, in each case, a slope close to -1.5. whereas that of sources lying in empty fields has the much steeper slope of about -2.2. The significance of this result depends,

of course, on the nature of the objects associated with empty fields:
whether these are a class of subluminous objects, or whether they are
galaxies or quasistellar objects beyond the limit of the Palomar Sky
Survey. The spectral index data (Section 4) indicate that the empty
field sources may be a mixture of the two latter classes in roughly
equal numbers, in which case the number-flux relation for both QSS
and radio galaxies has a slope similar to the overall slope found for
the S surveys. This would be surprising, in view of the different
volumes of space over which the two classes of objects are presumably
distributed, and deep optical searches for objects in the empty fields
are desirable.

4. THE SPECTRAL INDEX DISTRIBUTION

The distribution of the two-point spectral index, α, defined by
$\alpha = \log | S(\nu_2)/S(\nu_1)|/\log(\nu_2/\nu_1)$ and obtained from the flux densities
measured at 2.7 and 5 GHz, is shown in Figure 3 for the S survey sample.

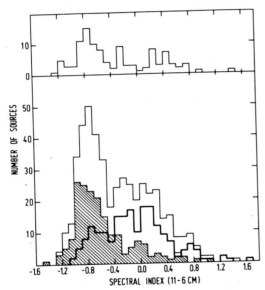

Figure 3. Spectral index distribution for the S survey sources. Top,
EF, bottom, all sources, QSS (heavy line), GAL (shaded).

The separate distributions for radio galaxies, quasistellar sources
and empty fields are also shown. The distribution is double-peaked,
the narrow peak near $\alpha = -0.8$ being due largely to the radio galaxies
and the broader peak near $\alpha = 0$ to the QSS. The median index, $\bar{\alpha}$, and
the fraction of sources which have $\alpha > -0.5$ are given in Table IV
for the whole sample, for the different identification classes and
for several ranges of the flux density.

Table IV
Spectral index distribution for the S surveys

Class	$\bar{\alpha}^+$	f	Number
All sources	-0.45±0.05	0.53±0.03	467
QSS	-0.11±0.04	0.75±0.06	187
GAL	-0.71±0.03	0.29±0.04	169
EF	-0.49±0.06	0.50±0.07	104
0.68>S>0.60	-0.56±0.14	0.48±0.08	73
0.80>S>0.68	-0.46±0.06	0.52±0.07	99
0.96>S>0.80	-0.55±0.10	0.47±0.07	95
1.30>S>0.96	-0.29±0.13	0.57±0.07	102
2.10>S>1.30	-0.31±0.08	0.59±0.09	68
S>2.10	-0.50±0.08	0.50±0.09	56

+The error in $\bar{\alpha}$ is the mean of the ranges on each side of the
median which contain $\sqrt{N}/2$ sources.

The QSS and radio galaxies have very different distributions: 75
percent of the former, but only 29 percent of the latter have the flat
or inverted spectra characteristic of compact, opaque radio sources.
The empty fields show an intermediate distribution which can be closely
reproduced by a mixture of radio galaxies and quasistellar sources in
a ratio of 5:4. The situation is different from that found for low-
frequency surveys, in which the empty field sources have spectra
similar to those of radio galaxies (Bolton 1966, Pauliny-Toth and Kellermann 1968).

There is no obvious dependence of $\bar{\alpha}$ on the flux density, even when
the spectral index data (Pauliny-Toth et al. 1974) for the D survey
sources are included, and consistent with this, there is no difference
in the slope of the number counts for sources with steep or flat
spectra. By contrast, the spectral index between 5 GHz and 408 or 318
MHz does show a dependence on flux density (Pauliny-Toth and Kellermann
1972, Condon and Jauncey 1974), in the sense that weaker sources tend
to have steeper spectra over this range of frequency. Such a dependence
is predicted by the model of Fanaroff and Longair (1973),or, more
simply, from the number-flux relation and spectral index distribution
at longer wavelengths (Pauliny-Toth and Kellermann 1972), but it
probably does not continue below S(5 GHz) $\sim$ 0.1 Jy (Section 2).

5. ANISOTROPY

The difference in the slope of the number-flux relation for the
stronger sources in the S surveys on the one hand, and the Parkes
surveys on the other, has been pointed out in Section 2. This difference is even larger when the counts in the northern and southern
galactic hemispheres are considered, as Table V shows.

Table V
Number flux relation for sources having S(5GHz)>0.6 Jy
in the two galactic hemispheres

Sample	$b^{II} < -10^{o}$			$b^{II} > +10^{o}$		
	K	x	N/sr	K	x	N/sr
All data	55.1±2.2	-1.54±.08	121±6	47.2±2.3	-1.80±.10	119±6
S surveys	60.7±3.5	-1.51±.11	131±10	46.0±2.9	-1.83±.12	117±7
Parkes	50.9±2.9	-1.56±.11	113±8	56.2±5.0	-1.56±.17	125±13

The difference between the values of the slope x in the two hemispheres for the whole sample is significant at the 2σ level, and is
due to the S surveys alone: the Parkes surveys give a slope which is
close to the Euclidean value in both hemispheres. When the S surveys
are considered separately, the anisotropy is present in each of them,
albeit with reduced significance, so that one may conclude, first,
that the effect is real, and second, that the angular scale of the
anisotropy is large.

The anisotropy appears to be present to some extent for all the
identification classes, but is most marked for the radio galaxies: of
the sources having S(5 GHz)>1.3 Jy, the number identified with radio
galaxies is 20.5±3.5 sr^{-1} and 8.6±1.9 sr^{-1} in the south and north
galactic hemispheres, respectively. The observed anisotropy in the
source counts is therefore at least partly due to a relative deficiency of bright radio galaxies in the north galactic hemisphere (within
the S survey regions). There is no significant difference in the
spectral index distribution between the two hemispheres, for the sample
as a whole. There is some indication that the spectral index distributions for the S surveys differ: the median indices are -0.55±0.05,
-0.30±0.10 and -0.40±0.08 for the S2, S3 and S4 surveys respectively,
and the S2 survey does not show the double-peaked distribution of α
seen for the other two surveys. Furthermore, the low-frequency spectral
index, α(5 GHz-408 MHz), shows anisotropy, in that the slope of the
number-flux relation for sources having α(5 GHz-408 MHz)>-0.5 is
steeper in the north galactic hemisphere than in the south. This last
effect may be connected with the previously mentioned deficiency of
bright galaxies in the north.

For the weakest sources in the survey, there is no evidence at
present for any anisotropy: the number counts from the two deep MPIfR
surveys agree almost exactly, and the numbers of sources in the
separate regions covered by the second of these surveys are consistent
with a random distribution.

6. SUMMARY

The 5 GHz source counts show the following features:
(i) a slope of -1.5 over a wide range of source density (20 sr^{-1} to

2×10^3 sr^{-1}),
(ii) a steeper slope, of -1.66 for the stronger sources,
(iii) a flattening of the slope to about -1.3 for source densities greater than 2×10^3 sr^{-1},
(iv) an excess of faint sources over the numbers given by the model of Fanaroff and Longair (1973),
(v) a marked anisotropy in the counts for the stronger sources, the slope being significantly steeper than -1.5 only in the area having $b^{II} > 10°$ and Declination >10°, and being partly due to a deficiency of bright galaxies in this region,
(vi) anisotropy in the spectral index distribution of the sources.

The anisotropy observed for the strong sources, the persistence of a slope close to the Euclidean value, and the source counts for the faint sources suggest that the present evolutionary models are not adequate to explain the high-frequency source counts.

7. ACKNOWLEDGEMENT

I wish to thank Drs. M. Davis, E. Fomalont, K. Kellermann, E. Preuss and A. Witzel, as well as Mr. H. Kühr, who have all contributed to the measurements on which this paper is based.

REFERENCES

Bolton, J.G. 1966, Nature 211, 917.
Bolton, J.G. and Shimmins, A.J. 1973, Australian J. Phys. Astrophys. Suppl. 30, 1.
Bolton, J.G., Shimmins, A.J., Wall, J.V. and Butler, P.W. 1975, ibid. 34, 33.
Condon, J.J. and Jauncey, D.L. 1974, Astron. J. 79, 1220.
Crawford, D.E., Jauncey, D.L. and Murdoch, H.S. 1970, Astrophys. J. 162, 405.
Davis, M.M. 1971, Astron. J. 76, 980.
Davis, M.M. and Taubes, C. 1974, Bull. American Astron. Soc. 6, 485.
Fanaroff, B.L. and Longair, M.S. 1973, Monthly Not. Roy. Astron. Soc. 161, 393.
Johnson, K.H. 1974, Astron. J. 79, 1006.
Kellermann, K.I., Pauliny-Toth, I.I.K. and Davis, M.M. 1968, Astrophys. Letters 2, 105.
Kühr, H. 1977, in preparation.
Longair, M.S. 1974, IAU Symposium No. 63, "Confrontation of Cosmological Theories with Observational Data", p. 93, publ. D. Reidel, Holland, ed. M.S. Longair.
Pauliny-Toth, I.I.K. and Kellermann, K.I. 1968, Astron.J. 73, 953.
Pauliny-Toth, I.I.K. and Kellermann, K.I. 1972, ibid. 77, 797.
Pauliny-Toth, I.I.K. and Kellermann, K.I. 1974, IAU Symposium No. 63, "Confrontation of Cosmological Theories with Observational Data", p.111, publ. D. Reidel, Holland, ed. M.S. Longair.

Pauliny-Toth, I.I.K., Kellermann, K.I., Davis, M.M., Fomalont, E.B.,
    and Shaffer, D.B. 1972, Astron.J. 77, 265.
Pauliny-Toth, I.I.K., Preuss, E., Witzel, A., Kühr, H., Davis, M.M.,
    Fomalont, E.B. and Kellermann, K.I. 1977, in preparation.
Pauliny-Toth, I.I.K., Witzel, A. and Baldwin, J.E. 1977, in prepara-
    tion.
Pauliny-Toth, I.I.K., Witzel, A. and Preuss, E. 1974, Astron. Astro-
    phys. 35, 421.
Pauliny-Toth, I.I.K., Witzel, A. and Preuss, E. 1977, in preparation.
Shimmins, A.J. 1971, Australian J. Phys. Astrophys. Suppl. 21, 1.
Shimmins, A.J. and Bolton, J.G. 1972a, ibid. 23, 1.
Shimmins, A.J. and Bolton, J.G. 1972b, ibid. 26, 1.
Shimmins, A.J. and Bolton, J.G. 1974, ibid. 32, 1.
Shimmins, A.J., Manchester, R.N. and Harris, B.J. 1969, ibid. 8, 1.
Wall, J.V. and Cooke, D.J. 1975, Monthly Not. Roy. Astron. Soc. 171,9.
Wall, J.V., Shimmins, A.J. and Merkelijn, J.K. 1971, Australian J.
    Phys. Astrophys. Suppl. 19,1.
Wall, J.V., Bolton, J.G., Wright, A.E., Savage, A. and Vander Hagen, J.
    1976, Australian J. Phys. Astrophys. Suppl. 39, 1.

DISCUSSION

P. Véron: You have said that in one of the instalments of the 5 GHz
survey, there is a deficiency of flat spectrum sources. Could you not
say just as well that there is an excess of steep spectrum sources?

Witzel: We feel that there is rather a deficiency of flat spectra
sources in S2, since in contrast to the double peaked distribution in
S3 and S4 there is only one peak (at the steep spectra side) with a
tail on the flat spectra side visible in S2.

Bolton: The difference between the 6 cm data and my own probably arises
from the fact that most sources with their flux maxima at cm wavelengths
have their maxima at wavelengths near 6 cm. Thus a positive spectral
index source between 11 and 6 cm becomes a negative spectral index
source between 6 cm and 3 cm. Thus the clear discrimination between
QSO's and Galaxies at 11 cm becomes somewhat obscured at 6 cm if the
spectra are based on the shorter wavelength data only.

THE EFFECT OF SOURCE VARIABILITY ON SAMPLE STATISTICS

G.D. Nicolson

At centimetre wavelengths a large fraction of radio sources have
flat radio spectra. Because these sources also exhibit large flux
density variations the statistical properties of complete samples of
radio sources will fluctuate with time. This effect has been investigated by constructing samples which are instantaneously complete above
2.4 Jy at 2.3 GHz by combining surveys at several epochs with variability data measured for individual sources.

The results show that between 1969.5 and 1975.5 sources with $b \leqslant -10°$ had a constant source count slope of 1.95 ± 0.26 while the slope for sources with $b \geqslant 10°$ varied between 1.49 ± 0.23 and 1.68 ± 0.24. This was caused by fluctuations in the number of sources with flat radio spectra at different times and in particular for the region $b \geqslant 10°$, $\alpha \geqslant 13^h$ the fraction changed from $29\% \pm 9\%$ to $9\% \pm 6\%$ over five years compared with a time averaged fraction of $26\% \pm 4\%$ for the entire sample. The statistical significance of the apparent anisotropies therefore changes with time and the reality of these anisotropies should be viewed with caution.

Mills: I believe that the question of variability is quite irrelevant. It merely gives different samples with the same statistics. Averaging samples taken at different times cannot help to beat the uncertainty.

Wall: Nelson Schuch (MRAO, Cambridge) has done a numerical experiment to examine the effects of intensity variations on the form of the high flux density and of Source counts at high frequencies. He has made the extreme assumption that the variability index distribution derived at Algonquin Radio Observatory is applicable to _all_ sources in high frequency surveys, and in a random number experiment has derived "source counts" at 200 different epochs, starting from a) invented samples with integral slopes of -1.5 and -1.8, and b) the NRAO 5 GHz survey list. In all cases the different epoch slopes lie within the $\pm$ 1 σ statistical errors obtained from maximum fitting to the original sample. Is this consistent with your observations?

Nicolson: Yes.

COSMOLOGICAL IMPLICATIONS OF VARIABILITY

Jeffrey D. Scargle

Two cosmological effects have been investigated using the systematic survey of radio source variability contained in the Algonquin Radio Observatory data. In any expanding cosmology, or any cosmology in which the redshift is Doppler or gravitational in origin, the time-scale for variability should vary as $1+z$. Although there is much scatter in the data, the effect has probably been detected, and the "tired light cosmology" can be ruled out at the 96% confidence level - unless evolutionary effects or observational selection have produced the observed relation

$$T = T_o (1+z)^{0.82 \pm 0.38}$$

The second cosmological effect is that some quantity derived from the luminosity variations of QSO's may possibly serve as a more reliable standard candle than does the luminosity at a randomly selected time, which has previously been used for radio-magnitude/redshift tests. Various such quantities do show reduced scatter in the magnitude/

redshift plane, but possible selection effects, and as shown by
Petrosian, the luminosity function and its possible evolution, prevent
useful cosmological information from being derived from any such
relation.   The detection of the wavelength dependence of the arrival
time of variations, for most radio sources studied, lends some weight
to the expanding plasma cloud model for the variations.

THE STATISTICAL ANALYSIS OF ANISOTROPIES

ADRIAN WEBSTER
Institute of Astronomy, University of Cambridge.

One of the many uses to which a radio survey may be put is an analysis of the distribution of the radio sources on the celestial sphere to find out whether they are bunched into clusters or lie in preferred regions of space. There are many methods of testing for clustering in point processes and since they are not all equally good this contribution is presented as a brief guide to what seem to be the best of them. The radio sources certainly do not show very strong clustering and may well be entirely unclustered so if a statistical method is to be useful it must be both powerful and flexible. A statistic is powerful in this context if it can efficiently distinguish a weakly clustered distribution of sources from an unclustered one, and it is flexible if it can be applied in a way which avoids mistaking defects in the survey for true peculiarities in the distribution of sources.

An ideal survey for statistical analysis is one made with infinitessimal telescope beam area, infinite receiver signal to noise ratio and absolutely constant gain sensitivity across the surveyed region. Measured against this standard every real survey is defective and every catalogue of sources is inaccurate to a greater or lesser degree. First, the finite beam area causes sources to be blended together or 'confused', the principal effect of which is to mistake a close pair of sources for one source: the catalogue thus exhibits an artificial absence of close pairs of sources. Second, the chief effect of noise is the reduction of sensitivity to sources at low galactic latitudes caused by the galactic background radiation. Third, the variation of gain can in principle give a variety of effects but in practice one effect predominates. Most surveys are drift-scan surveys in which the Earth's rotation sweeps a fixed beam across different declination strips each day: variations in receiver calibration from day to day then cause the sensitivity to depend on declination but not on right ascension and it is as a result prudent to be prepared for artificial variations of source density in the direction of increasing declination. Some statistics are better than others in discriminating instrumental effects from celestial effects,

D. L. Jauncey (ed.), Radio Astronomy and Cosmology, 75-81. All Rights Reserved.
Copyright © 1977 by the IAU.

as will be indicated below.

It is convenient to divide clustering statistics into two classes and to discuss the classes separately:

NUMBER DENSITY STATISTICS

In this class the measured flux density of a source is only used to decide whether the source is bright enough to be included in the analysis; all sources which are included are treated equally, with no further reference to their flux densities.   In this class two statistical methods stand out as being the best available.

In binning analysis the surveyed area is divided, somewhat arbitrarily, into a number of disjoint 'bins', and the number of sources in each bin is counted.   These numbers are tested by a straightforward application of chi-square (e.g. de Vaucouleurs 1971) or perhaps by a more oblique method such as statistical reduction (Zieba 1975) to find out whether they are consistent with the distribution of sources being a realisation of a Poisson process. Binning analysis is the best method for the very largest scales, such as are met in testing for differences between the north and south galactic hemispheres, because of its simplicity and because it copes with arbitrarily shaped bins.   The problem of galactic bachground noise can be met by excluding areas near the galactic plane, and the drift scan effect by choosing bins which have the same shape and size and differ only in right ascension.

Power spectrum analysis consists of defining a spiky function over the surveyed area by erecting a delta function at the position of each source, representing this function by a Fourier series and employing the squares of the values of the coefficients in this series as statistics (Bartlett 1964, Webster 1976a).   A spherical harmonic series may in principle be employed instead of the Fourier series (Peebles 1973) but in practice the extra computing involved makes this method less attractive.   Power spectrum analysis is powerful and flexible because each wave in the Fourier series contributes information which is practically independent of the information from every other wave, so a large number of waves may be investigated in order to maximise the statistical power, and if the coordinate system is carefully chosen any instrumental effect such as the drift scan effect only contaminates a very few waves which can be discarded from further consideration without significantly weakening the test.   The test thus beats binning analysis on all but the largest scales (even if many different binning configurations are tried in order to increase the power analogously to trying many waves) because it is not clear how independent the results of the different binning configurations are, and because most of the configurations are not immune from any given instrumental effect.

The family of neighbour-statistics, and in particular the method of nearest neighbour analysis, is well known but well worth avoiding. Compared with power spectrum analysis this method is weak, inflexible and full of pitfalls in its application (e.g. Webster 1976b).

LOG N / LOG S STATISTICS

In this class the slope of the log N / log S relation in one bin of the surveyed area is compared with that in other bins to find out whether the balance of bright sources and faint sources varies with direction. The log N / log S relation for the whole survey can usually be represented quite accurately by a straight line power law fit, so the statistics chosen are usually estimators of the slope of the power law which best fits the sources in each bin. The most powerful statistic therefore is the maximum likelihood estimator of the slope (Crawford et al. 1970) because of a general theorem (Kendall & Stuart 1967) that the ML estimator of a population parameter has a smaller sampling variance than any other estimator of that parameter. Almost as good are the least-squares and 'luminosity-volume' estimators (Pearson 1974) but the most obvious method of comparing the ratio of bright sources to faint in one bin with the ratio in other bins is weak because much of the information in the measured flux densities is wasted.

All of these tests are variants of the method of binning analysis and therefore suffer from the relative inflexibility and weakness of binning analysis mentioned above when comparing it with power spectrum analysis. A variant of power spectrum analysis called cross spectrum analysis (Peebles 1974) retains the power and flexibility of power spectrum analysis but seems never to have been applied to our problem.

Moving on now to mention the results of clustering analyses carried out to date, it seems to me that there is precious little good evidence in favour of significant clustering of the radio sources. Many investigators have indeed reported that they were unable to distinguish the actual distribution of sources in various catalogues from random distributions; my own power spectrum analyses of the 4C, GB, MC1, PKS 2700 MHz, B2 and 5C5 catalogues have led me too to this conclusion. Of the reports of significant clustering:
i) a few have been shown to be due to unanticipated instrumental problems or errors of analysis;
ii) some have not been supported by the results of comparable surveys of the same areas;
iii) many are analyses of surveys which have not been exhaustively shown to be of sufficiently high quality to put the possibility of instrumental error beyond doubt; and
iv) none has produced a result with a statistical significance of more than a few standard deviations anyway so the clustering has never been demonstrated beyond reasonable doubt.

It thus seems to me that there is no good evidence that the radio sources are distributed on the celestial sphere in any fashion other than uniformly, independently at random.   This lack of structure is of considerable fundamental significance quite apart from its bearing on whether the measured log N / log S curves are representative of the radio source population in the Universe as a whole.   In the first place it is direct evidence for the assumptions of isotropy and homogeneity of the Universe on large scales which underly the Friedmann cosmological models and the Robertson-Walker line element. For example, the power spectrum analysis of the Bologna B2 survey shows that the number of radio sources (and presumably also the density of matter) within a cube of side 1 Gpc or larger varies by less than about 3% as the cube is moved from place to place.   This information on the large scale homogeneity is better than that which can be had from the observed isotropy of the microwave background radiation for several reasons.   First it is more secure because the background radiation may have been scattered by free electrons after the epoch of recombination and this scattering, depending on circumstances, may make the surface brightness of the sky more or less patchy than it was at recombination.   Second, the density contrast of the large scale irregularities is expected to grow with time so an upper limit at a late epoch ($z \sim 1 - 3$ for the radio sources) is more valuable than one at an early epoch ($z \sim 1,000$ for the microwave background).   The homogeneity revealed by the radio sources confirms a point first made in connexion with the isotropy of the microwave background radiation: the Universe is more homogeneous than it has any known reason to be, in that the density of radio sources in widely separated regions is constant despite the fact that the radio sources formed before a light signal had time to travel from one region to the other.   Furthermore the lack of clustering is inconsistent with the local hypothesis for quasars if the quasars are expected to show the clustering and superclustering shown by galaxies in the same region of space.   Finally any model of radio sources in which the sources originate in pairs or higher multiples (such as Arp's 1967 model) cannot account for a significant fraction of the radio sources in the Universe because the multiplicity would show up as clear clustering.

REFERENCES

Arp, H.,1967. Astrophys. J., 148, 321.
Bartlett, M.S., 1964. Biometrika, 51, 299.
Crawford, D.F., Jauncey, D.L. & Murdoch, H.S., 1970. Astrophys. J., 162, 405.
Kendall, M.G. & Stuart, A., 1967. "The advanced theory of statistics", 2nd edition, volume 2. Charles Griffin & Co., London.
Pearson, T.J., 1974. Mon. Not. R. astr. Soc., 166, 249.
Peebles, P.J.E., 1973. Astrophys. J., 185, 413.
Peebles, P.J.E., 1974. Astrophys. J. Supp. Ser., 28, 37.
de Vaucouleurs, G., 1971. Publ. astr. Soc. Pac., 83, 113.

Webster, A.S., 1976a. Mon. Not. R. astr. Soc., 175, 61.
Webster, A.S., 1976b. Mon. Not. R. astr. Soc., 175, 71.
Zieba, A., 1975. Acta Cosmologica, 3, 75.

DISCUSSION

Peterson: How does the method of projection of a sphere on to a plane affect the amplitude of the Fourier coefficients? Does it dilute clustering that may be present?

Webster: Scarcely at all. Certainly, the clustering is not diluted. The chief effect is a small distortion of shapes, so that a circular cluster becomes elliptical, but this is an unimportant matter.

Jauncey: If you know what sort of clustering to look for, it seems that you can make a much stronger statement than just the general tests for anisotropy.

Miley: About how many bright radio sources could be haloes of widely spaced doubles which are not recognised as belonging together.

Webster: There cannot be more than about 5% of faint sources which are unrecognised wide doubles, or the power spectrum analysis would show it. This may or may not help decide about the bright sources.

Arp: The statement that steep spectrum sources seen at high frequency arise preferentially from relatively bright galaxies makes it seem natural that the Northern Hemisphere anisotropy is due to the greater number of local supercluster galaxies in the Northern Hemisphere. In that case it is unsophisticated to talk about North-South differences. The brighter galaxies actually are in the projected area of the super-cluster. That is a sharper, more sensible test of the anisotropy which would resolve the problem. Along that line, and in contradiction to what Adrian Webster claimed, if you plot the 3CR quasars between V = 17 and 19 Mag. you see they are missing in the 13 to 17^h region and fall in the 8^h - 12^h region with the bulk of the local super cluster galaxies.

Kellermann: I would like to make some historical and perhaps provocative remarks which may stimulate further discussion.

About twenty years ago, not too far away from this room a radio source survey was made at frequency of 81 MHz. Only a few of the approximately 2000 sources were identified (indeed as it turns out only a few more were ever real), but nevertheless profound cosmological conclusions were reached based on the unexpected large deviations of the N(S) relation from the "expected" -1.5 power law, and it was claimed that the observed isotropy excluded interpretations based on a local anomaly. Later surveys give results very much closer to the canonical -1.5 law, especially when differential source counts are used in place of the misleading integral counts used previously.

Today the experimental results are very much improved :   Surveys now
exist over a wide range of wavelengths which actually measured flux
densities down to very low values.   The data presented today by
Pauliny-Toth and Wall at 5 GHz are very different from the old data.
Except for the strongest 100 or so sources, the results agree quite well
up to $\sim 10^5$ sources sr^{-1} with the -1.5 slope corresponding to a random
distribution of sources in a hypothetical Eucledean Universe.   And for
the strongest 100 sources which do derivate from this law, the evidence
for isotopy is not clearly established.   Although the evidence for
anisotropy is only marginal, the important thing is that neither is the
evidence that these sources are isotropically distributed established.
Since the derivation from the -1.5 law is no greater than the apparent
anisotropy, it is not clear that this apparent steep slope is of any
cosmological significance.   The high degree of isotropy which is
observed for the weak sources is not relevant.

But although the experimental situation has changed drastically during
the past twenty years, the conclusions drawn from the source counts has
not!   What has changed is the argument, which now goes that because the
radio sources are so distant, the expected effect of the redshift is to
depress the counts below the -1.5 law, so that even the observed value of
-1.5 requires evolution.   But most of the identified radio galaxies are
not very distant, and there are still some (perhaps only one or two) who
question the cosmological interpretation of the redshift.   The question
which we really want to answer is the same as was originally posed years
ago :   What can we learn from the source counts alone, independent of any
assumption about the nature of the redshift?   After all, if the quasars
are cosmological, the great abundance of large redshifts is immediate and
obvious evidence without further analysis.

I cannot help but be impressed by the apparent coincidental agreement of
the 5 GHz source counts with that expected from a random distribution of
sources which are either relatively nearby or which are located in a
non-expanding universe.   I often wonder how the course of radio astronomy
and cosmology might have been changed, had the advance of radio technology
been reversed, and the 2C survey made at centimeter rather than meter
wavelength.

Rees: It seems to me that the anisotropy problem differs in one important
respect.   Whereas most of the other tests are controversial because
they involve both the physics of the sources and the cosmological model,
the question of large scale inhomogeniety is essentially independent of
the physics.   So it might be better to examine the extent to which it
can be improved.

Webster: To improve the isotropy tests one simply wants more and more
radio sources.   Give me a catalogue with 10^6 sources, then the accuracy
can be improved by a factor of 10 over the results from existing
catalogues.   I feel it won't get much better than that because a radio
telescope with sufficient resolving power to find a million sources
starts to split up the double radio sources.   Then we would not be able

to separate out multiplicity within radio sources from multiplicity within groups of sources. I suspect that the limit would not then come down very much farther.

Rees: I would like to question the usual assumption that the tests of large scale isotropy and homogeniety are more powerful when applied to the microwave background rather than to radio sources. When we look at the microwave background, we are looking back to redshifts of the order of 10^3. In the standard models for the evolution of density perturbations, these have had time to grow by a factor of 10^3 since then. So, in order to make a test as good as a 10% test of a scale of 1 Gpc you have to look for fluctuations of order 10^{-4} in the microwave background. Looked at this way, it is not so obvious that the radio source tests lose out.

Another point is that, particularly in low density cosmologies, you can look for fluctuations in the density that would not necessarily give rise to fluctuations in the velocity field. This is something that is not so easily done with the microwave background. So it seems to me that the Webster type tests on radio sources have certain advantages over tests on the microwave background.

II

SPECTRAL INDEX DISTRIBUTIONS

SPECTRA OF SOUTHERN RADIO SOURCES

J.G. Bolton
Division of Radiophysics, CSIRO, Sydney, Australia

Surveys of the sky between declinations +25° and -90° at 2700 MHz (11 cm) have been in progress for the past 10 years. Excluding some regions close to the galactic plane the whole sky south of +25° has been surveyed to a flux density limit of 0.6 Jy at 2700 MHz and within this area surveys to limits of 0.35, 0.25 or 0.1 Jy have been made covering 3.5 sr. Flux densities have been measured at 5000 MHz for all sources stronger than 0.35 Jy at 2700 MHz. The source positions have an average accuracy of 10" arc in both coordinates and the positions have been examined for optical identifications on Palomar, ESO or SRC sky survey plates, which now cover 95% of the area. The first part of this paper concerns the relationships between the spectral indices α(2700 to 5000 MHz) and the identifications of the 2300 sources with galactic latitudes greater than 10°. It is a statistically significant sample, since the sources stronger than 0.35 Jy cover 3.5 sr. It is also a representative sample, since no selection was made on the basis of spectral index or identification. It cannot however be claimed as a complete sample, for two reasons. A substantial fraction of sources found in radio surveys at high frequencies are variable - variations of up to a factor of three can occur on a time scale of a year - thus the various sections of the survey are complete only for the relevant epoch. Many of their optical counterparts are also variables - variations of up to a factor of 100 can occur on a time scale of one year. It is hoped to make some assessment of the effect of these two factors in the next two years, when second-epoch Parkes surveys will begin and SRC Schmidt plates will overlap the Palomar Sky Survey.

About half the sources in the sample can be identified and of these half are galaxies and half quasars or possible quasars. The identifications with galaxies are based solely on positional coincidence. About 70% of the identifications with quasars are supported by photometry, two-colour photography or spectroscopy; the remainder are based largely on positional coincidence, although for some there is additional support such as radio or optical variability. The small fraction of the latter objects which may be mis-identified is statistically unimportant. Figure 1 shows histograms of the number of sources

D. L. Jauncey (ed.), Radio Astronomy and Cosmology, 85-97. All Rights Reserved.
Copyright © 1977 by the IAU.

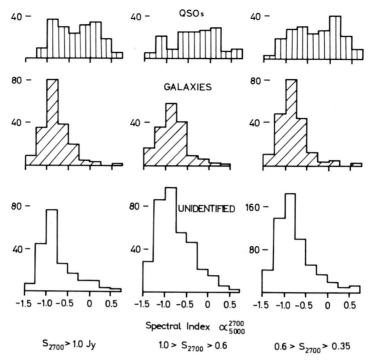

Fig. 1.   Distribution of spectral indices between 2700 and 5000 MHz for
sources in three identification classes and three ranges of flux density.

in nine spectral index bins for three flux-density ranges and three
classes of identifications - quasars, galaxies and unidentified sources.
The vertical scales on all these diagrams are the same except for the
weak unidentified sources at the lower right, where it is half that of
the rest.   Obvious from this figure are two results which I first
surmised 12 years ago from similar but much less data from the Parkes
408 MHz survey.   The spectral index distributions for the QSOs differ
markedly from those of the radio galaxies or the unidentified sources.
The close parallel between the two latter groups suggests that the
majority of unidentified sources are galaxies beyond plate limit.   The
distributions show little change in form with flux density;   however,
the fraction of unidentified sources rises rapidly with decreasing flux
density.   This is perhaps clearer in Figure 2, where the percentage of
sources in each identification class is shown in spectral-index bins
for the three ranges of flux density.   The percentage of identifications
- which are largely quasars - is always greater for sources with
spectral indices $\alpha > -0.25$.   In the strong source group the identified
quasars outnumber the unidentified sources by four to one.   Certainly
some of the latter are quasars which were temporarily subluminous at
the epoch of the Palomar Sky Survey;   second-epoch plates in the $+4°$
zone increased the number of quasar identifications by 10%.   Some of

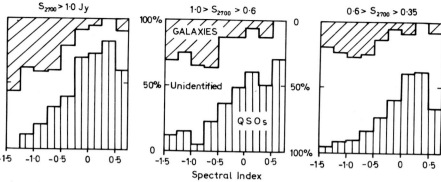

Fig. 2.   Percentage of identifications against spectral index for sources in three ranges of flux density at 2700 MHz.

the others are undoubtedly objects which were ruled out as quasars because they did not show an ultraviolet excess. Probably less than half the 'missing' identifications arise from this cause and amongst these are likely to be a high fraction of quasars with redshifts greater than 2.5.   However, colour is by no means an exclusive indicator for the redshift range of a quasar.   In a rather limited sample of ~80 emission-line quasars observed with the Anglo-Australian 4-m telescope in the past year two out of four with z > 2.3 have an ultraviolet excess.   This sample is however heavily biased towards ultraviolet excess objects.   Out of seven neutral stellar objects examined, four have emission-line spectra with z ≪ 2 and one (discussed later) has an absorption-line spectrum with z > 2.3.   The relationship between colour and redshift is probably influenced as much by the form of the continuum and the presence of absorption features as by the location of strong emission lines.

If the result for the quasars with α > -0.5 can be taken as characteristic of quasars with all values of spectral index, then for the strong source group the number of unidentified sources which are below plate limit or are neutral stellar objects is unlikely to be greater than, say, 20% of the identified quasars.   Though this percentage may increase with decreasing flux density it represents an insignificant fraction of all unidentified sources, which, as the similarity in overall spectral index distribution implies, must be galaxies beyond plate limit.

Luminosity Distributions

The luminosity distributions for the identified radio galaxies, divided into three ranges in flux density and three ranges in spectral index are shown in Figure 3.   It is expected that the numbers in the last magnitude range would be underestimated; near plate limit the ability to distinguish the image of a faint galaxy from that of a faint star is affected both by seeing conditions for the original plate

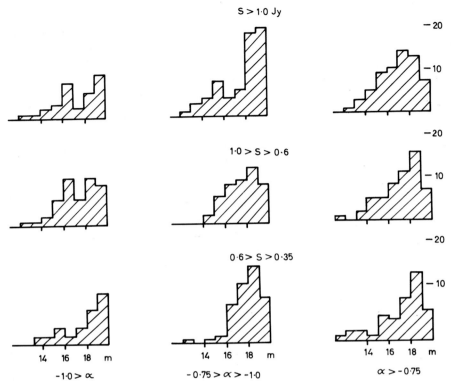

Fig. 3. Luminosity distributions for radio galaxies in three ranges of flux density and three ranges of spectral index.

and the quality of the copy plates or prints. Within the limits of statistical variations there are no significant differences in the luminosity distributions. They are consistent with a radio luminosity function which is extremely broad and a cut-off imposed by the limit of existing sky survey plates. The possible forms of the luminosity distribution beyond plate limit for the three flux density ranges are indicated in Figure 4. These diagrams were constructed using three constraints: (a) that the rise to within 1 magnitude of plate limit continues past the plate limit; (b) that the highest luminosity objects are comparable with the highest already known; and (c) that the number of sources to be accommodated is 80% of the unidentified sources in each range.

Currently the fractions of identified sources in the three flux density ranges are 0.53, 0.39 and 0.25. If the extrapolation of the luminosity distributions are realistic a gain of 1.5 magnitudes in plate limit, as may be expected from the SRC Schmidt IIIa-J survey plates, would raise these fractions to 0.83, 0.64 and 0.47; identification of such objects would require radio positions accurate to 1" or 2" arc. This forecast could be modified by the occurrence of a

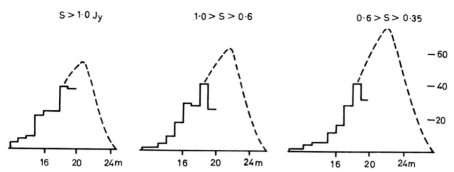

Fig. 4. Luminosity distributions for radio galaxies in three ranges of flux density with an indication of their extrapolation beyond existing sky survey plate limit.

population of faint sources which perhaps have to be associated with a cluster of galaxies rather than the brightest or any individual member. A preliminary examination of a limited number of SRC Schmidt plates of radio positions accurate to only 10" arc suggest that such an effect might be present.

The luminosity distributions for the quasars, again divided into three ranges of flux density and three ranges of spectral index, are shown in Figure 5. Here, although there is a trend for the peaks in the distributions to move towards fainter magnitudes with decreasing flux density, the peaks occur well before plate limit. The addition of a relatively small number of identifications missed because of a colour bias could not radically change the form of these distributions. The occurrence of a peak in the luminosity distribution of the radio quasars has been a result of almost all large-scale identification programs. In strong contrast, the luminosity distributions of quasars selected optically on the basis of ultraviolet excess and in some cases also infrared excess (for example, by Braccesi and his co-workers) appear to rise steeply beyond 18th magnitude. Unfortunately spectroscopic confirmation is not available for many of the faint optically selected quasars and a large fraction might be galactic stars; it is important that more be investigated spectroscopically. On the assumption that they are correctly identified some explanation is needed of the difference in the luminosity distribution of the radio quasars and the optically selected quasars. Qualitatively the difference could be understood if the radio emission was fairly closely related to the optical emission for one group of quasars and completely absent in another. There is good evidence for the latter but the situation on the first requirement is not so clear. For the variable quasars, changes in optical brightness appear to have related changes in radio emission and the ratio between radio and optical emission of the radio quasars observationally has a much smaller range than that for the radio galaxies; however, this may be an effect of the inherently smaller ranges in the individual observed quantities.

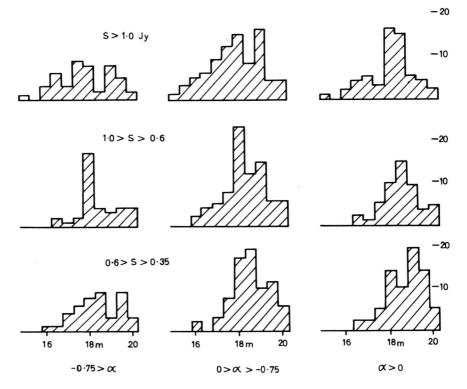

Fig. 5. Luminosity distributions for quasars in three ranges of flux density and three ranges of spectral index.

Radio and Optical Spectra

From their radio characteristics, the radio quasars can be divided into three classes. The first class contains objects which are in the majority amongst quasars found from low-frequency radio surveys. They have relatively steep power-law radio spectra and radio brightness distributions similar in form (and absolute size if redshifts are cosmological) to many bright radio galaxies. These sources occasionally develop weak high-frequency components with time-varying characteristics which agree well with the expanding opaque source models of Kellermann and Pauliny-Toth or Van der Laan. From the rather complete list of redshifts available for the 3C and 4C sources it is clear that most of these objects have emission-line spectra.

A second recognizable group of radio quasars have time-averaged spectral indices near zero and about half the sources found in the Parkes 2700 MHz surveys are of this type. Variations in the radio range suggest repeated outbursts of the Kellermann/Pauliny-Toth type, and this has been confirmed in some cases by very-long-baseline interferometry. Variations in flux density at 2700 MHz are typically 30% in a year and in visual brightness 1 magnitude. Collectively, at a given epoch 30% to 50% of these objects show no emission lines. Individually, the chance of detecting emission lines appears, rather perversely, to be best when the objects are faintest. Whether the rise in continuum simply swamps the line emission or substantially modifies the gaseous envelope of the quasar provides an interesting subject for future quantitative spectroscopy.

The third class of quasars comprise objects with radio spectra apparently characteristic of opaque synchrotron sources - but which show no variations to within an accuracy of 1% or 2% even on a time scale of 10 years. The high stability of these objects is very well documented: they have been used as position calibrators at Parkes for many years and their flux densities have been measured relative to Hydra-A several times each year. Optically these objects have very strong emission lines and it seems possible that the apparent synchrotron cut-off may be in fact due to free-free absorption. The cut-off is not steep enough to be due to a compact source at the centre of an isotropic sphere but could be modelled on the basis of a distributed source within an inhomogeneous medium. A rather logical extension of this idea would be to understand the non-radio quasars as having envelopes which are completely opaque to all radio wavelengths. It might help to explain the absence of compact central radio sources in galaxies with obvious nuclear activity. One such case is Pictor-A, where recent observations with the Fleurs synthesis telescope at 1410 MHz shows little or no trace of a central component. The optical spectrum of the nucleus of Pictor-A obtained by R.A.E. Fosbury with the Anglo-Australian 4-m telescope and the Wampler image dissector scanner (Fig. 6) has a spectrum in which the emission lines are extremely strong.

Optical Spectra of Quasars

Spectra of about 120 quasars have so far been obtained with the Wampler image dissector scanner and of these about 80 yield redshifts ranging from 0.2 to 3.3. A composite average spectrum for quasars formed from this data shows that the strongest permitted lines after Lyman-α are NV (1240), CIV (1549) and Hα, and that [OIII] (5007) is the strongest forbidden line. The strength of the NV line is somewhat surprising; its proximity to Ly-α would make it very difficult to distinguish on photographic spectra. One of the most interesting objects studied is 0528-250; a neutral stellar object was indicated as the identification by Parkes measurements and confirmed by observations with the NRAO interferometer. Part of its spectrum is shown in Figure 7; there are no recognizable emission lines, but the many prominent absorption lines give a unique redshift of 2.813. The absorption spectrum is very similar to that of 3C191 and further investigations with higher resolution should prove very interesting. Figure 7 is shown as log F_ν v log ν and was derived from the original data by scaling against a standard star and re-binning. This treatment allows determination of the optical spectral index of the continuum

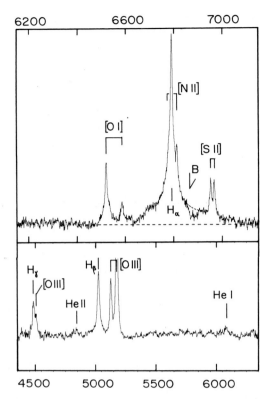

Fig. 6. Spectrum of the radio galaxy
Pictor-A obtained by R.A.E. Fosbury with
the 3.9-m Anglo-Australian Telescope.

where spectra were obtained at a moderate zenith angle and where they
are not obviously curved. A histogram of the spectral indices for
40 such cases is shown in Figure 8. (3C273 was observed to check the
validity of the reduction procedure.) The objects shown by the cross-
hatching are those with no discernible emission lines. For the objects
with redshifts the effects of a continuum K-term (correction for bandwidth compression and shift in wavelength with redshift) shows that on
the average the fainter objects are disadvantaged as regards detection
to a greater extent than the brighter objects. Similarly the higher
redshift objects are disadvantaged relative to the lower redshift
objects, with the notable exception of two very high redshift objects
with strong UV excess. The two effects appear to be independent.

Non-Cosmological Redshifts

     During the past 10 years circumstantial but not compelling
evidence has been accumulating that the redshifts of quasars and some
galaxies have a non-cosmological component. I will very briefly

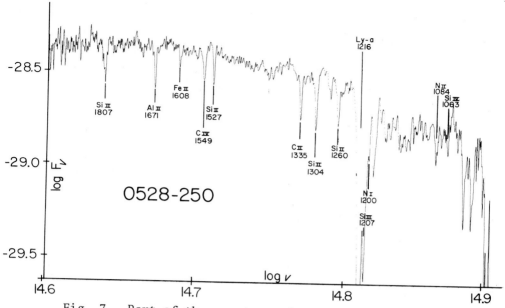

Fig. 7. Part of the spectrum of the quasar 0528-250.

discuss two. One concerns pairing of quasars with bright galaxies. Some years ago Burbidge, O'Dell and Strittmatter showed that there was an apparent relationship between the linear separation of a 3C quasar and a nearby galaxy and the redshift of that galaxy. At the same time it was pointed out that similar close associations did not appear in the published identifications of the first zones of the Parkes 2700 MHz catalogue. However, unless radio positions are really precise there is a tendency to associate a radio source with a nearby bright galaxy, since radio and optical centroids do not necessarily coincide. 3C275.1,

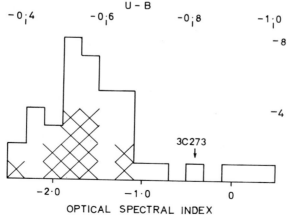

Fig. 8. Distribution of optical spectral indices for 40 quasars. Objects cross-hatched have no emission lines.

for instance, a quasar in the 3C sample, was originally identified with the nearby galaxy NGC 7413. A number of quasar-bright galaxy pairs are now known in the Parkes sources. For three of these, redshifts have

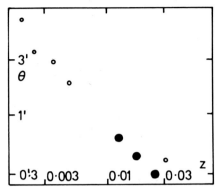

Fig. 9. Galaxy-quasar separation against galaxy redshift (from Burbidge, O'Dell and Strittmatter) with the addition of three new pairs (large circles).

been measured for the nearby galaxy. They are shown together with the original 3C sample in Figure 9. No comment appears necessary.

In addition to the quasar-galaxy pairs there are a number of quasar-quasar pairs. Most of these were found on two-colour image tube plates taken by B.A. Peterson and me in the positions of radio sources with flat or inverted radio spectra and thus likely to be quasars. Out of a sample of 100 plates, each 2' arc in radius and containing an object with ultra-violet excess at the position of the radio source, we found eight objects with a second ultra-violet excess object nearby. Spectra by Peterson and myself and by D. and B. Wills have proved that in five out of six cases investigated the non-radio objects are also quasars. Estimates of the surface density of quasars to the limiting magnitude of the plates (B=19-19.5) would predict only one object by chance in the 100 fields. Thus the result appears highly significant. For three of the pairs redshifts are known for both components, and in no cases are the redshifts similar, as is also true of the pair found by Stockton and the pair found by Wampler et al. If the pairing of quasars with discordant redshifts can be verified in a larger sample, the cosmological origin of the quasar redshifts will be open to question.

I thank my colleagues, B.A. Peterson, A.E. Wright, Ann Savage and D.L. Jauncey for unpublished material discussed in this paper.

DISCUSSION

Arp: I have reviewed the quasar-galaxy associations for the Paris Colloquium and find 7 of 8 new cases fall closely along the original Burbidge-Strittmatter-Odell relation with a constant distance of about 20 KPC between the quasar and the galaxy. The new points which Bolton has shown also substantiate this relation and, I think, are mostly in addition to the points I have added.

Wampler: What is the error in your spectral index measurements of the optical spectra of QSO's? In particular are you sure that you have several objects with positive continuum spectral index ($\alpha > 0$)?

Bolton: Yes, some objects have positive spectral index. Several of these objects have high z.

Menon: You mentioned that the ratio of radio to optical flux densities for QSO's is either constant or zero. Does this apply to both the flat and steep spectral index QSO's? In the case of steep spectra QSO's the total flux density need not be correlated with the optical flux density since the radio emission from the central component is smaller than the total flux density. However for the flat spectra QSO's all of the radio emission originates in the central optical object. Hence the ratio of the radio to optical flux density should be computed only when detailed structural information is available, especially for the steep spectra QSO's.

Bolton: I think that it must apply to both types. The relationship might be tighter in the case of the flat spectrum objects in spite of the large variations in both the radio and optical emission.

Webster: Could you please give some idea of any selection effects in your angular separation/redshift diagram?

G. Burbidge: The associations between the QSO's and bright galaxies were investigated statistically and were tested by several methods by Burbidge et al in their paper in 1971 and the results were confirmed using Monte Carlo methods by Kippenhahn in 1973. It was then found that a plot of $\Delta\theta$ against z_G gave the relation just shown by John Bolton and Burbidge, O'Dell & Strittmatter. We showed in that paper that selection effects would probably work against their observed effect.

Murdoch: You said only one new QSO was discovered on objective prism plates. Does this mean most QSO's discovered by objective prism are radio sources.

Bolton: No, the comparison was not with radio QSO but with objects selected as having uv excess on two colour plates. Only one emission line QSO was found which did not correspond to an object already picked out by its uv excess in a sample of 25, and this had no counterpart on the two colour plates - i.e. it had changed in brightness.

van der Laan: The purely positional identification with galaxies, given position of accuracies of ± 10", would seem hazardous beyond m ∿ 18. Do you not get a lot of spurious identifications near the plate limit?

Bolton: I don't believe so.   The 10" is a limit on the radio position rather than an rms error.

de Felice: Is there any significant difference in the QSO distribution between the Northern and the Southern Hemisphere?

Bolton: The subject was not considered but there is little evidence of significant difference.   Any difference is probably due to unexpected obscuration.

PAIRS OF QUASARS

B.J. Wills and D. Wills

The probability that the QSO pairs discussed by John Bolton could arise from chance associations of radio QSO's and background radio-quiet QSO's is 1 in 1000, if one uses the background QSO counts given by Setti and Woltjer (Annals N.Y. Acad. Sci., 224, 8 (1973)) from the search by Braccesi et al. (Astr. and Astrophys., 5, 264 (1970)).

With Alan Vomoto, we have made an independent search for QSO pairs under conditions as closely as possible identical to those applying to Bolton and Peterson's search.   Two colour (U,B) plates were obtained with the 2.1 m McDonald telescope, centred at the positions of flat-spectrum QSO's, most of them from the Green Bank 5 GHz (S2 and S3) surveys.   In nearly 100 fields, we found three UV-excess objects within 2' arc of the radio QSO's, and with B < 19.0.   Our spectroscopy showed two of them to be stars, and the third to be a QSO.   The expected chance number of QSO's expected is 0.6.   Searching to $\underline{B}$ = 19.5, the total is 2 QSO's found, compared with 1.3 expected by chance.   In neither case are the redshifts of the radio and non-radio QSO's the same.

Thus, our results are quite consistent with the number of objects expected on the basis of current estimates of the surface density of QSO's.   Our result can be made consistent with Bolton and Peterson's result, at the expense of increasing current estimates of the surface density of QSO's by a factor of about four.

G. Burbidge: What is the numerical value of the background density of QSO's that you used?

B. Wills: We used Setti and Wotjer's numbers derived from the Braccesi et al. counts, which gives 2.1 or 4.3 per square degree to B = 19.5 respectively.

Sanitt: By what factor could the background density of uv excess quasar be increased before there is a contradiction with your results.

B. Wills: If the background density (to $19^{m}.5$) were increased by a factor of 5, our 2 observed QSO's would have a 3% probability of occurrence.

SPECTRAL INDEX STUDIES OF EXTRAGALACTIC RADIO SOURCES

P. Katgert, L. Padrielli and J.K. Katgert
Laboratorio di Radioastronomia, Bologna

A.G. Willis
Sterrewacht, Leiden

1. INTRODUCTION

Spectral index distributions can be indicated conveniently by $g_S^{\nu_1}(\alpha_{\nu_1},\nu_2)$, i.e. the distribution of the two-point spectral index between frequencies ν_1 and ν_2 for a sample of radio sources complete to flux density S at the selection frequency ν_1 (flux densities will be expressed in Jy and frequencies in GHz). Such a detailed specification is necessary because $g(\alpha)$ has been found to depend on all three parameters. The effect of varying ν_1 is well-known: the fraction of flat-spectrum sources increases with selection frequency. Variations of $g(\alpha)$ with S have also been found: both the fraction of flat-spectrum sources and the mean spectral index of the steep-spectrum sources depend on S, at least at high frequencies. The magnitude of this dependence on S as well as $g(\alpha)$ itself appear to depend on the choice of ν_2, or rather on the relation between ν_2 and ν_1. This complex behaviour of $g(\alpha)$ is not unexpected if one considers that the redshift and luminosity distributions of a source sample vary with selection frequency and flux density limit. Intrinsic or induced correlations between spectral index and redshift or luminosity, or different redshift distributions of flat-and steep-spectrum sources may cause variations of $g(\alpha)$ with ν and S. An understanding of the behaviour of $g(\alpha)$ requires observations over large ranges of frequency and flux density. Even then, it will be difficult to interpret these data without information about the composition of the various samples, which can be obtained only through optical work. Here we describe recent spectral index information for weak sources selected at 1.4 GHz, as well as work on spectral index/optical identification correlations in 5 GHz samples.

2. SPECTRAL INDICES OF WEAK SOURCES SELECTED AT 1.4 GHz

Spectral index work with the Westerbork Synthesis Radio Telescope has concentrated so far on reobservation at 0.6 GHz of source samples defined earlier at 1.4 GHz. The first result of this work, an estimate of $g_{0.007}^{1.4}(\alpha_{1.4},0.6)$ from a sample of 36 sources (Katgert and Spinrad, 1974) was rather unexpected. About half of the sources appeared to have flat

D. L. Jauncey (ed.), Radio Astronomy and Cosmology, 99-106. All Rights Reserved.

spectra (i.e. $\alpha < 0.5$) resulting in an unusually low mean spectral index
of about 0.5. At present, 0.6 GHz observations are available also for
parts of the other 1.4 GHz Westerbork surveys, viz. the 1st survey (Katgert et al. 1973), the 3rd survey (Katgert,1975) and the survey of background sources (Willis et al. 1976). The main questions to be answered
by these new observations are whether the above-mentioned result of the
second survey is representative for this selection frequency and flux density level, and whether there is real evidence for variations of $g_S^{1.4}(\alpha)$
with S.

From the 1.4 GHz samples - which are all complete to at least 0.01
Jy - we selected a complete sample of 183 sources with S > 0.02 Jy, for
174 of which a 0.6 GHz flux density is known at present. The choice of
this rather high completeness limit should ensure that corrections for
flux density dependent selection against flat-spectrum sources (among
other things due to the primary beam attenuation) are small. In fact,
we found that such corrections are negligible for these samples and
therefore did not apply any. In Table 1 we give the parameters of the
distribution $g_{0.02}^{1.4}(\alpha_{1.4,0.6})$ for the total sample and for the various
subsamples. There appear to be rather large differences between the
four surveys, and especially the 2nd survey is anything but representative.
The sample of background sources probably comes nearest to being representative because of its size and distribution on the sky. Because of the
observed differences between surveys it is not clear to what extent the
total sample is representative, but it may be noticed that it is not
very different from the sample of background sources.

A comparison with Gillespie's (1975) result based on 1.4 GHz surveys
of the 5C regions (made with the Half-Mile Telescope) is interesting because the 3rd Westerbork survey and Gillespie's survey of the 5C2 region
do have some overlap. The mean spectral index of the two subsamples is
found to differ by 0.21±0.08, a result which can be explained completely
by differences between the zero-points of the two spectral index scales.
For 17 sources from the original 5C2 survey (with a 0.4 GHz attenuation
between 2.0 and 5.0) and detected also at 0.6 and 1.4 GHz in our observa-

Table 1

Parameters of Spectral Index Distributions $g_S^{1.4}(\alpha_{1.4},\nu_2)$

Sample	S	ν_2	n	$<\alpha>$	σ_α	$\%(\alpha < 0.5)$
BDFL/B2	2.00	0.4	50	0.61±0.05	0.37	0.24±0.07
All Wbk	0.02	0.6	174	0.64±0.03	0.34	0.25±0.04
1st Wbk	0.02	0.6	46	0.73±0.04	0.27	0.22±0.07
2nd Wbk	0.02	0.6	19	0.44±0.09	0.38	0.47±0.16
3rd Wbk	0.02	0.6	45	0.59±0.05	0.34	0.31±0.09
BGS Wbk	0.02	0.6	64	0.66±0.04	0.34	0.17±0.05
All 5C	0.01	0.4	140	0.71±0.03	0.36	0.21±0.04
5C2	0.01	0.4	31	0.80±0.06	0.33	0.19±0.08

tions, we find $<\alpha_{1.4,0.4}-\alpha_{1.4,0.6}> = 0.16\pm0.03$. The difference between the two 1.4 GHz flux density scales (Katgert, 1976) accounts for an additional spectral index difference of about 0.08. Although the discrepancy can be explained empirically, its origin remains to be investigated.

Also shown are the parameters of the spectral index distribution of a strong source sample taken from the 1.4 GHz BDFL catalogue. For the spectral indices we used 0.4 GHz flux densities from the Bologna catalogue, hence the sample only covers declinations between 24° and 40°. The 0.4 GHz flux densities have been corrected for the effects of partial resolution using the structure information of the BDFL catalogue. Before comparing the strong and weak source samples, it is again necessary to establish the relation between the two spectral index scales. In the Westerbork samples there are nine unresolved sources for which we have flux densities at 0.4 (Bologna), 0.6 and 1.4 GHz (Westerbork). For these sources we find $<\alpha_{1.4,0.4}-\alpha_{1.4,0.6}> = -0.02\pm0.05$. Since the Westerbork and BDFL 1.4 GHz flux density scales are identical to within the errors (Fomalont et al. 1974) we find a formal difference of 0.01 ± 0.09 between the mean spectral indices of the weak and strong source samples. Note that the uncertainty in this result is only partly due to limited statistics. The fraction of flat-spectrum sources is also practically the same for both samples.

Given the rather large uncertainty of this result we will discuss only very briefly some of its possible implications. It is clear that the two samples differ markedly with respect to average luminosity and redshift. Because direct redshift information is not available, we have computed redshift and luminosity distributions of both samples (see Table 2) on the basis of a conventional evolutionary model which reproduces the observed 1.4 GHz source count satisfactorily. Of course, these distributions are approximate, if only because the luminosity function (and its dependence on redshift) may be different for e.g. flat- and steep-spectrum sources. In view of the reported correlation between spectral index and luminosity for extended radio sources identified with elliptical galaxies (see e.g. Véron et al. 1972), the change in the luminosity distribution might produce variations of $g(\alpha)$, because the intrinsically strong sources (presumably with steep spectra) are almost absent from the weak sample. In order to account for the apparent absence of a change in the spectral index distribution it may appear necessary to postulate, e.g. a redshift dependent ratio of flat- and steep-spectrum sources as a function of luminosity.

Table 2

Hypothetical Redshift and Luminosity Distributions of 1.4 GHz Samples

Sample	z										log $P_{1.4}$			
	0.0	0.5	1.0	1.5	2.0	2.5					21	23	25	27
S > 2.00	46	11	8	7	6	6	5	4	4	3	1	3 11 18 19	27	21
S > 0.02	12	10	8	8	8	8	9	11	14	12	1	3 10 25 55	6	0

3. SPECTRAL INDEX / IDENTIFICATION CORRELATION FOR 5 GHz SAMPLES

The most convincing evidence for variations of $g(\alpha)$ with S comes from high-frequency samples. In particular, $g_S^{5.0}(\alpha_{5.0}, \nu_2)$ has been studied in considerable detail on the basis of the NRAO 5 GHz surveys (see e.g. Condon and Jauncey, 1974 and Pauliny-Toth et al. 1974). The main result appears to be that the mean low-frequency spectral index (i.e. $\nu_2 < 1$ GHz) increases significantly between a sample complete to 0.6 Jy and one complete to 0.067 Jy. However, the high-frequency spectral index (i.e. $\nu_2 > 2$ GHz) remains practically constant. Because the identification percentage of high-frequency samples is rather high, one has a good opportunity to study possible spectral index/identification correlations. The first analyses of this kind made use of identifications based on radio positions of moderate quality (see e.g. Fanti et al. 1974). At present, radio positions of arc-second quality allow much more reliable identifications to be made. For a sample of 135 sources with S > 0.6 Jy, Johnson (1974) obtained accurate identifications using radio positions obtained with the RRE interferometer, while Condon et al. (1975) reidentified part of the weak 5 GHz sample (i.e. S > 0.1 Jy) on the basis of NRAO interferometer positions.

We have made 5 GHz observations with the Westerbork telescope of essentially all sources in the weak 5 GHz sample to obtain accurate radio positions and structures. For a small number of sources, not detected at 5 GHz due to strong resolution effects, additional 1.4 GHz observations were made. About 40 per cent of the sources were found to be appreciably extended (with sizes of up to 4-5 arc minutes). Identifications were carried out on the PSS prints, on the basis of positional coincidence only. The number of spurious identifications is estimated to be less than three. For a comparison with the strong source sample (S > 0.6 Jy), we limit the weak sample to the 91 sources with flux densities between 0.067 and 0.6 Jy. The spectral separation of the weak sample is based on $\alpha_{5.0,0.4} \gtrsim 0.50$; for the strong sample (of 118 sources) we used $\alpha_{5.0,0.3}$ (Condon and Jauncey, 1974), increased by 0.05 to account for the difference between the 0.3 and 0.4 GHz flux density scales.

The percentage of flat-spectrum sources changes from 56% in the strong sample to 36% in the weak sample, in good agreement with earlier results. In Table 3 we compare the identification content of the flat- and steep-spectrum subsamples in the strong and weak surveys.

Table 3

sp.	sample	n	QSO	GAL	EF	NI
flat	strong	66	64%	23%	13%	--
	weak	33	55%	21%	24%	--
steep	strong	52	17%	52%	29%	2%
	weak	58	19%	28%	50%	3%

The percentage of identified flat-spectrum sources changes from 86±11 in the strong sample to 76±15 in the weak sample, which is not significant. On the other hand, the percentage of identified steep spectrum sources changes from 69±12 to 47±9. This change is wholly due to the loss of galaxy identifications in the weak sample (from 52±10 to 28±7 per cent).

The fraction of flat-spectrum sources identified with galaxies apparently does not vary with flux density limit. One might think that this indicates a low value for the ratio between radio and optical luminosity of flat-spectrum galaxies. However, in that case, most flat-spectrum galaxies in the strong sample should have optical magnitudes well above the PSS limit, which appears not to be true. The constancy of the fraction of galaxy identifications among flat-spectrum sources is probably a result of the very flat source count for the flat-spectrum subsample (which is determined largely by the quasars). The percentage of quasar identifications is more or less independent of flux density limit for both flat- and steep-spectra quasars, i.e. the counts of flat- and steep-spectra quasars also differ considerably.

It would seem that the optical identification information does not really enable one to choose between possible explanations for the increase of the fraction of steep-spectrum sources with decreasing flux density. For instance, the validity of the model proposed by Fanaroff and Longair (1973) cannot be checked without redshift information. Observations at lower flux densities are needed to see whether the steepening continues (as predicted by the "cosmological" model), or whether the variation in spectral index distributions is due to relatively local irregularities in the density of radio sources.

REFERENCES

Condon, J.J. and Jauncey, D.L., 1974, Astron. J. 79, 1220.
Condon, J.J., Balonek, T.J. and Jauncey, D.L., 1975, Astron. J. 80, 887.
Fanaroff, B.L. and Longair, M.S., 1973, Monthly Notices Royal Astron. Soc. 161, 393.
Fanti, R., Ficarra, A., Formiggini, L., Gioia, I. and Padrielli, L., 1974, Astron. Astrophys. 32, 155.
Fomalont, E.B., Bridle, A.H. and Davis, M.M., 1974, Astron. Astrophys. 36, 273.
Gillespie, A.R., 1975, Monthly Notices Royal Astron. Soc. 170, 541.
Johnson, K.H., 1974, Astron. J. 79, 1006.
Katgert, J.K. and Spinrad, H., 1974, Astron. Astrophys. 35, 393.
Katgert, P., Katgert-Merkelijn, J.K., Le Poole, R.S. and Laan, H. van der, 1973, Astron. Astrophys. 23, 171.
Katgert, P., 1975, Astron. Astrophys. 38, 87.
Katgert, P., 1976, Astron. Astrophys. 49, 221.
Pauliny-Toth, I.I.K., Witzel, A. and Preuss, E., 1974, Astron. Astrophys. 35, 421.
Véron, M.P., Véron, P. and Witzel, A., 1972, Astron. Astrophys. 18, 82.
Willis, A.G., Oosterbaan, C.E. and Ruiter, H.R. de, 1976, Astron. Astrophys. Suppl., in press.

DISCUSSION

Wall: If the "flat spectrum" galaxies are weak emitters and hence optically bright, would this not explain the lack of change in identification rate for these objects as flux density decreases?

Katgert: Certainly, but one would then expect to find the flat spectrum galaxies in the strong sample well above the sky survey limit, which is found not to be the case.

Jauncey: How do you distinguish flat spectrum galaxies from QSO's?

Katgert: Basically on their optical appearance and colour, but I admit that near the sky survey limit the classification is rather uncertain.

Shaffer: Did you say 40% of the weak survey sources were resolved? At what resolution?

Katgert: Yes, at 6".

THE SPECTRAL INDEX DISTRIBUTION OF A SAMPLE OF VERY FAINT SOURCES FROM A SURVEY AT 4.8 GHz

M.M. Davis

Yesterday Dr. Pauliny-Toth described a survey using the Bonn 100m telescope at 4.8 GHz in selected small areas. This is the faintest sample we are likely to have at the shortest wavelengths for some time to come. It is of particular interest to investigate the spectral content of this sample, as some evolutionary models predict a rapid decrease in the flat spectrum population of radio sources at low flux densities.

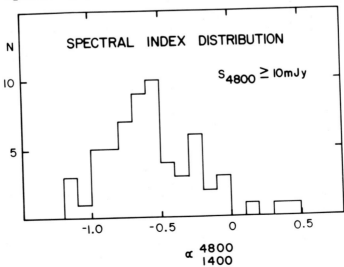

The spectral index distribution shown in the figure was determined with the Arecibo telescope using a 3.3 arcminute beam at 1.4 GHz. The sensitivity was sufficient so that only sources with spectral index flatter than +0.5 might be undetectable. For 61 sources > 9 mJy at 4.8 GHz, 34 percent are flat spectrum ($\alpha > -0.5$). The median spectral index is -0.60, with a quartile range of 0.46, from -0.32 to -0.78. For the brighter half of the sample, cutting off at 20 mJy at 4.8 GHz, the fraction of flat spectrum sources is essentially unchanged (36%). Hence the large ratio of steep to flat spectrum population of about 5 to 1 predicted by the Fanaroff and Longair model is found to be too large; the actual decrease is from a ratio of about 1 to 1 at high flux density levels to 2 to 1 at this very faint level.

Ekers: How does your result on the lack of change in $\langle\alpha\rangle$ compare with the change reported yesterday in $\langle\alpha\rangle$ with flux density for the 5 GHz surveys?

Davis: $\langle\alpha\rangle$ does change, down to about 0.5 Jy, but stays constant from there down to about 0.01 Jy at a flat spectrum population of about 35%.

SPECTRA OF RADIO SOURCES SELECTED AT 408 MHz

H.S. Murdoch

A complete set of "normal" spectrum sources selected at 408 MHz to a limit of 0.97 Jy have a mean spectral index steeper by about 0.1 than a comparison set from an all sky catalogue of sources > 10 Jy at 408 MHz, at a significance level > 3σ. The spectral index distribution of QSO's in the sample is remarkably compact but the distribution for galaxies is broader due to a correlation between spectral index and optical apparent magnitude. For further details and references, see Mon. Not. R. Astr. Soc. (1976) 177, 441.

Conway: In calculating spectral index using 178 MHz did you correct the flux densities for finite angular diameter, since the raw 4C flux densities contain a downwards bias from this effect?

Murdoch: 178 MHz pencil beam flux densities, where available, were used and increased by 10%. 4C interferometer values were increased by 15% but I regard the 178 MHz scaling as somewhat uncertain.

Condon: Since most of the sources in a complete sample lie within a factor of 2 of the lower flux-density limit, each sample describes only a small portion of the number-flux density diagram. The mean spectral index can easily be determined from the number counts at different frequencies. See, Fomalont, E.B., Bridle, A.H., Davis, M.M., 1974, Astron. Astrophys., 36, 273.

Longair: There has been considerable discussion about the validity, or otherwise, of the models which Bernard Fanaroff and I developed to account for spectral index distributions at different frequencies and flux densities. I think we should distinguish clearly the two types of variation in spectral index distributions reported today and yesterday. Hugh Murdoch has shown that there is a shift in the mean of the spectral index distribution of the sources observed at low frequencies. This is a small but significant effect. On the other hand at high frequencies there is a very major change in the fraction of flat spectrum sources with flux density. This can be seen very clearly by comparing the results of I. Pauliny-Toth and Mike Davis. At low flux densities the fraction of flat spectrum sources decreases markedly with flux density.

The point of our work was to show that this second type of variation occurs very naturally in evolutionary world models of the types developed to account for the source counts. The point is that if one supposes that the same form of evolution is valid for sources with $\alpha \simeq 0$ and $\alpha \simeq 0.75$ and that the spectral index distribution observed at low frequency is applicable to all sources at all epochs, the powerful flat spectrum sources are observed preferentially at high flux densities in high frequency surveys; the steep spectrum sources are observed at relatively smaller redshifts. Therefore, the sources with $\alpha = 0.75$ and $\alpha = 0.0$ observed in high frequency surveys give information about the evolution over different redshift ranges. Because the evolution eventually "saturates" for all classes of source, the flat spectrum sources fall out first whilst the relative proportion of steep spectrum sources continues to increase with decreasing flux density. This is the essence of the models and my comparison of our models with the results shown today suggests that the models are in remarkable agreement with the observations - certainly not as bad as suggested by the speakers In these models one can obtain large changes in the spectral index distribution at high densities at high frequencies - which is not possible in non-evolutionary models. There is plenty of scope for improving the agreement of the models and the observations using all the new information.

SOME EVIDENCE FOR LARGE SCALE CLUSTERING OF RADIOSOURCES

G. Grueff and M. Vigetti

A complete sample of 526 radiosources with $S_{408} > 0.9$ Jy, $24°$ < DEC < $34°$, $23^h 30^m$ < R.A. < $02^h 30^m$, and $07^h 30^m$, and $07^h 30^m$ < R.A. < $17^h 30^m$ has been optically identified on the Palomar Sky Survey, and all the spectral indices α_{408}^{5000} were also obtained. The distribution of such sources was investigated for possible anisotropies. By dividing the sky strip in intervals half an hour long in R.A. no evidence for anisotropy was found in the density of sources. However when computing the average spectral indices in each box, their distribution was found to be non-random for Quasars, at the significance level of 4%, and for Empty Fields, at the significance level of 0.3%. Any possible systematic error in flux measurement has been excluded by checking that the variations of average spectral index for Quasars and Empty Fields are completely uncorrelated. An attempt to decide whether the steep spectra or the flat ones were producing the anisotropy indicates the latter as more probably responsible.

III

ANGULAR DIAMETER-REDSHIFT AND FLUX DENSITY TESTS

RADIO SOURCE ANGULAR SIZES AND COSMOLOGY

R.D. Ekers
Kapteyn Laboratory, Groningen

G.K. Miley
Huygens Laboratory, Leiden

1. INTRODUCTION

The main cosmological tests using radio sources as probes are summarized below.

Test	Advantages	Difficulties
Source counts[1] N(s)	Only requires information from a radio source survey.	i) Survey must be complete and unbiased with s. ii) Interpretation depends on the radio luminosity function (RLF).
Hubble relation[2] for radio sources s(z)	Independent of the RLF.	i) Knowledge of z requires complete optical identifications to avoid bias on z with s. ii) RLF is broad and very nearly critical.[3]
Angular size – redshift[4,5,6] $\theta(z)$	i) Complete sample is not critical unless its incompleteness is a function of θ. ii) Independent of the RLF.	Requires a large sample of objects with both θ and z measured.
Angular size – flux density[7,8] $\theta(s)$	i) Complete sample in s is not critical. ii) Only requires a radio catalogue of angular sizes.	i) Interpretation depends on both the RLF and the linear size distribution function.
Angular size[8] distribution $N(\theta,s)$	i) Includes N(s) relation but θ gives additional constraints.	Will be affected by any correlation between radio power and linear size.

D. L. Jauncey (ed.), Radio Astronomy and Cosmology, 109-117. *All Rights Reserved.*
Copyright © 1977 by the IAU.

Footnotes to table:

1. Jauncey (1975); 2. Bolton (1966); 3. von Hoerner (1973); 4. Hoyle
(1959); 5. Wardle and Miley (1974); 6. Hewish et al.(1974); 7. Swarup
(1975); 8. Kapahi (1975a).

There are three tests involving angular sizes of radio sources which
appear to be at least as good as the traditional N(s) test. These have
not been used extensively until now, probably because of a lack of high
resolution angular size measurements over a wide range of radio source
flux densities. This situation is changing rapidly especially with the
increasing resolution and sensitivity of aperture synthesis telescopes.
In this review these methods and some of the recent results are
discussed.

2. MEASURES OF THE ANGULAR SIZE OF RADIO SOURCES

When using photographic observations of galaxies the simplest
measure is an isophotal diameter - in contrast to this the radio
astronomer can usually measure directly the more fundamental metric
diameter. However, there is still considerable choice regarding which
characteristic of the angular distribution of emission to use. General
considerations to keep in mind are that the measure of diameter
i)   should not depend on relative resolution,
ii)   should not depend on relative sensitivity,
iii) should be independent of frequency and
iv)   can be used on the largest possible fraction of the data.
The most common measure is the separation of the components of a double
source and since the majority of strong extragalactic radio sources are
double this is a simple and effective definition. In order to satisfy
points i) and iv) it is necessary to extend it to partially resolved
and unresolved sources by modelling them on the basis of the characteristics of the more resolved sources. To satisfy point ii) a cut off
must be set on the acceptable ratio of intensities for the two components. These conditions will not introduce bias providing the brightness
distribution or the fractions with different morphology are not themselves a function of distance. Other possibilities are to use some
characteristic in the visibility function, e.g. the spacing for which
the visibility first falls to 0.5, or the first moment of the source
distribution which can be derived either from the visibility function
or the brightness distribution (Burn and Conway, 1976). These have the
advantage that they can easily be applied to partially resolved sources.
Another estimate sometimes used is the distance between outermost contours. This is a bad measure since it depends on receiver sensitivity
and resolution, and is not a purely metric diameter.

There is a general problem with complex sources. Is the correct
measure for a tail source like NGC 1265 (e.g. Wellington et al., 1973)
the separation of the two components of the tail or the length of the
tail? A general answer is that it does not matter so long as it is done
consistently. A less physical choice will increase the dispersion in the
linear size distribution and hence increase the scatter in the angular
sizes but this is not as bad as introducing a bias in diameter with flux

density or distance. For this example the more physical measure is the separation of the two components of the tail but this will introduce a bias unless the total sample is restricted to those objects of sufficiently large angular size that this distinction could, in principle, be made for all of them.

In the following discussion we have used the separation of components of double sources, or an equivalent measure.

3. LINEAR SIZE DISTRIBUTION FUNCTION

The distribution of linear sizes for the 3CR radio sources with $z < 0.25$ and $b > 10^\circ$ are shown in Figure 1.

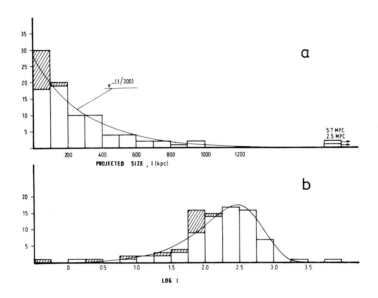

Figure 1.   Distribution of projected linear sizes of 3CR radio galaxies from Kapahi (private communication), assuming H = 50.   a) linear scale with equal interval bins, b) logarithmic scale with logarithmic binning. In each case the same exponential of form k exp(-ℓ/300) is shown.

Jacobs (this symposium) has pointed out that the use of logarithmic binning and logarithmic scales as in (b) has led to the erroneous notion that the number of radio sources peak at a characteristic separation of about 300 kpc. This has not only confused the development of radio source theories but has led to an overdependence on the largest angular size (LAS) statistics. Figure 1(a) shows the same data in a linear plot with equal bins and it can be seen that an exponential distribution of projected sizes is a better fit. It is clear that for samples drawn from such a distribution the mean or the median of the sample will have much

smaller sampling uncertainty than the largest member of the sample. The
effects of random projection on this distribution are relatively unimportant.

4. ANGULAR SIZE - REDSHIFT RELATION

Miley (1968) first showed that there was an angular size-redshift
relation for quasars and further work on this has been reported by Legg
(1970), Miley (1971) and Wardle and Miley (1974). These observations
showed that the upper envelope of both the angular separation distribution and component sizes of quasars decreased approximately as z^{-1} up
to z ~ 2.5. This is expected in a Euclidian universe but is a faster
decrease than is expected for most cosmological models unless there is
a linear size-redshift relation of the form, $\ell \propto \ell_0 (1 + z)^n$ with n ~ -1,
which just cancels the geometrical effects (Kellermann, 1972). Such a
linear size-redshift relation is not however unexpected (e.g. van der
Kruit, 1973). Figure 2 summarizes the angular size-redshift data as
given in Wardle and Miley (1974) by estimating the median of the angular
size distribution for four ranges of z.

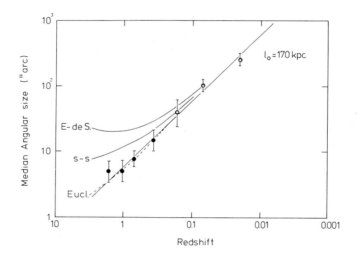

Figure 2.   Median angular sizes of quasars, ●, Wardle and
Miley (1974), and 3CR radio galaxies, o, plotted against
redshift.   N.B.: Selection effects discussed in text. The
result for three world models are indicated by the continuous
line. The broken line is for an Einstein-de Sitter model with
$\ell \propto \ell_0 (1 + z)^{-2}$.

A similar statistic is given for 3C radio galaxies. The median is a
convenient estimate to use in this case since sources which only have
upper limits can also be included. This diagram clearly shows the relation and also shows a continuity between radio galaxies and quasars if

the quasar redshifts are assumed to be at cosmological distances. The models shown are for the linear size distribution function for radio galaxies given in Figure 1. A good fit can be obtained for a Euclidian model and also for other cosmologies provided $\ell \propto \ell_0(1 + z)^n$ with n between −1 and −2.

At this stage a strong word of warning is still required. The data available in the literature, especially for the quasars, is still very heterogeneous and contain unknown frequency dependent effects so a more detailed analysis of these results should not be attempted until a more systematic analysis of diameters subject to the points raised in Sections 2 and 3 has been made.

One such investigation is now in progress as a follow up to the work which led to the discovery of the (θ, z) correlation. Hartsuijker and Miley have mapped 117 quasars with the Westerbork telescope at 5 GHz with a resolution of 6" x 6" cosec δ. The sample includes all those quasars whose redshifts were published up to 1972, with LAS > 7" or with unknown structure. Together with the smaller diameter sources this gives structural information on 211 quasars with known redshift. 30% of these are 3C and 35% are 4C sources so several complete sub samples can be isolated. These measurements give both the intensity and polarization distribution. Some preliminary results are:
i) Central components are detected in 41 out of the 43 cases with sufficient resolution to make this separation. For 24 of these the central components comprise more than 10% of the total flux density of the source. Comparison with the results of Fanti (this symposium) shows that this fraction is much larger than is the case for the radio galaxies.
ii) For 75 sources that are sufficiently resolved, 85% are symmetrical doubles (D1) or triples and only 15% are asymmetric doubles (D2) like 3C273.
iii) The number of sources on the $\theta(z)$ diagram has been almost doubled from Miley (1972) but its form remains substantially unaltered. There is no apparent difference in the behaviour of 3C, 4C, Parkes and Bologna sources. More detailed analysis is still in progress.

An interesting variation on this test is to use the interplanetary scintillations as a measure of the angular sizes of compact components in extragalactic sources (Hewish et al., 1974). Some further results from this technique are described by Hewish (this symposium).

5. ANGULAR SIZE FLUX DENSITY RELATION

Early attempts (Longair and Pooley, 1969) and (Fanaroff and Longair, 1972) to use this test were not very promising mainly because of the small samples of extended sources available at low flux densities. Now that large amounts of information on the angular sizes of radio sources can be routinely obtained it is possible to exploit this test more fully. The hope is that the very large data samples which are available will offset the double smoothing of the $\theta(s)$ relation by the RLF and the linear size distribution function.

The most complete published analysis of θ(s) data comes from the
lunar occultation sample with the Ooty telescope combined with the 3C
data (Swarup 1975, Kapahi 1975b). A sample of data at lower flux densities has been obtained by analysing the angular size information on
background sources found during routine mapping observations with the
Westerbork Synthesis Radio Telescope (Ekers, Hummel and Jacobs - in
preparation). These two sets of data are summarized in Figure 3.

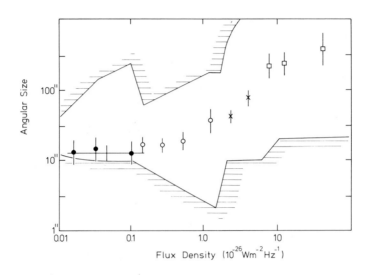

Figure 3.   30 percentile points in the angular size
distribution of radio sources plotted against flux density.
Data is from Westerbork (Ekers et al., in preparation), •,
and Katgert (1976), +, Ooty (Kapahi, 1975b), o, BDFL (Bridle
et al., 1972), x, and all sky (Robertson, 1973),□. The
hatched area is excluded by selection effects.

The 30 percentile points of the angular size distribution (i.e. 30% of
the sources are larger than the size given) for a number of flux
density bins are plotted against flux density. Since the Westerbork
data is all obtained at 1400 MHz, the BDFL catalogue (Bridle et al.,
1972) has been used instead of the 3C sample. No other data is available
in the intermediate flux density range so the Ooty sample has been
shifted using a mean spectral index of -0.75 between 327 and 1400 MHz.
Similarly, the all sky sample (Robertson, 1973) has been corrected from
408 to 1400 MHz. A further measurement of the angular sizes in the
0.01 - 0.1 Jy* range at 1400 MHz can be obtained from the Westerbork
observations of the 5C2 regions (Katgert, 1976). The 30 percentile

----

* $1 \text{ Jy} = 10^{-26} \text{ W m}^{-2} \text{ Hz}^{-1}$

points derived from this data are also shown in Figure 3. Error bars indicate statistical uncertainty only.

The hatched area in Figure 3 is a rough indication of the area excluded by selection effects. The upper limit results from the use of peak rather than integrated flux densities in radio source surveys and the lower limits result from the finite resolution and sensitivity of the radio telescope used to measure the angular sizes. The separation between these regions becomes uncomfortably small for some ranges of flux density and indicates that further attention to these effects is warranted. The fact that the lower angular size limit for the Westerbork data is very close to the measurements is just a consequence of using the 30 percentile points since this percentile was specifically chosen to avoid this selection effect.

A very suggestive comparison of the (θ,s) and (N,s) results are illustrated in Figure 4 in which we show the data normalized by the Euclidian values. There is a striking similarity between these two plots.

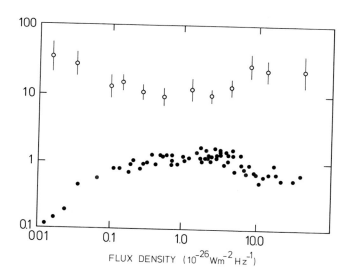

FLUX DENSITY $(10^{-26} Wm^{-2} Hz^{-1})$

Figure 4. The normalized (θ,s) relation, o, and normalized differential (N,s) relation, •, plotted against flux density.

Both relations are close to the Euclidian result over the same flux density range. In both cases there is a much steeper slope in the 3-10 flux unit range and at low flux densities both relations become less steep than Euclidian. A few remarks on this result which indicate the potential of the (θ,s) test can be made.
i) The (θ,s) results are independent of the (N,s) results since the measurement of θ for the sources is independent of the density of these sources in a catalogue. Hence an explanation of the steep $N(s)$ relation in the 3-10 flux unit range as a statistical fluctuation in the number of strong sources would not explain the (θ,s) result.

ii) If the deficiency of strong sources were due to selection against strong extended sources such as DA240 (Willis et al., 1974) this would have given smaller rather than larger sizes in the (θ,s) data in this density range.

iii) A more likely explanation of this similarity is that the relation between average flux density and distance, which depends on the RLF and its evolution, will be the same for both relations.

Model calculations including the RLF and the linear size distribution and their evolution are discussed in the following contribution by Kapahi.

6. SUMMARY

Good information on the angular sizes of radio sources over a large range of flux densities and redshifts is now available. This information can be used to place additional constraints on world models and to give more information on the evolution of radio source sizes and intensities. A schematic illustration of the way in which angular size information could in principle be used to determine results of cosmological significance is illustrated below.

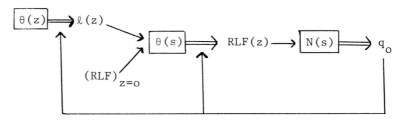

The θ(z) relation depends only on the cosmology and the size evolution function ℓ(z) so for a given cosmological model, quantified by q_0 in this simple outline, we can deduce the ℓ(z) relation. This plus the local RLF can be used to determine the evolution of the RLF from the θ(s) relation. Once we have determined RLF(z) independently of the N(s) relation we can put it in the N(s) relation to determine the cosmology. If this differs from the original assumption we could iterate to search for a consistent solution.

REFERENCES

Bolton, J.G. 1966, Nature 211, 917.
Bridle, A.H., Davis, M.M., Fomalont, E.B. and Lequeux, J. 1972, Astron. J. 77, 405.
Burn, B.J. and Conway, R.G. 1976, Mon.Not.R.Astr.Soc. 175, 461.
Fanoroff, B.L. and Longair, M.S. 1972, Mon.Not.R.Astr.Soc. 146, 361.
Hewish, A., Readhead, A.C.S. and Duffet-Smith, P.J. 1974, Nature 252, 657.
von Hoerner, S. 1973, Astrophys. J. 186, 741.
Hoyle, F. 1959, Paris Symposium on radio astronomy, p.529, ed. R.N. Bracewell, Stanford University Press.

Jauncey, D.L.   1975, Ann.Rev.Astron.and Astrophys. 13, 23.
Kapahi, V.K.   1975a, Mon.Not.R.Astr.Soc. 172, 513.
Kapahi, V.K.   1975b, Ph.D.Thesis, Tata Institute of Fundamental Research, Bombay.
Katgert, P.   1976, Astron. and Astrophys. 49, 221.
Kellermann, K.I.   1972, Astron. J. 77, 531.
van der Kruit, P.C.   1973, Astrophys. Lett. 15, 27.
Legg, T.H.   1970, Nature 226, 65.
Longair, M.S. and Pooley, G.G.   1969, Mon.Not.R.Astr.Soc. 145, 121.
Miley, G.K.   1968, Nature 218, 933.
Miley, G.K.   1971, Mon.Not.R.Astr.Soc. 152, 477.
Swarup, G.   1975, Mon.Not.R.Astr.Soc. 172, 501.
Wardle, J.F.C. and Miley, G.K.   1974, Astron. and Astrophys. 30, 305.
Wellington, K.J., Miley, G.K. and van der Laan, H.   1973, Nature 244, 502.
Willis, A.G., Strom, R.G. and Wilson, A.S.   1974, Nature 250, 625.

DISCUSSION

Baldwin: Is it possible to improve the error bars on the angular diameters in the θ - s diagram at high flux densities?

Ekers: Yes. We could obtain a complete sample of angular sizes for the Southern hemisphere sources.

Wampler: How does one correct for the fact that the sample contains objects such as 3C 273, with one component of the double source on the QSO and the other off the QSO?   Does one double the angular separation?

Ekers: One possibility would be to give any difference and just say that this class is one of the contributions to the linear size distribution function.   Providing the relative number of these objects does not change with redshift this will not introduce any bias.   If one has sufficiently high spatial resolution to recognize these objects for the whole sample then the scatter in the (θ,z) relation may be slightly decreased by including this correction.

McCrea: There is one effect that would account qualitatively for the effects in numbers and sizes at larger redshifts, but I do not know how well it would do so quantitatively.   If two sources are confused and counted as one, the result is to get the size too large and the number too few.

THE ANGULAR SIZE - FLUX DENSITY RELATION

Vijay K. Kapahi
Netherlands Foundation for Radio Astronomy, Dwingeloo
and
Tata Institute of Fundamental Research, Bombay, India

The relation between angular size and flux density depends on the world model and more strongly on evolution in radio source properties with epoch. I shall consider here the simplest forms of evolution that explain the observed θ-S relations (Swarup 1975; Kapahi 1975a,b; Ekers, this symposium).

In order to calculate the expected N (S,θ) relations in any world model, one needs to know the generalized Luminosity Size Function, $\Phi(P,\ell,z)$, that gives the number of sources per unit volume at redshift z, in the luminosity range P to P + dP, with projected linear sizes in the range ℓ to ℓ + dℓ. For simplicity let us assume that the local Φ can be factorized in terms of the Radio Luminosity Function (RLF) and the Radio Size Function (RSF), i.e.

$$\Phi(P,\ell) \ dP \ d\ell = \rho(P) \ dP. \ \psi(\ell) \ d\ell \ ; \qquad \text{where} \int_0^\infty \psi(\ell) \ d\ell = 1$$

This requires that P and ℓ be completely uncorrelated. The observational evidence is presented in Figure 1, where we have plotted P_{178} against ℓ(with H = 50 km/s/Mpc) for a complete sample of 87 3CR galaxies with z < 0.25 (redshift known for 84 and estimated from optical magnitudes for 3 galaxies). The low luminosity part of the figure has been filled in by plotting the B2 sample of radio sources identified with bright ellipticals (Colla et al. 1975) and a sample of giant ellipticals of low radio luminosity (Ekers, private communication). It is clear from Fig. 1 that for $P_{178} \gtrsim 10^{24}$ W/Hz/sr there is little correlation between P and ℓ. While the sizes do appear to be somewhat smaller at lower luminosities, it must be remembered that it is difficult to observe large sources of low luminosity. We shall therefore assume that P and ℓ are independent over the entire range of P, but will keep in mind the possibility that the low P sources may be physically smaller.

We now need to know the RLF and RSF. The local RLF is fairly well determined at low and intermediate P. If we scale the recent results of Perola et al. (this symposium) from 1400 MHz to 178 MHz (using α=-0.75)

D. L. Jauncey (ed.), Radio Astronomy and Cosmology, 119-123. All Rights Reserved.
Copyright © 1977 by the IAU.

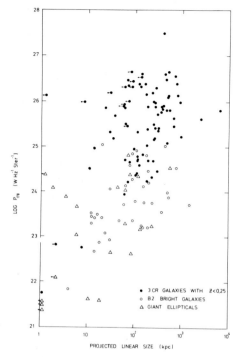

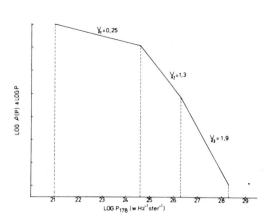

Figure 2. Radio Luminosity Function.

Figure 1. Luminosity-size diagram.

the RLF can be well approximated by two power laws of the form $\rho(P) \, dP \propto P^{-\gamma} \, dP$; with $\gamma_1 = 1.25$ in the range $10^{21} < P_{178} < 4 \times 10^{24}$ W/Hz/sr and $\gamma_2 = 2.3$ for $4 \times 10^{24} < P_{178} \lesssim 2.10^{26}$ W/Hz/sr, as shown in Fig. 2. It is not possible to determine the local RLF for $P \gtrsim 10^{26}$ W/Hz/sr independent of world model and evolution. There is evidence however that the slope of the RLF steepens considerably at higher P. The steepening is in fact required to explain the angular size data in terms of simple evolution (Kapahi 1975a,b). We shall take $\gamma_3 = 2.9$ for $P > 2.10^{26}$ W/Hz/sr and assume that no source has $P > 2.10^{28}$ W/Hz/sr.

The distribution of projected linear sizes for the sample of 87 3CR radio galaxies (shown by Dr. Ekers in the previous talk) is fitted reasonably well with an exponential. We therefore take the local RSF to be given by

$$\psi(\ell) \, d\ell = (1/\ell_o) \exp(-\ell/\ell_o) \, d\ell, \qquad \text{with } \ell_o = 300 \text{ kpc.}$$

Note that this gives a better fit to the observed distribution of sizes than the function used by us earlier (Kapahi 1975a,b).

The $N(S,\theta)$ relations are now readily computed. The expected relations between median values of θ and S in the Einstein-de Sitter Universe,

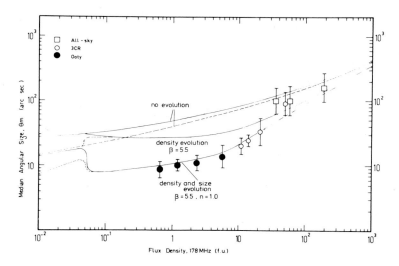

Figure 3. The observed $\theta_m(S)$ relation compared with predictions of evolutionary model.

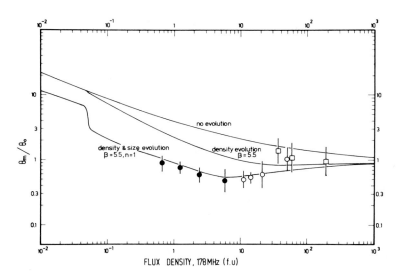

Figure 4. Same as Fig. 3, but with θ_m normalized with respect to the Euclidean expectation, θ_o.

with and without evolution, are compared with observations in Figure 3. This is done also in Figure 4, where θ_m has been normalized with respect to that expected in a static Euclidean Universe ($\theta \propto S^{\frac{1}{2}}$). A few remarks concerning the results in Fig. 3:

(a) The data cannot be explained without evolution because even at high flux densities one is seeing to appreciable z, and since the θ-z relation in most models is fairly flat at high z, going to lower flux levels (higher z) does not reduce θ_m very much.

(b) Values of θ_m at small S can be made smaller by having a larger contribution from more distant sources, as in density (or luminosity) evolution, or by requiring that distant sources have smaller physical sizes (size evolution). Either of the two types of evolution is not enough by itself; both are necessary to fit the data. By requiring a fit with the observed logN-logθ relation (Kapahi 1975b) for the 3CR sample in addition to the θ_m-S data, the amount of density and size evolution can both be determined. The simple forms of evolution assumed and the best fit parameters are listed below.

Density evolution:

$\rho(P,z)=\rho(P,z=0)(1+z)^\beta$ for $P_{178}>2.10^{26}$W/Hz/sr; $\beta=5.5$

$\rho(P,z)=0$ for $z > z_c$; $z_c=3$

Size evolution:

$\psi(\ell,z) = \{1/\ell_o(z)\}$ exp $\{-\ell/\ell_o(z)\}$ dℓ

$\ell_o(z)= \ell_o(z=o) (1+z)^{-n}$; $n = 1.0$

(c) The sharp discontinuity in the θ_m values at $S_{178} \simeq 5$ x 10^{-28}Jy, seen in Figures 3 and 4 is somewhat artificial, resulting from limiting the density evolution to sources with $P>2.10^{26}$W/Hz/sr and from the cutoff in P and z. An extension of the θ_m-S relation to such low flux densities may throw light on the evolutionary properties of sources of low and intermediate luminosity.

(d) The effect of the low P sources having smaller physical sizes is shown by the dashed curves in Fig. 3, where we have assumed that the RSF for $P<4$ x 10^{24}W/Hz/sr has the same exponential form as for higher luminosities but only one third as large an e folding size, i.e. $\ell_o=100$ kpc. Such a correlation of P and ℓ affects the θ_m values only at very low flux densities because low P sources make a negligible contribution at higher flux levels.

The predictions of the evolutionary model are compared in Fig. 5 with the distribution of angular sizes obtained by Katgert (1976) from Westerbork observations of the 5C2 region down to 13 mJy at 1400 MHz.

To summarize, it seems clear that the angular size data provide independent evidence of evolutionary effects similar to those inferred from the analysis of source counts, the V/V_m tests and the θ-z relation

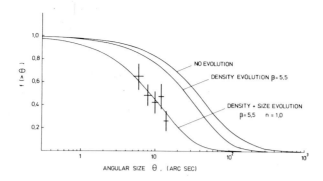

'Figure 5. The fraction of sources with angular size >θ
in the 5C2 region (Katgert 1976) compared with predictions
of evolutionary model.

for QSOs. The models we have considered are admittedly oversimplified.
It should be possible to refine these when detailed distributions of
angular sizes specially at low flux levels become available.

I thank R.D. Ekers for useful discussions.

REFERENCES

Colla, G., Fanti, C., Fanti, R., Gioia , I., Lari, C., Lequeux, J.,
     Lucas, R. and Ulrich, M.H.: 1975, Astron.Astrophys. 38, 209.
Kapahi, V.K.: 1975a, Ph.D. Thesis, Tata Institute of Fundamental Research,
     Bombay.
Kapahi, V.K.: 1975b, Monthly Notices Roy. Astron. Soc. 172, 513.
Katgert, P.: 1976, Astron.Astrophys. 49, 221.
Swarup, G.: 1975, Monthly Notices Roy. Astron. Soc. 172, 501.

DISCUSSION

Conway: A sample of 72 sources from 4C has been mapped at Westerbork
by Conway, Burn and Vallee.   The distribution $N(\theta)$ of the component
separations can be compared with the $N(\theta)$ for the Mackay 3CR sample.
The agreement is excellent apart from the presence in the Mackay sample
of 12% of sources with $\theta > 100"$, which is virtually absent from the 4C
sample.   We have shown that this is not a selection effect introduced
by the WSRT, but it is quite consistent with the result of Caswell and
Crowther that 4C is missing 15% of sources with large θ.   If one
corrects for this, there is no difference between 4C and 3CR in the
statistic $N(\theta)$.   (Flux range between 3.7 Jy and $\sim$ 12 Jy at 178 MHz.

THE FLUX DENSITY-ANGULAR SIZE DISTRIBUTION
FOR EXTRAGALACTIC RADIO SOURCES

G. SWARUP and C. R. SUBRAHMANYA
Radio Astronomy Center, Tata Institute of Fundamental
Research, P.O. Box 8, Ootacamund, 643001, India

SUMMARY

The median values of angular sizes of weak extragalactic radio
sources, the flux densities of which lie in the range of about 0.3 to
5 Jy at 327 MHz, have been determined for a new sample of 119 sources observed during 1973-74, and agree well with the value of about
10 arc sec determined earlier by Swarup (1975). For 8 different flux
density ranges, the angular size distribution for the All-sky, 3CR and
Ooty radio sources have been compared with theoretical predictions
based on the evolutionary model by Kapahi (1975) and show a remarkable agreement with his model except that the best fit is found for a
linear size evolution proportional to $(1+z)^{-1}$.

1. INTRODUCTION

In this paper are presented some new data on the flux density-
angular size distribution, $N(S, \theta)$, for extragalactic radio sources
and these data are used for cosmological investigations. Recently,
Kapahi (1975) presented an independent evidence of evolution in source
properties with cosmological epoch by examining simultaneously (a)
the angular size distribution $N(> \theta)$ for the 3CR radio sources and
(b) the relation between the median value of maximum angular extent,
θ_m, and flux density S derived by Swarup (1975). Here we have
examined all the available data, i.e. angular size distributions $N(S, \theta)$
for 8 different flux density ranges of the All-sky, the 3CR and the
Ooty occultation surveys. A more general function $N(S, \theta, z)$ or
$N(S, \theta, m)$, where z is the redshift and m the apparent magnitude
is also discussed briefly - this function broadly represents the map
of the radio universe and combines several relations $N(S)$, $N(\theta)$,
$N(z)$, $S(\theta)$, (z), $z(S)$ discussed by earlier authors.

D. L. Jauncey (ed.), Radio Astronomy and Cosmology, 125-132. All Rights Reserved.
Copyright © 1977 by the IAU.

2. OBSERVATIONS AND RESULTS

Angular size statistics of 163 weak extragalactic radio sources observed at 327 MHz by the method of lunar occultation at Ootacamund during 1970-71 were presented recently by Swarup (1975) who compared them with the angular extents of the stronger sources from the 3CR and the All-sky catalogues. The open triangles in Fig. 1 show the median values θ_m for an independent sample of 119 radio sources observed at Ootacamund during 1973-74. The new values show a good agreement with the values derived by Swarup (1975) from the earlier sample of 163 sources (closed triangle). It is also seen that median values based on the recent high resolution maps of 3CR sources by Pooley and Henbest (1974), Riley and Pooley (1975) and other recent Cambridge papers, as shown by crosses in Fig. 1, are almost the same as the values derived from the earlier data by Swarup (filled circles).

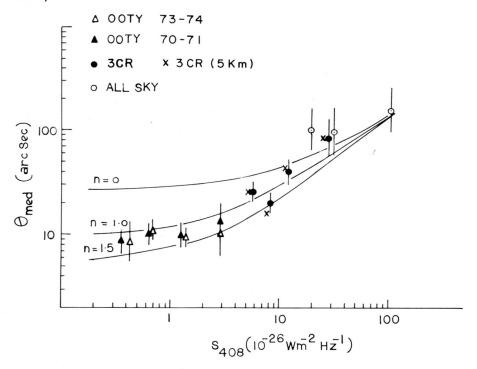

FIG. 1: The median values of angular size θ_m for the new sample of 119 sources observed during 73-74 are compared with the earlier values found for 70-71 data. Also the θ_m values for the 3CR sources found by Swarup (1975 - filled circles) have been revised to include recent high resolution observations. The vertical lines show statistical errors. The 3 curves are theoretical predictions from Fig. 8 by Kapahi (1975).

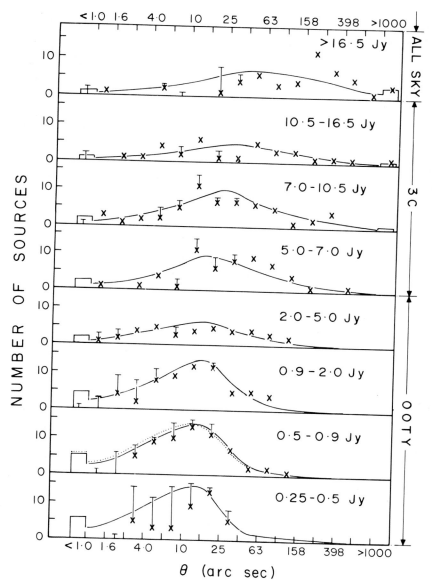

FIG. 2: The observed number of sources and theoretical predictions with angular size in the intervals shown at the top and bottom are plotted for 8 different flux density ranges of the All-sky, 3C and Ooty sources. The observed definite values are shown by crosses and upper limits by bars and theoretical predictions by full line curves. Some 3C sources have flux below 5 Jy but are included in the range 5 to 7 Jy and a few Ooty sources with flux 5 Jy are included in 2 to 5 Jy range.

In Fig. 2 are given distribution of angular sizes of 513 radio sources comprising of (a) 62 radio sources with $S_{408} > 16.5$ Jy from the All-sky Catalogue (Robertson, 1973), (b) 197 sources of the complete sample from the 3CR survey, including 28 sources with $S_{408} > 16.5$ Jy on Wyllie's scale which form part of the above All-sky sample and (c) 282 Ooty occultation sources observed during 1970-71 and 1973-74. The definite values of number of sources N are shown by crosses and upper limits by bars. The full line curves are theoretical predictions for the evolutionary model as defined in Table I of the paper by Kapahi (1975), except that we have taken the value of the exponent n = 1.1 for the linear size evolution proportional to $(1 + z)^{-n}$ because it provides a better fit to the data.

Since the area of the sky covered by Ooty Survey is not known accurately, we have normalized the total number of sources for the theoretical curve in each flux density range to be equal to the observed number. It may be noted anyway that Kapahi's luminosity function (1975) gives a reasonable fit to the source counts, N(S), observed by the Cambridge group. For a statistical comparison of the theoretical calculations (solid lines) and experimental data of Fig. 2 we have considered 72 bins with $\log \theta \leqslant 0.6$, 0.6 to 0.8, ..., 1.8 to 2.0 and > 2.0 for each of the 8 flux density ranges. There are, however, only 64 degrees of freedom as we have normalized for the total number of sources in the 8 ranges as noted above. The standard chi-square test gives values of 74, 71, 73 and 104 for n = 1.0, 1.1, 1.2 and 1.5 respectively, corresponding to a probability of about 30 percent for the first three and about 0.02 percent for the last. Kapahi's 3-slope luminosity function (1976) also gives a chi-square of 71 for n = 1.1 and density evolution parameter $\beta = 6.0$.

It should be noted that the observed values of the maximum angular size θ for the Ooty sources are smaller than the true values because these sources have been scanned along two or three position angles only. To minimize these projection effects, only those sources have been included in the sample of 282 Ooty sources for which the difference between the position angles of the scans is between 30° to 150°. Further, since the major axes of the radio sources are expected to be randomly distributed with respect to the scanning direction, it can be shown that the projection effects essentially shift the theoretical curves towards lower values of θ by a small amount as shown by the dotted line for the range 0.5 to 0.9 Jy. With these corrections a value of n = 1.0 is favoured.

The theoretical curves in Fig. 2 predict that there are less than 8% sources with $\theta > 40$ arc sec around 1 Jy. In order to verify

TABLE I. Comparison of observed and calculated values of n(S, θ, z) for different intervals of S, θ and z

S-range	*z range:	< 60"			30" : 60"			10" : 30"			≤ 10"			n(S)
		1	2	3	1	2	3	1	2	3	1	2	3	
>16.5	obs	37	11	0	4	7	4	0	4	4	0	4	26	27
	cal	38	11	1	4	8	5	3	7	9	2	5	8	
10.5 : 16.5	obs	21	19	3	5	5	5	5	10	8	3	3	13	38
	cal	23	11	1	3	8	6	2	8	16	1	5	16	
7.0 : 10.5	obs	6	10	1	1	6	10	7	12	15	1	6	22	67
	cal	17	10	1	2	7	7	2	7	21	1	5	21	
5.0 : 7.0	obs	3	14	3	5	6	14	2	19	17	3	6	18	64
	cal	13	8	1	2	7	6	1	7	24	1	4	27	
2.0 : 5.0	obs	4	0	11	0	0	14	1	4	21	0	4	43	28
	cal	6	5	0	1	4	6	1	4	30	0	3	39	
0.9 : 2.0	obs	0	0	2	2	4	0	4	4	35	2	2	43	46
	cal	3	3	1	1	2	6	0	2	34	0	2	45	
0.25 : 0.9	obs	0	0	2	0	0	1	1	4	34	1	6	50	90
	cal	1	2	0	0	1	5	0	1	37	0	1	52	

*z range -- 1: ≤ 0.1;   2: 0.1 to 0.4;   3: > 0.4

NOTE: n(S) = total number of observed sources in the corresponding S-ranges. For the 3C sample, area of the sky covered is 4.2 steradian but is not known accurately for the Ooty sample.

whether the occultation method has excluded detection of any large
diameter sources of low flux densities, we have formed a sample of
43 sources from Molonglo and Bologna catalogues with $S_{408} > 0.5$ Jy
(median value ~ 1.2 Jy) and which were expected to be occulted by
the moon during the period of observations at Ootacamund, but exclu-
ding sources with $|b| < 10^{\circ}$ and also if the difference of position
angles of the observed strip scans by the occultation method was
$< 30^{\circ}$.   Occultations were clearly seen for 41 sources and only 6 or 7
out of 43 sources have $\theta > 40$ arc sec, which agrees reasonably
with the expectation from Fig. 2.

3. DISCUSSIONS

It is seen from Fig. 2 that the evolutionary model provides a good
fit to the observed N(S, θ) distribution.   The Steady State model has
a poor fit with chi-square values corresponding to probabilities less
than 10^{-6} and the fit is much poorer if the source counts are also
considered.

In making chi-square test we have given equal weight to all the
flux density ranges.   In fact one should give equal weight to equal
volumes of space.   Also, other observable parameters such as
spectral index distribution, redshift and/or apparent magnitude distribution need to be considered simultaneously.   As shown in Table I,
we have made a preliminary comparison of the theoretical predictions
for a 3-slope luminosity function by Kapahi (1976) with the experimental data of N(S, θ , z) and find a reasonably good agreement.
Unidentified sources were assumed to have $z > 0.4$ and for sources
with no redshifts it was assumed that the absolute magnitude of the
associated optical object was -23 for radio galaxies and -25.8 for
QSOs.   For the weaker sources, any spread in the optical luminosity
function may be important.

To conclude, the flux density-angular size distribution provides
independent evidence for the evolutionary model of the universe.   With
improved data it may be possible to derive the parameters of the
model more accurately.

4. ACKNOWLEDGEMENT

One of us (G. S.) was a recipient of a Jawaharlal Nehru Fellowship
during this period.

REFERENCES

Kapahi, V. K. 1975, Mon. Not. R. astr. Soc., 172, 513.
Kapahi, V. K. 1976, Private Communication.
Pooley, G. G., Henbest, S. N. 1974, Mon. Not. R. astr. Soc., 169, 477.
Riley, J. M., Pooley, G. G. 1975, Memoirs R. astr. Soc., 80, 105.
Robertson, J. G. 1973, Austr. J. Phys., 26, 403.
Swarup, G. 1975, Mon. Not. R. astr. Soc., 172, 501.

DISCUSSION

P. Véron: You have used in your model a redshift cut-off of z = 3 which may be a little unrealistic. What would happen if you were using instead z = 4 or 5?

Kapahi: At the flux levels of the Ooty data the value of the cut-off redshift is not very critical.

Readhead: There are some difficulties associated with the detection of radio sources of angular size greater than a few hundred arc seconds by lunar occultations. Could you describe briefly how your analysis procedure takes account of these, and could you give us a figure for the limiting surface brightness detectable with the Ooty telescope? How many large sources (if any) might have been missed in the survey?

Swarup: The occultation observations are made using the "phase-switched" mode of the Ooty radio telescope, in which the two halves are multiplied together, so that the total drift due to movement of the moon in declination is less than 5° K (~ 2 Jy) over a period of several hours. However, over 2/3rd of the moon's disc the drift is less than about 1° K over periods of few hours. Thus the number of sources of diameter more than about 100 arc second and flux $\gtrsim$ 1 Jy which have been missed are likely to be insignificant, particularly because the measured distributions cut off at a much lower value of the angular size.

Conway: Is it possible that the "angular size" quoted is a mixture in some cases of "Largest Angular Size", i.e. isophotal diameter, and in other cases of Component Separation. Would it be possible to restrict the data, so as to be homogeneous, only to include Component Separations?

Swarup: The "largest angular size" θ defined by us refers not to the separation of the outermost components for the double or complex sources but to the half-power width for unresolved or single sources (see Swarup, MNRAS, 172, 501, 1975). But since more than 80% of the sources are found to be double, the effect of the mixture is considered small. In any case, with the availability of the detailed maps made with the 5 km telescope for the 3CR complete sample and higher resolution observations at Ooty, it is planned to investigate finer features of N(s,θ) distribution.

A STATISTICAL ANALYSIS OF THE ANGULAR SIZE FLUX DENSITY RELATION

J.V. Narlikar

This is a brief description of the investigation by Dr. S.M. Chitre and myself, of the θ-s relation from the Ooty and 3CR surveys.   The θ-s diagram has considerable scatter of the wide range of power and size distributions of radio sources, as well as the projection effects. Indeed, the scatter in the θ-s is worse than that in the optical m-z relation, even for the QSO's.   Just as it is impossible to determine the Hubble constant reliably from the m-z plot for the QSO's alone, so it appears ambitious to arrive at unambiguous cosmological conclusions from the present θ-s data.   A single statistic like θ_m (s) used by Drs Swarup and Kapahi cannot do justice to the information content of the θ-s scatter diagram.

To show this we have used two independent tests of the data.   The first one is a modified form of the familiar χ^2-test which takes into account the variable (and unknown) areas of the sky covered by the Ooty survey.   The second test uses ranking techniques which properly take into account the dispersion of the median θ_m (s).   The tests are used to compare the observed plot with that predicted by evolutionary and non-evolutionary models.   A wide range of models of both types are consistent with the data at 1% level.   The Kapahi model for n = 1.5 appears to be ruled out by the χ^2-test while a non-evolutionary model with a mild power-size correlation survives.   Data with considerably reduced scatter is needed to draw any meaningful cosmological conclusion.

Kapahi: It seems to me that in comparing the predictions of the Steady State and the evolutionary model with the observed θ-s correlation, Dr. Narlikar has chosen rather unrealistic parameters for the luminosity function and the size function which minimize the discrepancy between the steady state and the observations.   For the evolutionary model however, he has used a value of n = 1.5 which is not the best fit value.

Narlikar: In choosing the parameters described by me other checks besides the θ-s relation were performed.   For example, the N-θ relation was taken into account in the 3C-R part in the same way as done by you. Also, a predicted nearby sample on the basis of this model does not seem to disagree with what we know about the luminosity-size distribution of radio sources.

As regards the second point, the cases n = 1, 1.1 etc. have not been examined in this way.   Prof. Swarup tells me that he finds the minimum χ^2 for these cases is better than for n = 1.5.   In that case, it illustrates the power of the proposed minimum χ^2 test in distinguishing between different values of n.   My own interest in the problem was not to look for the best n but to show that at present the data does not rule out non-evolutionary models.

THE ANGULAR DIAMETER—REDSHIFT TEST FOR LARGE REDSHIFT QUASARS

J.M. Riley, M.S. Longair and A. Hooley
Mullard Radio Astronomy Observatory, Cavendish Laboratory,
Cambridge, England.

Complete samples of quasars in the 3CR and 4C catalogues have been observed with a resolution of 2" arc in RA and 2" cosec δ in dec with the Cambridge 5-km telescope at 5 GHz (Riley & Pooley 1975; Jenkins et al 1977), from which it has been possible to determine the overall angular sizes of their radio structures (LAS). The angular diameter-redshift test for quasars has been reexamined using these data; details are described fully elsewhere (Hooley et al 1977).

The samples of quasars used were selected as follows:
a) the 40 quasars in the complete sample of 166 3CR sources which have flux densities $S_{178} \geqslant 10$ Jy using the values given by Kellermann et al (1969), $\delta \geqslant 10^{\circ}$ and $|b| \geqslant 10^{\circ}$.
b) the 19 quasars with $z > 1.5$ and flux densities $S_{178} \geqslant 2.5$ Jy in the complete samples of 4C quasars of Lynds & Wills (1972) and Schmidt (1975).

1. THE ANGULAR DIAMETER-REDSHIFT DIAGRAM

The angular diameter-redshift ($\theta - z$) plot is shown in Fig. 1. It includes all the radio galaxies and quasars with measured redshifts in the sample of 166 3CR sources and the 19 4C quasars with $z > 1.5$. It can be seen from this plot that the well-defined upper bound noted by Miley (1971) still remains. Comparison of this plot with the expectations of homogeneous world models, and its interpretation, depend on whether the quasars and radio galaxies are treated separately or not. If they are considered together, the upper bound is best described by the line corresponding to $\Omega = 0$, though the line $\theta \propto z^{-1}$ is also a reasonable fit. For the quasars alone, however, the upper bound is more or less independent of z and a large value of Ω would fit the data well; this result is similar to that of Hewish et al (1974) which is based on the distribution of the angular diameters of the compact structure within radio components as determined by the scintillation technique, a sample which is also dominated by quasars.

D. L. Jauncey (ed.), Radio Astronomy and Cosmology, 133-138. All Rights Reserved.
Copyright © 1977 by the IAU.

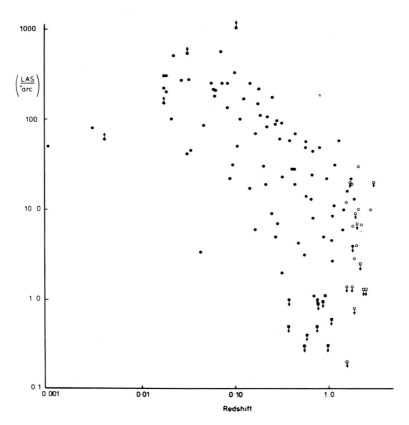

Fig. 1.   The angular diameter-redshift diagram for all the radio galaxies and quasars with redshifts in the 166 sample of 3CR sources (filled
circles and squares) and the 19 4C quasars with z > 1.5 (open circles
and squares).

2.   VARIATIONS IN THE OVERALL PHYSICAL SIZES OF QUASARS WITH REDSHIFT

     The present data may be used to test whether the overall physical
sizes of radio sources of given luminosity are smaller at large redshifts.
A plot of the radio luminosity at 178 MHz emitted frequency, P, against
overall physical size, D, for all the quasars in the present samples is
shown in Fig. 2; it is assumed that $\Omega = 0$ and $H_0 = 50$ km s^{-1} Mpc^{-1}.
The test can be made by inspection of Fig. 2 in the region in which there
is a mixture of sources of large and small redshifts in the same range
of intrinsic luminosity i.e. $6 \times 10^{27} \lesssim P \lesssim 3 \times 10^{28}$ W Hz^{-1} sr^{-1}.   In
this region, the 4C quasars with z > 1.5 span the same range of physical
sizes as the 3CR quasars with z < 1.5 and, in particular, the source
4C 28.40 with z = 1.989, is almost as large as the most extended sources
at small redshifts.   The significance of this result is not great in
view of the very small sample.   It is interesting to note from Fig.2

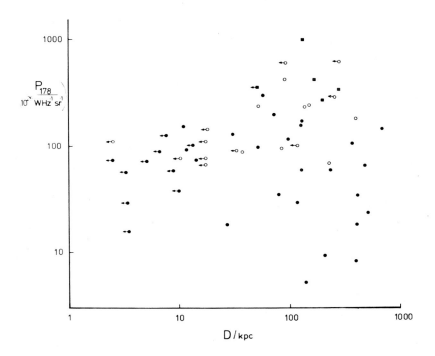

Fig. 2. The luminosity–linear size plot for the 40 quasars in the 3CR 166 sample and the 19 4C quasars with z > 1.5 for $\Omega = 0$. The symbols used are : ● 3CR quasars with z < 1.5, ■ 3CR quasars with z > 1.5, ○ 4C quasars.

that the most luminous sources in the samples are not the most compact.

3. THE STATISTICS OF QUASARS IN THE 3CR AND 4C SAMPLES

Because the 4C quasars are drawn from complete samples it is possible to compare the actual numbers of quasars in different angular size and redshift ranges with the numbers predicted from the 3CR sample of quasars. In this way the overall distributions of physical size in the samples can be used rather than just the upper bound in the θ – z plot or the crude distribution in the P – D diagram.

In order to predict from the 3CR sample the numbers expected in the 4C samples, it is necessary to take account of the effects of cosmological evolution on the comoving space density of quasars at large redshifts. The evolution laws derived by Jackson (1974) from the combined 3CR sample and the 4C sample of Lynds & Wills (1972) were adopted. It was then assumed that the 3CR quasar sample is complete down to 10 Jy, and hence the space density of quasars, with physical properties similar to those of a given 3CR quasar, can be derived directly from the limiting redshift at which that 3CR quasar could be observed and still remain in the 3CR sample.   The angular size distributions expected for quasars in the 4C samples similar to each 3CR quasar were calculated taking account of the optical ($m_v \leqslant 19.5$) and radio limits ($S_{178} \geqslant 2.5$ Jy) of the 4C samples.   The predicted distributions for $\Omega = 0$ and $\Omega = 1$ were very similar.

There is excellent agreement between the predicted and observed angular size distributions for 4C quasars with $z > 1.5$ for both the $\Omega = 0$ and $\Omega = 1$ cases.   This indicates that at least down to a flux density level of 2.5 Jy at 178 MHz the observations are consistent with the hypothesis that the physical sizes of the most powerful quasars do not change with cosmological epoch.   It is important to note that this method of using each 3CR quasar individually means that any correlations between the radio luminosity and overall linear size of quasars are automatically incorporated in the predictions.

This result is in marked contrast to that of Swarup (1975) and Kapahi (1975) who find that evolution in physical size is required to explain their angular size–flux density relation.   However, as the fraction of quasars in their samples is small, exclusion of known quasars from their samples has little effect on the degree of physical size evolution required to explain their data, and there is little evidence from their data alone for physical size evolution in quasars.   The apparent contradiction between the two results could therefore be reconciled by a difference between the overall physical size evolution in radio galaxies and quasars.

4.   CONCLUSIONS

The results of this investigation are consistent with the hypothesis that the overall physical sizes of the radio structures of quasars of the highest radio luminosity do not change with cosmological epoch.

REFERENCES

Hewish, A., Readhead, A.C.S. & Duffett-Smith, P., 1974.   Nature, 252, 657.
Hooley, A., Longair, M.S. & Riley, J.M., 1977.   Mon.Not.R.astr.Soc., in preparation.
Jackson, J.C., 1974.   Mon.Not.R.astr.Soc., 166, 281.
Jenkins, C.J., Pooley, G.G. & Riley, J.M., 1977.   Mem.R.astr.Soc., in preparation.
Kapahi, V.K., 1975.   Mon.Not.R.astr.Soc., 172, 513.
Kellermann, K.I., Pauliny-Toth, I.I.K. & Williams, P.J.S., 1969.   Astrophys.J., 157, 1.
Lynds, R. & Wills, D., 1972.   Astrophys.J., 172, 531.
Miley, G.K., 1971.   Mon.Not.R.astr.Soc., 152, 477.
Riley, J.M. & Pooley, G.G., 1975.   Mem.R.astr.Soc., 80, 105.
Schmidt, M., 1975.   Astrophys.J., 195, 253.
Swarup, G., 1975.   Mon.Not.R.astr.Soc., 172, 501.

DISCUSSION

THE LARGEST ANGULAR SIZE–REDSHIFT DIAGRAM FOR A COMPLETE SAMPLE OF QUASARS

J.F.C. Wardle and R. Potash

The complete sample of quasars from the 3C and 4C catalogues, recently published by M. Schmidt, has been mapped using the NRAO three element interferometer, and a new LAS–z diagram has been constructed. There is an apparent lack of large redshift large linear size quasars. In a q_o = 0 cosmology, which minimises this effect, it is still significant at a level of about 2%. The distribution in redshift and luminosity of the quasars has been analysed using Lynden Bell's C^m method, for three groups of linear size. There is no evidence of a difference in their spatial distribution between the largest and the smaller quasars. However, the optical luminosity junctions appear to be significantly different, in the sense that the luminosity function of the largest sources cuts off at a lower maximum luminosity than that of the smaller sources. The average optical luminosity, $F(2500 \, \overset{o}{A}_{emitted})$ is a factor of 2 less for sources larger than 200 kpc. The distributions of the ratio of radio to optical luminosity show no significant dependence on linear size, implying that on the average the radio luminosities are also smaller for the largest sources.

We suggest that as a source reaches large linear sizes and becomes older, its optical and radio luminosities both decline, leading to an under representation of large sources in any sample which is limited in flux density and apparent magnitude. There is no evidence that the space density of large sources evolves differently from the smaller ones at large redshifts.

Wardle: Several speakers have suggested that the linear size distribution of quasars changes with redshift. I want to stress that our results are in complete agreement with those presented by Mrs. Riley, and we both find no evidence that the linear size distribution is different at high redshifts.

Readhead: Julia Riley has shown that in the complete sample of 3CR and 4C quasars the upper envelope of the LAS distribution does not show the decrease with redshift which is found in the complete sample of 3CR quasars and radio galaxies. As she mentioned, the scintillation results show an apparent lack of component angular size with redshift. I wish to point out that if one looks at the scintillating radio galaxies in the complete 3CR sample, there is good continuity of their LAS with those of the quasars, and that for these galaxies too there is no variation with redshift in the upper envelope to the LAS distribution.

Condon: I would like to point out an important selection effect on the QSO angular diameter–redshift test. This test for QSO's is affected by any incompleteness which depends on source diameter. The spectra of sources identified in samples found at 11cm (Condon, Balonek and

Jauncey, Astrophys. J. 79, 1220, 1975) and 6cm (Condon, Balonek and
Jauncey, Astrophys. J. 80, 887, 1976) indicate that the larger diameter
QSO's have relatively low optical-to-radio luminosity ratios.   All radio
selected QSO's belong to optically incomplete samples (in the sense that
empty fields remain), and the high redshift large diameter QSO's are the
most likely to be missed.   The result of this selection is to make it
appear that we are in a static Euclidean universe (Miley, Mon. Not. Roy.
Astron. Soc., 152, 477, 1971, Wardle and Miley, Astron. & Astrophys.,
30, 305, 1974).

        A closely related selection effect explains the apparent excess of
QSO's larger than 7 arc sec in the northern galactic hemisphere relative
to the southern galactic hemisphere (Miley 1974).   The radio surveys
which found the sources in the two hemispheres were made at different
frequencies; the southern sources being found at the higher frequency.
A chi-squared test shows that the division of sources into the steep
and flat spectrum groups in the northern (50 steep, 23 flat) and
southern (25 steep, 24 flat) galactic hemispheres is nearly as signific-
and as the anisotropy in angular sizes.   Thus statistical QSO diameter-
redshift tests should not be used to determine either the geometry of
the universe or the evolution of QSO's unless the QSO sample is both
optically complete and taken from a single, homogeneous radio survey.

Ekers: I wish to emphasise that the effects of optical selection on
the $\theta(z)$ relation do not apply to the $\theta(s)$ relation.   Perhaps the fact
that the linear size evolution effects required to explain the $\theta(s)$
relation are weaker than those apparently required to explain the $\theta(z)$
results from this optical solution effect in the $\theta(z)$ data.

THE ANGULAR DIAMETER-REDSHIFT RELATION FOR SCINTILLATING RADIO SOURCES

A. Hewish, A.C.S. Readhead and P.J. Duffett-Smith
Mullard Radio Astronomy Observatory, Cavendish Laboratory,
Madingley Road, Cambridge CB3 OHE.

Observational evidence on the angular diameter-redshift relation using radio sources has, until recently, been confined to studies of the overall angular size. In this paper a different approach is described in which the very compact hot spots in radio sources are used as standard measuring rods. Such hot spots appear to be especially suitable for cosmological studies for the following reasons.

(a) Hot spots which contain one half or more of the total flux density are generally found only in the most powerful sources $(P_{178} \gtrsim 10^{27}$ W Hz^{-1} sr$^{-1})$.

(b) Hot spots appear to show a smaller dispersion in physical size than the overall source dimensions.

Hot spots typically have steep radio spectra ($\alpha > 0.6$) and are found near the outer extremities of powerful double sources (Readhead & Hewish, 1976). They should not be confused with the central components of radio sources which have much flatter spectra and have angular diameters in the milliarcsec range.

The only powerful source which is near enough for the hot spots to be mapped in detail is Cygnus A. A small amount of data is available from intermediate baseline interferometry at low frequencies, but for information on a large sample we must, for the present, rely upon the technique of interplanetary scintillation which provides an angular resolution in the range 0".1 - 2".0.

In the method developed by Readhead and I the angular diameter is derived from observations of scintillation over a large range of solar elongation. Scintillation versus elongation curves of this type are found to exhibit good repeatability from year to year and the angular diameter is derived from the slope as measured between elongations of $35^{\circ} - 90^{\circ}$. The method is believed to have random errors of about ± 0".1 for strongly scintillating sources although there is a possibility of some systematic error due to uncertainties in the adopted model of the

D. L. Jauncey (ed.), Radio Astronomy and Cosmology, 139-147. All Rights Reserved.

solar plasma. Cross-checking of scintillation and interferometric measurements at the same frequency should remove any systematic error. In the meantime the limited data available suggest that this error is not large.

The information provided by scintillation is roughly equivalent to that given by the fringe visiblity of an interferometer taken across the largest dimension of a scintillating component. If a source contains several hot spots within an overall diameter of $\lesssim$ 1", or if a hot spot lies within a halo of < 1".0, an equivalent diameter of intermediate size is derived and the apparent flux density in the scintillating component is reduced. If several hot spots are separated by more than about 1" they scintillate almost independently and a mean diameter, strongly weighted towards the smallest component, is derived. A study of blending effects has shown that errors in the case of core-halo type sources only become serious when the halo has an angular diameter of $\lesssim$ 0".5 (Hewish & Readhead, 1976).

In a study of identified sources in the complete sample of extragalactic 3C sources away from the galactic plane a strong correlation has been found between the occurrence of hot spots and the overall linear size. The scintillation measurements were made at 81.5 MHz with the 4.5 acre telescope (Readhead & Hewish, 1974). These results show that hot spots which contain at least half of the total flux density are a common feature of powerful sources whose overall extent is less than 200 kpc, but are rarely found in sources with an overall size larger than 300 kpc. This is illustrated in Fig. 1.

The linear diameters of the hot spots themselves are scattered in the range 1 - 10 kpc and show no correlation with overall size. As shown in Fig. 2, however, there is evidence for an increase of size with increasing luminosity. A selection effect which acts against the detection of the larger hot spots in the weaker, and therefore nearer, sources certainly exists, but it is unimportant when the luminosity P_{178} exceeds 10^{26} W Hz^{-1} sr^{-1}. It should also be stressed that the upper limit of about 10 kpc for the hot spots cannot be explained by observational selection. There are only 8 sources in the sample classified as weak scintillators having P_{178} > 10^{27} W Hz^{-1} sr^{-1}. All of these have been observed with the 5 km telescope and in only one instance (3C 300) is there evidence for a hot spot having a diameter in the range 10 - 20 km. It follows that hot spots are a distinct feature of the most powerful sources and that there is not a continuous distribution of component sizes between hot spots and more extended features having sizes greater than 20 kpc.

To study the angular diameter-redshift relation requires the most accurate scintillation measurements and for this purpose a restricted sample was chosen for which the angular diameters are believed to be accurate to ± 0".1. All the sources in this sample contain at least 40% of their total flux density in hot spots. A few sources in this category having spectral indices of less than 0.6 were rejected to

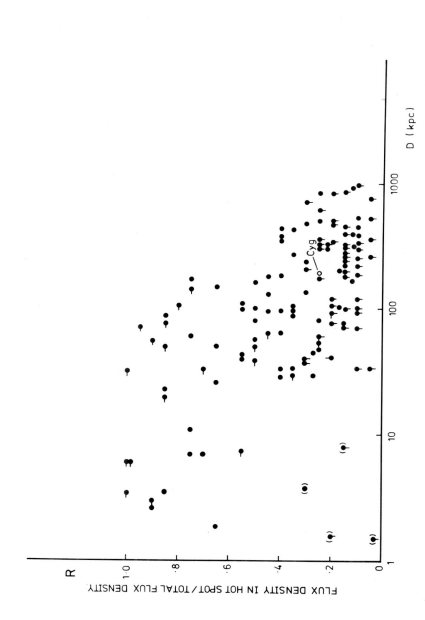

Fig. 1 The fraction of the flux density in hot spots versus overall physical size.

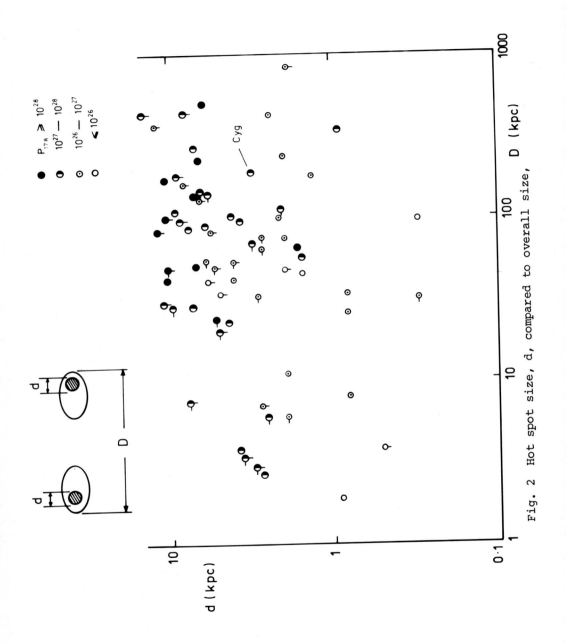

Fig. 2 Hot spot size, d, compared to overall size, D (kpc)

eliminate any possible contribution to scintillation from central components. The results are shown in Fig. 3.

The angular diameters are substantially independent of redshift in the range $0.2 < Z < 2.0$. The absence of sources near the lower resolution limit at large redshifts is particularly significant since the scintillation method is strongly weighted towards the detection of sources of the smallest angular size. These results cannot be seriously influenced by interstellar scattering since many of the sources have been studied at both 151 MHz and 81.5 MHz and the angular diameters do not scale as (frequency)$^{-2}$.

The possibility of blending between hot spots and more extended features which also scintillate has been suggested by Swarup & Bhandari (1976) to account for the absence of small angular diameters at large redshifts. We have shown that this explanation would require a large fraction of the sources to contain hot spots of angular diameter $\lesssim 0\rlap{.}''2$ within more extended components of diameter $\sim 0\rlap{.}''5$ (Hewish & Readhead, 1976); another possibility would be that hot spots are clustered within an area of diameter $\sim 0\rlap{.}''5$. In the light of present evidence suggesting a low value for the deceleration parameter q_o, the simplest model appears to be that the hot spots actually are physically larger in sources which have large redshifts.

In conclusion, brief mention should be given to a new type of scintillation study being carried out by Duffett-Smith. The sensitivity of our radio telescope at 81.5 and 151 MHz should be adequate to detect a background level of scintillation due to sources of low flux density which are too numerous to be resolved individually. The method is essentially an extension of the statistical P(D) analysis introduced by Scheuer long ago. It is difficult to apply in practice since great care is needed to eliminate spurious effects which might arise from, for example, ionospheric scintillation or terrestrial interference. Weather balloons over the continent are already causing difficulties. One result which can be stated now is that the maximum scintillating flux density at an elongation of 15° does not exceed 0.23 ± 0.03 Jy at 151 MHz. This value is appreciably lower than would be expected if the background sources contained similar hot spots to the 3C sample. The hot spots in these weak sources must either have a larger angular size, or contain a smaller fraction of the total flux density, than the stronger sources. The cosmological implications of these conclusions will depend, of course, upon the intrinsic luminosity of the sources. If the sources have comparable luminosities to the 3C sample, but are at larger redshifts, the result could indicate an increasing trend for hot spots to have large angular size at large redshifts.

References

Hewish, A. & Readhead, A.C.S., 1976, Astrophysl Letters (in press)
Readhead, A.C.S., & Hewish, A., 1976, Mon.Not.R.astr.Soc.(in press)
Swarup, G. & Bhandari, S.M., 1976, Astrophys. Letters, 17, 31.

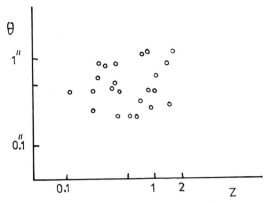

Fig. 3 The θ-z relation for hot spots in strongly scintillating sources.

DISCUSSION

THE S - θ RELATION AT 408 MHz

H.P. Palmer

Radio linked interferometers have been used at 408 MHz to investigate this relation.   (Baselines 24 and 127 Km, $\gamma_{2.7}$ > 0.5 if θ < 2.7" and γ_{127} > 0.5 if θ < 0.57").   Anderson and Richards made drift observations at constant hour-angle in the Dec. 30° (Bologna) area, and found that the fraction of sources having γ_T > 0.5 increases from 29% to 41% between S = 14 Jy and S = 3 Jy for the 24 Km baseline.   However the fraction remains approximately constant at 20% ± 4 for the same source observed with the 127 Km baseline.   There are approximately 200 sources per point for the drift observations on figure 1.

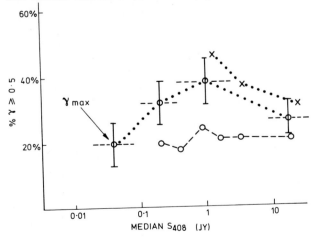

Speed, Spencer and Warwick have made tracking observations at 408 MHz using the 127 Km baseline. They found that the maximum value of γ observed on each track was the best indicator of the size of source components. The fraction of sources having $\gamma_{max} > 0.5$ was found to increase from 21% at $S_{medium} = 14$ Jy to 38% at 1 Jy and then to fall again to 20% by 80 mJy (for observations in the 5C2 and 5C3 areas). When the scintillation data of Readhead, A.C.S. and Hewish, A., Mem. R. Astr. Soc., 78, 1-49, (1974) and Swarup, G. and Bhandari, S.M., Ap. Letters, 18, 31, (1976) is expressed in the same form, it shows reasonable agreement on the rise between 14 and 3 Jy.

Speed has shown that these observations can be predicted from the properties of a complete sample of 57 3CR sources, considered to be moved to progressively larger redshifts, only if a density evolution index $\beta \simeq 6$ is assumed.

Scheuer: Prof. Hewish and Dr. Palmer have both made measurements of the sizes of components (as opposed to LAS). Would either of them care to say whether their results agree or disagree?

Palmer: A source by source plot of 45 sources showing R_{scint} vs $\gamma_{Def\ 408}$ shows good correlation along a 45° line. However the scintillation data do not show a clear variation of R with flux. Both techniques show that R or $\gamma_{T\ Def\ 408} \geqslant 0.5$ for 20% ± 5 for sources with $S \simeq 1$ Jy at 408 MHz.

Concerning the components smaller than 0.2 arcsec absent from the scintillation data at 81 and 151 MHz, this interferometer data does not yet extend smaller than 0.3 arcsecs. Components showing partial resolution on this scale are seen at 1660 MHz, and by VLBI at lower frequencies.

Hewish: I do not see any inconsistency until the intrinsic luminosities of the sources have been measured. Our data indicate a significant decrease in linear size of hot spots with decreasing absolute luminosity and this might explain your correlation with flux density at 327 MHz. The interferometric data do not yet indicate that a large fraction of the high luminosity sources are less than 0.3 arcsec. It will be interesting to see what results are obtained from low frequency interferometry at higher resolution.

Swarup: According to Dr. Hewish, most of high luminosity radio sources have compact components with size at 81 MHz $\gtrsim 0.3$ arc sec. But, according to the Jodrell Bank observations presented by Dr. Palmer, the angular size of compact components decreases with flux density up to ~ 1 Jy at 408 MHz and these refer to sources of high luminosities at redshifts beyond about one, according to our present understanding. Decrease of angular sizes of the scintillating components with decreasing flux density is also indicated by scintillation measurements at 327 and 408 MHz, e.g. of half of the sources around 2 Jy at 327 and 408 MHz have visibility greater than 0.4, indicating that their sizes should be appreciably less than 0.3 arc sec. Thus there seems to be some inconsistency in the observations at 81 MHz when compared to those at the higher frequencies.

Webster: Could not the lack of small scintillation size at large red-shifts also be explained by taking a more realistic source model?

Hewish: One model which could explain this result would be if the bulk of the large redshift sources actually consists of one or more hot spots of size 0ʺ.1 superimposed upon a halo of size about 0ʺ.5.   So far there seems to be little evidence for the existence of steep spectrum sources having sizes as small as 0ʺ.1.

ON THE INTERPRETATION OF $(\theta - z)$ Data

R.C. Roeder

Dyer and I have investigated effects which occur when a source of radiation at a large distance in a universe with mean density ρ_u, on some sufficient scale, is examined by a line of sight along which the average density, $\langle\rho\rangle$, differs from ρ_u.   (Dyer, C.C. and Roeder, R.C., Ap. J. (Lett), 1972, 174, L115; 1973, 180, L31; 1974, Ap. J., 189, 167.) If we define $\alpha = \langle\rho\rangle/\rho_u$, and select lines of sight which pass sufficiently far from intervening galaxies that we can ignore the shear (Weyl focussing) introduced by them, we need only consider any Ricci focussing $(\alpha > 1)$, on each of it $(\alpha < 1)$, relative to the usual homogeneous models $(\alpha = 1)$.

If $\alpha < 1$, angular diameters at a given z will in general be smaller than those given by the usual $(\theta-z)$ relations for homogeneous models. In the limit $\alpha = 0$, $\theta(z)$ is given by

$$\theta^{-1} \propto \int_0^z (1 + x)^{-3} (1 + 2q_0 x)^{-\frac{1}{2}} dx,$$

where q_0 is the cosmological acceleration factor, assuming that Weyl focussing can be neglected.   In order to evaluate situations in which shearing cannot be neglected, numerical integrations must be done.

Thus one cannot properly interpret $(\theta-z)$ data without knowing something about the small scale structure of the universe.

de Felice: Would the determination of a minimum in the observed $\theta-z$ relation give a direct indication of the degree of anisotropy along the line of sight?

Roeder: It is believed that the parameter α is nearly zero.   In order to distinguish the shearing effect (Weyl focussing) from Ricci focussing, you have to be able to resolve sources.   If there were no Ricci focussing $(\alpha = 0)$, then a minimum in $(\theta-z)$ would provide information about the anisotropy.

Oort: Have you made an estimate of how large α would be on the basis of what we know concerning the distribution of galaxies, assuming that the distribution of intergalactic matter is the same as that of the galaxies?

Roeder: I have not done so myself, but it has been suggested to me by various people that $\alpha < \frac{1}{2}$.   However, the apparent detection of an intervening galaxy in the direction of 3C286 suggests that α will vary from one line of sight to another.

Narlikar: When $\alpha \rightarrow 0$, the minimum occurs at $z \rightarrow \infty$.   Is it a finite minimum or a zero minimum?

Roeder: Finite, very similar to the case for $q_o = 0$.

IV

OPTICAL SPECTRA AND IDENTIFICATIONS

THE PRESENT STATUS OF 3CR IDENTIFICATIONS

Jerome Kristian
Hale Observatories, Carnegie Institution of Washington,
California Institute of Technology

This paper is a partial and preliminary report on a review of the optical and radio properties of 3CR sources (Kristian, 1977). It will treat the overall status of the identifications and a few of the optical properties of the identified sources, including evidence for a faint magnitude cutoff in the apparent brightness of 3CR quasars. A summary of the status of 3 CR identifications as of early 1976 has been prepared by Smith, Spinrad, and Smith (1976). The present work includes new data on several dozen sources, as well as a systematic reevaluation of previous identifications.

Progress in 3CR identifications has quickened in the last few years. On the radio side, there has been a wealth of new high-resolution, high-accuracy radio maps and positions. The non-negligible number of earlier misidentifications that have been found point up the need for caution in assessing the reality of a given identification without precise radio and optical positions. On the optical side, new instruments enable spectroscopy and photometry to be done on objects that were inaccessibly faint five years ago. Much of the new optical work has been done by Spinrad, Smith, Burbidge, and their collaborators at Lick, and by Katem, Kristian, Sandage, and Westphal at Palomar. The technical means are now available to complete the 3CR survey optically, including spectroscopy and photometry, to a uniform optical limit of V = 22, although the faintest observations are time-consuming, and each one still a small triumph.

At present, one-third of all 3 CR sources are not identified. This includes most low-latitude sources, for which good radio maps and positions are generally not yet available. The identifications of high-latitude ($|b| > 10^o$) sources are summarized in the bar graph on the next page. Here "EF" (empty field) refers to sources with good radio positions which are blank to the limit of the best available photographs. "No ID" refers to sources without identification for a variety of reasons (no good positions, complex source, crowded field, etc.). One-fourth of all high-latitude sources are in these two categories and are still unidentified. The symbol "?" refers to good identifications, on the basis of radio-optical position agreement, which are too faint to tell whether their images are galaxies or starlike. Most of the identified QSOs have redshifts and photometry, although there are still a few candidates

D. L. Jauncey (ed.), Radio Astronomy and Cosmology, 151-155. All Rights Reserved.
Copyright © 1977 by the IAU.

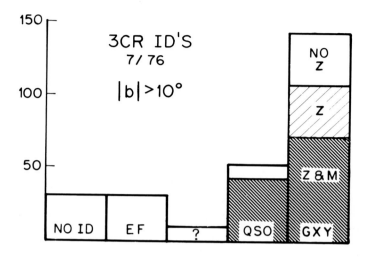

without spectral verification. One-half of the galaxies have redshifts and magnitudes, one-quarter have redshifts only, and one-quarter have not yet been measured optically, although the identifications seem secure from position agreement. Only about the 30 radio-brightest high-latitude sources are optically complete (identifications, redshifts, photometry).

The second figure (see next page) shows the frequency distribution of absolute V magnitudes for all sources with redshifts and photometry. The galaxies have been $\underline{K}$-corrected, assuming q_o (formal) = 1, with no evolutionary correction. This has been found (Kristian, Sandage, and Westphal, 1977) to be a good representation of the data for red magnitudes of galaxy clusters to redshift 0.75, including 3 CR sources that are bright cluster members. The galaxy absolute magnitudes are symmetrically distributed between -22 and -24. Three of the four sources fainter than -22 are suspect identifications on other grounds. Of the three galaxies brighter than -24, one is a variable N galaxy and two have redshifts greater than 0.5, where the $\underline{K}$-correction is large and uncertain. Ignoring these six objects, the mean absolute magnitude of the galaxies is -22.94 $\pm$ 0.42. Including the six increases the dispersion but does not change the mean. This may be compared with the bright cluster galaxy average of -23.30 $\pm$ 0.38 (Sandage, 1973). The quasar absolute magnitudes overlap the bright end of the galaxy distribution and extend to -28. The formal average is -25.95 $\pm$ 1.14.

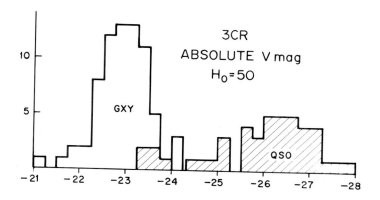

The final figure is the Hubble diagram for all extragalactic 3CR sources with measured redshifts (next page). Magnitudes are uncorrected, except for an aperture correction for the galaxies. Vertical lines are objects with magnitudes estimated from plates. The figure shows perhaps the most interesting results so far: namely, a strong decrease in the number of 3CR quasars fainter than apparent magnitude 19.5-20. Such an effect has already been suggested by Bolton (1969) from Sky Survey identifications of Parkes sources, and by Grueff and Vigotti (1975) from deep 48-inch plate material on B2 sources. There has been some reluctance to accept the effect because of the coincidence with the Sky Survey cutoff, but the present sample contains a good deal of deeper plate material.

Almost all quasars lie in the magnitude range 15-19.5, and show a frequency distribution peaked near magnitude 18. In this magnitude range, 40% of all 3CR identifications are quasars. By contrast, in the magnitude range fainter than 19.5, the 3CR Hubble diagram shows 10 galaxies with measured redshifts and measured or estimated magnitudes and only one quasar. Also, there are now 28 additional sources fainter than 19.5 with identifications based on position agreements of 1-3 arc sec. Of these 12 are galaxies, 6 are probable galaxies, 5 cannot be classified from the available plates, and 5 are quasar candidates. Most of the latter two classes are near the bright end of the range. If the cutoff is taken at 20 mag, there are 14 galaxies (6 with spectra), 6 galaxy suspects, 3 unclassified, 1 quasar, and

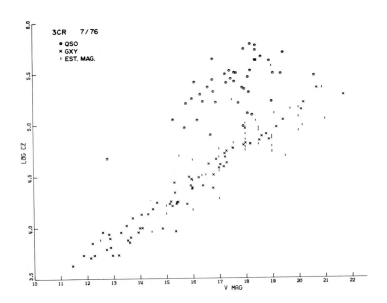

no additional quasar suspects. These frequencies are in sharp contrast to those at brighter magnitudes, and offer strong support for the reality of the decrease of quasars at faint apparent optical magnitudes.

References

Bolton, J.G. 1969, Astron. J., 74, 131.
Grueff, G. and Vigotti, M. 1975, Astron. Astrophys. Suppl., 20, 57.
Kristian, J. 1977, to be submitted to Astron. J.
Kristian, J., Sandage, A.R. and Westphal, J.A. 1977, submitted to
    Astrophys. J.
Sandage, A.R. 1973, Astrophys. J., 183, 731.
Smith, H.E., Spinrad, H. and Smith, E.O. 1976, Minkowski Symposium,
    submitted to Publ. Astron. Soc. Pacific.

This work was supported by a grant from the National Science Foundation.

DISCUSSION

D. Wills: Which 3CR QSO's do you no longer consider to be correctly identified, and is their number consistent with the estimated surface density of radio-quiet QSO's?

Kristian: There are 2 previously known cases, and perhaps one or two

more that are questionable, pending better data.   The numbers I think
are consistent with estimated radio-quiet densities.   These cases were
mentioned as extreme examples - the QSO's generally are in better shape
than the galaxies because they have been worked on harder and longer
and are mostly above the Sky Survey limit.

G. Burbidge: What about the problem of misidentification.   While radio
positions may be good to 1" of arc the intrinsic sizes are 10" - 20" or
greater, so how certain can one be that the identification is correct?
Second, if the galaxy is misidentified, may not the true identification
be a fainter galaxy in the same cluster, or a much more distant galaxy
with a much larger redshift.   Also isn't there a problem with 3C 123 -
the radio and optical positions not being in very good agreement.

Kristian: For individual extended sources I have tried to restrict the
definition of certain identifications to geometrically clean cases, say
a double width 10" - 20" separation with the ID within 1" of the
geometrical center, of which there are many examples.   The chances of
a spurious coincidence to even deep plate limits are very small, and the
confirmation has so far been 100% for those objects with spectra.   This
holds also for galaxies in clusters.   For cluster sources with good maps
and positions, the identifications have been unambiguously with the
brightest cluster galaxy.   3C 295 is a good example, in spite of 2 quite
nearby companions to the source.   With regard to 3C 123, I have just
been told by Pooley (this conference) that the Cambridge map shows a
weak central component exactly (< 1") coincident with the galaxy.

H.E. Smith: Three comments about the possibility of misidentifications
in the 3CR.

1)   The background density of galaxies even to V = 23 is only about 1
     per sq minute of arc; thus the possibility of chance coincident is
     not great.

2)   Many objects have spectral pecularities which confirm their
     identification with active objects.

3)   Specifically with respect to 3C123, the radio structure is a bit
     asymmetrical which may account for the apparent discrepancy.

QSO IDENTIFICATIONS FROM A MOLONGLO RADIO SURVEY

C. HAZARD
Institute of Astronomy, Madingley Road, Cambridge, England.

THE RADIO SURVEY

For the past few years a co-operative programme involving astronomers at the Universities of Cambridge, Sydney and Groningen and Lick
and Steward observatories has been in progress aimed in part at obtaining a complete sample of QSO's in a selected region down to given limits
of optical and radio flux. The radio sample from the 408 MHz Mills
Cross at Molonglo covers two declination strips each 2° wide and centred
at declinations $\sim$11°N and $\sim$17°N respectively. Part of the survey has
been published as the MC2 and MC3 catalogues (Sutton et al. 1974) covering the 11° strips in the R.A. interval 11^h28^m-01^h23^m and the 17° strip
in the interval 13^h30^m-04^h13^m and this section is complete to a flux
limit (S) of 0.45 Jy. The rest of the survey consists of an extension
of the 11° strip around the rest of the sky and to a lower flux limit of
about 0.15 Jy.

IDENTIFICATION PROGRAMME

The regions of sky around the sources are being examined on the
Palomar Sky Survey prints and all objects above the plate limits which
can be considered identifications on the basis of their positional
agreement are listed and their optical positions measured to an accuracy
of better than 1 arc sec. The search areas around each source are
determined mainly by the estimated errors in the radio positions ($\pm$ 5"
for S$\gtrsim$1 Jy increasing to $\pm$ 15" for S$\sim$0.45 Jy) but allowance has been
made also for real optical-radio position differences. Identifications
in general have been suggested only for BSO's brighter than $19^m_.5$ and
galaxies <18.5 but a few fainter objects have been identified for some
of the stronger sources. From random counts in the area of the survey
we estimate that these identifications will have a reliability of at
least 90%. With the relatively large search areas there should be
little discrimination against even large real optical-radio displacements arising due to extended asymmetric structure. It must be emphasized that this is the first stage of the identification procedure.

D. L. Jauncey (ed.), Radio Astronomy and Cosmology, 157-163. All Rights Reserved.
Copyright © 1977 by the IAU.

For objects not identified at this stage more accurate positions are
being obtained at Westerbork and NRAO and structures and spectra are
being obtained at Westerbork and Cambridge for at least the more interesting of the confirmed identifications. Even in the early stages of
the spectroscopic work attention is not being confined to the suggested
identifications (although these naturally have the highest priority)
but many of the neutral and red stellar objects close to the radio
positions are being observed. So far these observations suggest that
such objects will not significantly increase the number of QSO's over
that obtained by concentrating on blue objects alone although of course
they may include some of the interesting objects at very high redshift
(z). The majority have turned out to be either galactic stars or BL Lac
type objects although in the extension of the 11° survey an interesting
high redshift red QSO with z = 3.27 has been found (Beaver et al. 1976).
This, however, is an unusual object with the red colour due to a steep
optical continuum and strong CIV emission which at z = 3.27 falls near
the peak response of the Palomar E-print. A second interesting identification this time with a neutral stellar object is 1400+162, identified
with a BL Lac object in a cluster of galaxies. Weak narrow emission
lines give a redshift of 0.24, the same as the nearest cluster galaxy,
thus confirming it as a cluster member. Its radio structure is particularly interesting, consisting of a flat spectrum compact component and
steep spectrum extended structure some 25" in extent symmetrically
placed about the central component. At low frequencies it would appear
as a symmetrical type double source. This is a typical QSO structure
and links the BL Lac objects to the QSO's and suggests that fundamentally
they are similar objects.

OPTICAL STUDIES

The optical work is most complete for the sources in the MC2 and
MC3 catalogues and particularly in the R.A. interval $11^h30^m-17^h$. Table 1
shows the distribution in R.A. of the suggested BSO identifications in
this part of the survey and also the number so far confirmed as identifications, mainly by spectroscopic studies but in a few cases by accurate
Westerbork positions. The results for the $11^h30^m-17^h$ region confirm
that the reliability of the identifications is indeed of the order of
90%.

It is noticeable that the BSO density appears particularly high
for the intervals $11^h30^m-12^h30^m$ and 17^h-18^m. While the high density in
the latter region must be suspect because of its proximity to the galactic plane all but one of the identifications between 11^h30^m and 12^h30^m
have now been verified. In this interval 36% of the sources are QSO's
compared with a 20% BSO content for the survey as a whole. In fact the
majority of the identifications in this early part of the survey lie
between 11^h46^m and 12^h21^m where there are 10 QSO's out of a total of
20 sources. In contrast between 13^h35^m and 14^h15^m, a much larger area
since it includes both declination strips, there is only one suggested
BSO identification. While the data are too scanty to reach any definite

R.A.	$11°$			$17°$		
	N_s	BSO's	QSO's	N_s	BSO's	QSO's
$11^h30^m - 12^h30^m$	33	12*	11	—	—	—
12 30 - 13 30	36	8	7	—	—	—
13 30 - 14 30	24	3	1	30	7	4
14 30 - 15 30	41	8*	6	43	8*	6
15 30 - 16 30	28	5	5	40	4	4
16 30 - 17 00	15	3*	1	18	6	3
Totals	177	39	31	131	25	17
17 00 - 18 00	38	13	1	45	17*	1
20 20 - 21 20	34	6	—	29	11*	1
21 20 - 22 20	22	5	3	32	6	4
22 20 - 23 20	32	3	2	26	2	1
23 20 - 00 00	17	3	2	14	2*	1
Totals	143	30	8	146	38	8

Table 1. Summary of MC2 and MC3 identification data. An asterisk indicates that one of the BSO's has been shown to be a galactic star.

conclusions they do at least suggest the possibility that the QSO distribution may be non-uniform on a scale of $20°-30°$.

Figure 1 shows the magnitude distribution for all BSO's suggested as identifications up to 17^h R.A. and for comparison the magnitude distribution for the 3C sample of Schmidt (1970) and the 4C sample of Lynds and Wills (1972). All three show the familiar peak in the distribution well above the plate limits. I emphasize that this peak is a real feature of the distributions since when galaxy identifications are taken into account there are just insufficient sources for omitted identifications to greatly change the results although no doubt the steepness of the decrease to fainter magnitudes is exaggerated. This is an important feature since it means that at least for sources >0.45 Jy at 408 MHz the counts of BSO's <19^m5 in different areas will not be greatly affected by either high galactic latitude obscuration or variations in the print sensitivities. There is, however, clear evidence of a shift of the peak to fainter magnitudes with decreasing radio flux.

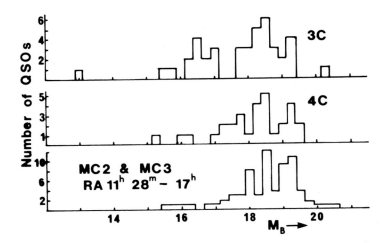

Figure 1. Magnitude distributions of 3C, 4C and MC2 and MC3
          QSO's up to 17^h R.A.

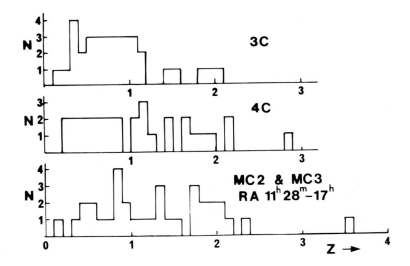

Figure 2. Redshift distributions of 3C, 4C and MC2 and MC3
          QSO's up to 17^h R.A.

      The redshift distributions shown in Figure 2 also show evidence of
a shift to higher values of z between the strong 3C sources and the
weaker 4C and Molonglo samples. The difference between the latter

samples, however, is scarcely significant. Neither is there a significant difference between the z distributions for Molonglo sources with
S>1 Jy and those with S<1 Jy.

The scarcity of objects with z≳2.2 is obvious and again it must be
emphasized that this is almost certainly a real effect. There are no
colour selection effects which would prevent the detection of high z
objects at least up to z∿3.5 except in the case of unusual objects such
as 0938+119. Furthermore once identified the redshifts of such objects
are generally simple to determine because of the presence of strong Ly-α
in the majority. The relative absence of high z objects is shown more
clearly in Figure 3(a) which gives the z distribution for all objects for
which redshifts are available.in the MC2 and MC3 catalogues. Our failure to detect high z objects (z≳3) after our initial successes some
years ago has been disappointing since we then confidently expected
that at least among the neutral St objects we would find many further
examples. However, most such objects have turned out to be continuum
spectrum objects although one or two low redshift QSO's have also been
found among them.

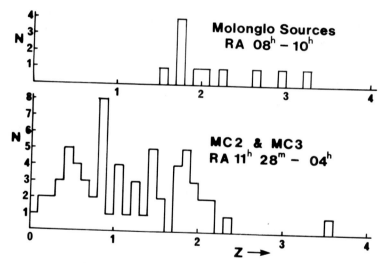

Figure 3.   Redshift distributions for all MC2 and MC3 QSO's
and for a sample of weak source around 08^h-10^h R.A.

We have recently started to examine the extension of the 11°
survey in the region of R.A. 08^h-10^h and the present results shown in
Figure 3(b) are surprising. In contrast to the MC2 and MC3 regions
where only one third of the QSO's have z≳1.5 all of the objects so far
examined in the 08^h-10^h region have z>1.5. It does not appear that
this result can be attributed solely to the lower flux limit of the
new survey since all of the objects with z>2 have S>0.45 Jy and could
have been included if present in the MC2 and MC3 sample. However, it

may be that there is a bigger dependence of redshift on flux than appears from the early results. Alternatively, the result may provide further evidence for a clustering of QSO's.

While it is clear that with the relatively scanty and incomplete data available it is premature to take seriously the evidence for QSO clustering it seems to me that if the paucity of QSO's beyond z∿2.2 and the steep density evolution proposed by Schmidt are real such clustering would not be too surprising. Nor would regions of high z QSO's. The paucity of QSO's with z∿2.2 can be interpreted as representing a real switch in time of QSO's or at least an epoch of maximum QSO formation whose onset is determined by the properties of the Universe at that time. Density perturbations and quite small differences in the expansion velocity about the mean would then imply different switch on times on different regions. When followed by steep density evolution this would lead to drastic differences in the density of QSO's in different regions the QSO's being a particularly sensitive indicator of the conditions in the Universe. Nearby regions which switched on early would now be deficient QSO's while high z objects would be found only in those distant regions where the QSO formation occurred earlier than the average. A completely uniform density of QSO's would certainly require a remarkable uniformity of the effective age of the Universe in unconnected regions. The uniformity of the 3° black body radiation of course argues for this high degree of uniformity and any observed density fluctuations in the QSO distribution must be reconciled with the isotropic background and also the apparently isotropic distribution of the radio sources in general.

Acknowledgements

The identifications discussed in this paper have been carried in collaboration with H. S. Murdoch. The author also acknowledges the contributions of all the other participants in the programme and the support of NATO grant No 647.

References

Beaver, E. A., Harms, R., Hazard, C., Murdoch, H.S., Carswell, R. F. and Strittmatter, P. A. : 1976, Astrophys. J. (Letters), 203, L5.

Lynds, R. and Wills, D. : 1972, Astrophys. J., 172, 531.

Schmidt, M. : 1968, Astrophys. J., 151, 393.

Sutton, J. M., Davies, I. M., Little, A. G. and Murdoch, H. S. : 1974, Austr. J. Phys. Astrophys. Suppl. No. 33.

DISCUSSION

Grueff: I would like to comment on the lack of optically faint quasars present in radio-catalogues. It is certainly true that the distribution of optical magnitudes for quasars in the 3C catalogue is tapering down well before the P.S.S. plate limit, with a peak value at about $m_V = 18$. However, for B2 sources stronger than about 1 Jy, the same distribution is peaked at about $m_V = 19$. Further, there is evidence that at $S_{408} \sim 0.2$ Jy a large fraction of quasars can have $m_V > 20.5$.

Hazard: I agree that the peak of the distribution shifts to fainter magnitudes with decreasing radio flux. This shift is clearly apparent between the 3C sample and the Molonglo sample.

Miley: What is the flux ratio of the components of your 60" quasar with a redshift of 2?

Hazard: Roughly equal.

Miley: So in this case the identification wouldn't have been missed by merely using the position of the centroid?

Hazard: No.

M.G. Smith: Kristian and Hazard have re-emphasised Bolton's point that there is a shortage of optically faint QSOs in the radio catalogues. However, according to unpublished work by Hoag and Weedman, faint QSOs do exist in large numbers at faint magnitudes. Even in the restricted redshift range from 1.7 to 3.3, they find more than a dozen QSOs per square degree in the magnitude range 18.5 to 21. Can Margaret Burbidge or Gene Smith comment on the magnitude distribution in this sample?

Hazard: The radio samples I discussed have a radio flux limit of about 0.5 Jy at 408 MHz compared with a limit of 9 Jy at 178 MHz for the 3C Survey, and show a significant shift of the peak in the magnitude distribution towards fainter magnitudes. No doubt as the flux limit is reduced, fainter QSOs will become more frequent among the radio selected objects and indeed there are already many around 20^m in the Molonglo sample. My point was that irrespective of the reasons for it, above 0.5 Jy the peak must be genuine although the steepness of the fall at fainter magnitudes will be somewhat exaggerated. Thus, provided counts are complete to around $19^m.5$ the numbers of BSOs counted in different regions cannot be greatly affected by high galactic latitude obscuration or print to print variations in the sensitivity of the Palomar Survey, certainly not by factors of the order of 2.

IDENTIFICATIONS FROM THE WSRT DEEP SURVEYS

J.K. Katgert
Laboratorio di Radioastronomia CNR, Bologna

H.R. de Ruiter, A.G. Willis
Sterrewacht, Leiden

1. INTRODUCTION

The WSRT yields samples of radio sources which are well suited for systematic identification work for the following reasons:
a) The surface density is high: at 1415 MHz an average of 10 to 30 sources
can be detected within $0°55$ from the field centre; at 610 MHz the numbers
are 40 to 100 within $1°0$. In any given field, most sources will fall
within the boundaries of plates for the present generation of optical
telescopes.
b) The positional accuracy is, on the whole, rather high. One can thus
propose identifications based on positional agreement only, avoiding
selection effects present when selecting "on type".

The purpose of the present paper is to discuss the identification
data available at present, either already published or in press. Completeness and reliability of the various samples will be reviewed, together with the statistical properties. The basic parameters of the
various surveys are listed in Table 1, which should be largely self-
explanatory. The percentage of identifications in Column 8 is the raw
percentage; Column 9 gives the contamination which will be discussed in
Section 2. The numbers given here do not always agree with those given
by the authors, because we have included only those sources for which
the map flux density, S, $> 6\sigma$, and the distance from the field centre is
less than $0°55$ at 1415 MHz or less than $1°0$ at 610 MHz.

2. COMPLETENESS AND RELIABILITY

High positional accuracy is now becoming more easily obtainable; also
efforts are being made to find non-blue QSO's. Identification work has
therefore been put on a much more quantitative basis than it was
until recently. Descriptions of possible methods for doing this have
been given by Richter (1975) and Condon et al. (1976). De Ruiter et al.
(in press), using WSRT data, use these methods to derive a likelihood
ratio LR as a measure of the reliability of an identification. The LR

D. L. Jauncey (ed.), Radio Astronomy and Cosmology, 165-170. All Rights Reserved.
Copyright © 1977 by the IAU.

Table 1

Surveys with Identifications

No.	Authors	Freq. (MHz)	N	S_{lim} (mJy)	Plates	m_{lim}	ID %	CP %
1	De Ruiter et al., 1976	1415	462	3 - 9	III a-J	22.5	27	4
			39	6 - 9	III a-J	24.0	49	6
2	Katgert & Spinrad, 1974	1415	53	7	PSS	20.5	26	1
			32	7	098-02	22.0	43	9
3	Jaffe & Perola, 1975 $\delta > 20$	1415		4 - 5	III a-J	22.5		
			40				28	6
4	$\delta < 20$		34				29	12
5	Katgert et al., 1973	1415	166	6 - 10	PSS	20.5	18	1
6	Valentijn et al., 1977	610	88	6	III a-J	22.5	35	17
7			38	16			26	4
8	Katgert, 1977	610	226	16 - 29	PSS	20.5	17	3

is defined as the ratio of the probabilities of an object at a given distance from the source position being either the correct identification or the first contaminating object.

These methods are of great use per survey, but they are less practical when comparing surveys, especially as some of the surveys to be discussed have been done at 610 MHz and therefore have much lower positional accuracy. In this case a simpler approach seems useful: by reducing each survey to a single number, one can compare at a glance the identification reliability of the various surveys. The number chosen is the average search area:

$$A = (2.7)^2 \pi/n \sum_n (\sigma^2_{\alpha_{rad}} + \sigma^2_{\alpha_{opt}})^{1/2} (\sigma^2_{\delta_{rad}} + \sigma^2_{\delta_{opt}})^{1/2}.$$

Including all objects within 2.7σ of the source position means that in this respect the sample will be 98.6% complete. σ_{opt} has been taken to be 1 arc sec in both coordinates; as quoted by all authors. This mean search area for each survey has been plotted against the mean density of objects on the plates used for the identifications (Fig. 1). Lines of equal percentage of contamination CP (in terms of the total radio sample) have also been drawn in. The densities of objects for some representative plate types have been marked on the right. The numbering of the surveys corresponds to that of Table 1.

The main survey (1) has been identified on III a-J plates with a limiting magnitude of 22.5. De Ruiter et al. quote an a posteriori reliability of the identifications of 85%, corresponding to a contamination rate of ca. 4.5% (their lower limit to the LR, 1.8, implies this value). This rate is in good agreement with the expected CP of 4% obtained from Fig. 1. For a subsample of survey (1) ($m_{lim} = 24.0$), CP

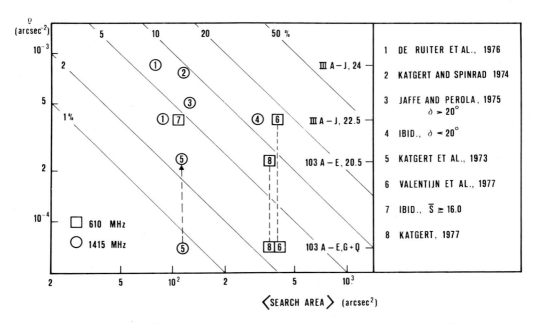

Fig. 1. Contamination for the various surveys. For a description see the text.

is 6% (3-4 sources). De Ruiter et al. find a contamination of 2% (1 source) at the cost of some (slight) loss in completeness: they set the same limit to the LR as for the total sample; due to the higher density of objects this effectively narrows down the search area. It should be noted that, as for this subsample only differential identifications have been done (i.e. m > 20), ρ in Fig. 1 here is really $\Delta\rho$; in Table 1 integral values have been listed for homogeneity. We have subjected the stellar and unclassified objects from survey (1) to a number of tests to see if they could be mainly responsible for the contamination, but they behaved in all respects like the galaxies and objects known to have blue-excess. We will henceforth assume they are good identifications, i.e. have a reliability of 85% as a class.

For survey (2), insofar as deep plates have been used for the identification work, CP is disappointingly large (9%). As the identification percentage is also high, the total reliability should still be acceptable (ca. 80%).

The difference in CP between (3) and (4) reflects the declination difference, as the error in declination is proportional to cosec δ.

In survey (5), Katgert et al. attempted to include normal colour objects also (lower point in Fig. 2). However, the identifications are only complete for galaxies and blue QSO's. It now seems that identification on position would have been sufficiently reliable (upper point).

On the other hand, surveys (6) and (8), when identified on position only, have an uncomfortably high CP (upper points in Fig. 1). Both surveys can usefully be re-defined to include only PSS type-identifications (lower points). As survey (8) was not yet finalized, this was done, and number 6 in Table 1 refers to the restricted case. For survey (6), the numbers in Table 1 refer to the published material.

Fig. 1 proves to be useful in deciding beforehand if it is worth while to obtain deep plates for any given survey. For WSRT 610 MHz surveys this will generally not be the case; for 1415 MHz surveys, it will be. Fig. 1 also indicates how to bring down CP, which one can do in two ways: a) confine oneself to lower densities, and b) go to smaller mean search areas. One can normally do this by setting higher limits in S, as on an average the faintest sources have the largest search area, at least at 610 MHz. Survey (7) is such a subsample of survey (6). At 1415 MHz, for surveys (2) and (5), this does not work, for here the extended sources have the largest errors and mostly a high map flux.

It should be stressed that either re-definition can be applied without loss of positional completeness. Of course, going back to identification by type one loses the neutral or red QSO's, but this seems unavoidable.

3. IDENTIFICATION STATISTICS

We have used the identifications for various statistical investigations. Survey (1) has been used as the primary source of statistics because of its size, and the other surveys have been used for confirmation. The most interesting results come from the magnitude distribution for stellar objects. This is given by De Ruiter et al. for all objects; they note that there is no clear indication of a cut-off at faint magnitudes, viz. that any lack of stellar objects may be caused by the difficulty of classifying identifications and therefore compensated by ?-objects. However, one obtains an interesting result by dividing the stellar objects into two classes of flux density, $S < 60$ mJy and $S > 60$ mJy. One obtains two congruent curves, the one for low S being shifted towards fainter magnitudes by ca. 1^m relative to the one for high S. Keeping in mind the limited numbers (22 in each sample), one can tentatively conclude:
a) that the presence of a considerable number of faint-magnitude QSO's is a reflection of the bi-variate luminosity function; the nearness of bright galaxies (most of the sources concerned lie within $0\overset{\circ}{.}5$ of an NGC galaxy) need not be invoked to explain this effect.
b) that there is some indication that the cut-off is also present at lower flux densities, but at fainter magnitudes. This cut-off would be real if the unclassified objects are faint galaxies.

The QSO's from the other surveys tend to follow an intermediate curve except for some incompleteness for $m > 20$; the numbers are too small to draw any conclusions. The magnitude distribution for galaxies in all surveys is normal; a relative lack of galaxies at m = 20 in survey (1) is not found for other surveys and must therefore be a chance fluctuation.

For survey (1), the identification percentages of the various types are: Galaxies, 10%, QSO's, 10%, Unclassified, 7%. These proportions are confirmed by the other surveys. For surveys (1a) and (2), with deeper plates, the percentages become: Galaxies, 28%, QSO's, 11%, Unclassified, 7%. Clearly the increase in identification rate is due to galaxies only. Again dividing the sources into low S and high S classes, it can also be noted that the brighter sources tend to be identified more frequently with QSO's: of all the sources with S > 60 mJy, 17% are stellar objects; at S < 60 mJy, 7% are. The difference is significant at the 2.6σ level; the identification content of the other 1415 surveys tends in the same direction. The difference could be made up if most or all of the low S unclassified objects are QSO's. No difference in identification content has been found between resolved and unresolved sources.

We have also looked at the colours of all stellar objects with $m \lesssim 20.5$. There are 31 such objects; 21 are blue, 9 are neutral or red, 1 is invisible (possibly variable). All of the brighter (m < 16.5) objects are neutral-to-red and at least two of them have been found to have normal stellar spectra. Assuming that only the fainter ones are QSO's, we estimate that of the order of 20% of all QSO's at the flux density levels concerned should be indistinguishable from normal stars.

4. CONCLUSIONS

WSRT surveys at 1415 MHz are well-suited for identification work on deep plates; at densities of about $4 - 8 \times 10^{-4}$ objects arc sec^{-2} the reliability should still be of the order of 80%. At 610 MHz this reliability is only reached when one confines oneself to the Palomar Sky Survey and uses colour and type information. There is evidence for a dependence on 1415 MHz flux density of the QSO mean magnitudes. There may also be a (flux-dependent) cut-off in magnitude. The fraction of sources that can be identified with QSO's appears to decline with flux density. An estimated 20% of QSO's at present flux density levels have stellar colours.

REFERENCES

Condon, J.J., Balonek, T.J. and Jauncey, D.L., 1975, Astron. J. 80, 887.
Jaffe, W.J. and Perola, G.C., 1975, Astron. Astrophys. Suppl. 21, 137.
Katgert, J.K. and Spinrad, H., 1974, Astron. Astrophys. 35, 393.
Katgert, P., Katgert-Merkelijn, J.K., Le Poole, R.S. and Laan, H. van der, 1973, Astron. Astrophys. 23, 171.
Richter, G.A., 1975, Astron. Nachr. 296, 65.
Ruiter, H.R. de, Willis, A.G. and Arp, H.C., 1976, submitted to Astron. Astrophys. Suppl.
Valentijn, E.A., Perola, G.C. and Jaffe, W.J., 1977, submitted to Astron. Astrophys. Suppl.
Willis, A.G., Oosterbaan, C.E. and Ruiter, H.R. de, 1976, Astron. Astrophys. Suppl., in press.

DISCUSSION

Arp: I would like to comment that of the 21 Quasars noted as blue on
the Sky Survey Prints, my multichannel scans showed them to be noticeably
redder, on the average, than 3CR Quasars.

Jaffe: Are the 3 stars you identified actually radio sources, or is it
reasonable to assume that they are chance contaminations?

J. Katgert: Probably contamination.

Condon: Are the radio sources identified with stellar objects more compact than average?

J. Katgert: No.

LUMINOSITY FUNCTIONS FOR EXTRA GALACTIC RADIO SOURCES

R. Fanti
Istituto di Fisica and Laboratorio di Radioastronomia, Bologna

G.C. Perola
Istituto di Scienze Fisiche dell'Università, Milano

1. DEFINITIONS

The monochromatic luminosity function of radio sources (RLF) is the number of sources per unit volume as a function of the luminosity P at a frequency ν and of the cosmic epoch (z). Symbol: $n(P(\nu),z)$. It is often given per interval of log P, or M_r, the absolute radio magnitude. This function is determined only for sources associated with optical objects (galaxies and QSO's). It can be given for all kinds of associations, or for sources associated with a specific type of object. In this case the normalized, or fractional, RLF is sometimes used, $F_i(P,z) = n_i(P,z)/\rho_i(z)$, where ρ_i is the space density of type i objects. The word "bivariate" is used for the RLF defined per interval of the optical luminosity (or magnitude M). A RLF can be determined using either a radio-optically complete sample of identified sources, or the radio observation of an optically selected sample. The merits of methods used to estimate a RLF from a complete sample are discussed by Felten (1976). Translation of a RLF from one frequency to another must be done with care if, at the two frequencies, different radio components (like the extended and the compact) would be preferentially sampled. We shall review the estimates of local (z = 0) RLF's using $H_o = 100$ Kms^{-1} Mpc^{-1} and the unit WHz^{-1} for P.

2. OVERALL RLF

Wall, Pearson and Longair (this volume) use a sample of 87 3CR sources with S(408 MHz) $\geq$ 10 Jy, of which 74 are identified with galaxies and QSO's having a measured redshift. Thus, they obtain a luminosity distribution that extends down to log P(408) = 21 with the help of Cameron (1971) and Caswell and Wills (1967) (see Sections 3 and 4). The n(P,z) is then estimated by fitting both this distribution and a distribution of source counts down to S(408) = 12 mJy. The best fit of n(P,0) in a Friedmann universe with q_o = 1/2 is shown in Fig. 1. This procedure must be followed for sources with log P > 26, whose local densities are too low for a direct determination. Below this power a local n(P) for galaxies can be constructed using sources within z = 0.1, where evolutionary effects can be neglected and the choice of the universe geometry is not critical. We shall consider separately the contribution of the elliptical and SO galaxies, (E + SO), and of the spiral and irregular galaxies, (S + 1).

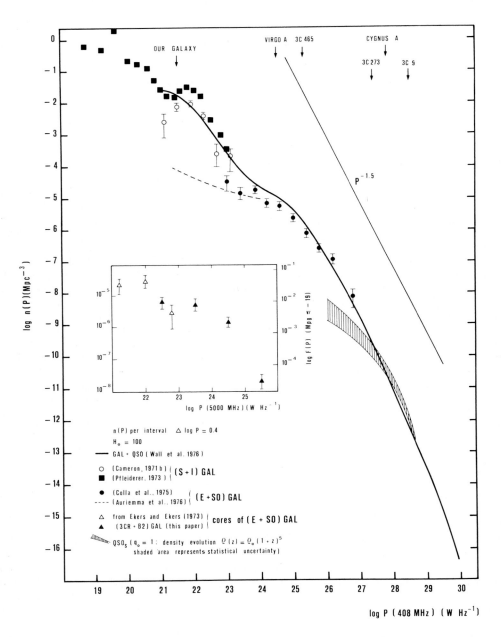

Fig. 1. The overall Radio Luminosity Function.

3. (E + SO) GALAXIES

Colla et al. (1975) use a sample of 53 B2 sources (S(408) > 0.20;0.25 Jy) and 53 3CR sources (S(408) > 5 Jy) identified with E and SO galaxies brighter than m_p = 15.7 and m_v = 17, respectively (the redshift available for all galaxies). Their RLF is given in Fig. 1 along with an extension down to log P = 21.5 by Auriemma et al. (1976; see Section 5) (error bars here and later are equal to $N^{-1/2}$ percent). No attempt is made to separate the E from the SO. Note the rather sharp change of slope at about log P = 25. Earlier estimates of n(P) for log P > 23, where the spiral contribution is negligible (Caswell and Wills, 1967; Merkelijn, 1971), are not as reliable because the distances were mostly estimated indirectly using a fixed absolute magnitude for radio galaxies (see Section 5). However, the flattening below log P = 25 seen by Caswell and Wills but not by Merkelijn is confirmed.

4. (S + I) GALAXIES

Cameron (1971a,b) observed an optically complete sample of bright galaxies (m < 11.0; 12.5) at 408 MHz and detected a total of 85 galaxies with known distance. 72 of these are (S + I) and have log P < 23.5; the corresponding RLF is shown in Fig. 1. The contribution of this type of galaxy peaks at log P = 22 and drops to a density comparable to that of the (E + SO) type at about log P = 23.5. The results of a similar work by Pfleiderer (1973) at 1.4 GHz on a larger sample of galaxies (twice as many detections with known distance) are also given in Fig. 1 (the RLF has been shifted to 408 MHz using a mean spectral index = 0.75). The discrepancy with Cameron's result is probably due to a rather uncertain correction for incompleteness. However, the peak at log P = 22 is confirmed. The extension to lower luminosities shows a minimum at about log P = 21, followed by a new rise. Pfleiderer's data indicate that this structure may be due to a different form of the RLF of the S and I galaxies, the irregulars becoming dominant at log P < 21. This part of the curve, however, is very uncertain due to the limited number of detections available.

5. CORRELATIONS BETWEEN OPTICAL AND RADIO LUMINOSITY IN GALAXIES

To investigate this property, the (E + SO) bivariate RLF has been studied by Ekers and Ekers (1973) at 5 GHz (see Section 7) and by Colla et al. (1975) at 408 MHz. Auriemma et al. (1976) extend the latter work by combining the B2 and 3CR samples with a sample of 17 radio galaxies in rich clusters (Jaffe and Perola, 1976) and a sample of 23 non-cluster radio galaxies (Ekers et al. 1976) to obtain a bivariate F(P) at 1.4 GHz for $M_p \leq -18$ per interval of one magnitude. Fig. 2a shows the result of fitting power laws to the data. Note the break in the slope for each interval of M_p at about the same power log $P^*_{1.4}$ = 24.5; at larger P the slope is equal to 1.3 ± .1 independent of M_p up to log P = 26; at smaller P the slope decreases with M_p. The latter property reflects the approach to the saturation (100%) of the integral value of F(P). Above P^* the

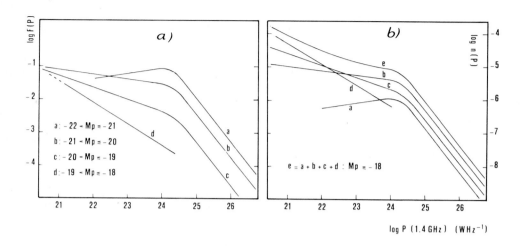

Fig. 2.   a) bivariate fractional RLF and b) bivariate RLF.

$F(P)$ depends on the optical luminosity L as $L^{-1\cdot5}$.   From the bivariate $n(P)$ (Fig. 2b) a most probable value of M_p = -20.3 for radio galaxies is derived for $P > P^*$, independent of P, while this value increases as P decreases for $P < P^*$.   The first result corresponds to that obtained by Sandage (1972) for a sample of 3CR sources, $\langle M_v \rangle$ = -21.48, and shows that the use of a constant $\langle M \rangle$ to construct a distance modulus for radio galaxies is justified locally only for 24.5 < log $P_{1.4}$ < 26.   Beyond log P = 26 a purely local determination of $\langle M \rangle$ is not possible for statistical reasons, while evolutionary effects might change its value at z > 0.1.

For the (S + I) galaxies, using the data from Cameron (1971a), one finds that the bivariate peaks at about log P_{408} = 22 for -21 < M_p ≤ -20 and -20 < M_p ≤ -19, and log P_{408} = 21.4 for -19 < M_p ≤ -18, showing that the average radio luminosity is larger for the brighter galaxies.

6. GALAXIES INSIDE AND OUTSIDE CLUSTERS

The results of the observations of the (E + SO) in 5 rich clusters by Jaffe and Perola (1976) at 1.4 GHz are compared by Auriemma et al. (1976) with those obtained also at 1.4 GHz by Ekers (1976) on a sample of (E + SO) galaxies outside rich clusters (but still mostly in aggregates of various sizes).   In Fig. 3 the two $F(P)$ for M_p ≤ -19 are shown.   To take care of the effects of the dependence on M of the bivariate F, in constructing the $F(P)$ from the second sample its optical luminosity distribution has been normalized to that of the cluster sample.   From the comparison there seems to be no systematic difference in the two function below log $P_{1.4}$ = 24.   Above this power, Riley (1976), from a critical

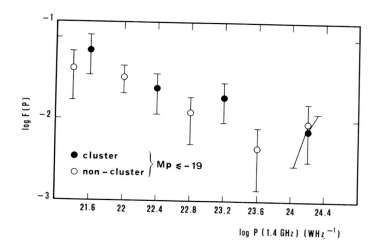

Fig. 3. RLF for cluster and non-cluster radio galaxies.

analysis of the identification of sources in various surveys with Abell clusters, concludes that a) the RLF of those sources is similar in form to the general one from the 3CR, and that between 1/10 and 1/4 (this gives a measure of the uncertainties) of all radio sources up to $\log P_{1.4} = 26.5$ lie in Abell clusters; b) galaxies brighter than $M_p = -20$ have the same probability of being a radio source inside as well as outside Abell clusters; and c) no correlation is evident between the richness of a cluster and its probability of containing a source. On the other hand Owen (1975), on the basis of observations of over 500 Abell clusters with the 300 foot NRAO telescope at 1.4 GHz, finds a form of the RLF that peaks at $\log P_{1.4} \simeq 24.5$. He tentatively explains this feature as being due to the summation in its beam (10') of the radio emission of more than one galaxy per cluster (see also Jaffe in this volume). McHardy (1974), using 3CR and 4C sources in clusters, finds that powerful radio sources tend to be more common in Bautz-Morgan class I clusters, a property which does not seem to be explicable only with the difference in the bright end of the optical luminosity function between class I and the other BM classes. Owen (1975) finds an analogous trend (higher detection probability of clusters dominated by a giant central galaxy) using the Rood-Sastry classification.

For the (S + I) Jaffe and Perola (1976) (see also Jaffe et al. 1976) find evidence that the spirals in Coma (a spiral poor cluster) are more powerful relative both to the nearby spirals and to those in a spiral rich cluster like Hercules. More clusters need to be studied to see if this is a general property differentiating spiral poor from spiral rich clusters.

7. STATISTICS ON COMPACT SOURCES IN ELLIPTICAL GALAXIES

High resolution, high frequency observations show the existence of small
diameter (< 1-2 arcsec) radio sources (cores) in the central region of
elliptical galaxies. These cores are generally found either by high
frequency observations of nearby elliptical galaxies (e.g. Heeschen, 1970;
Ekers and Ekers, 1973) or by high frequency, high resolution observations
of previously known radio galaxies (e.g. Pooley and Henbest, 1974; Riley
and Pooley, 1975; Colla et al. 1975).

The physical sizes of these cores, as derived from VLBI observations,
range from ~ 1 parsec (e.g. Cyg A, Kellermann et al. 1975) to about 1 Kpc
(e.g. 3C 236, Fomalont and Miley, 1975). However, the actual size distribution is not known due to the incompleteness of the sample studied with
the VLBI technique. Typically core sources of "Cyg A" type have a flat
spectra, while core sources of "3C 236 type" have normal radio spectra.
These cores are often considered as a manifestation of the explosive
phenomena occurring in the nuclei of elliptical galaxies, which ultimately
supply energy to the extended component of the radio source. A statistical study of the properties of cores could shed light on the problem
of how extended radio sources are formed and then powered during their
evolution.

With the data presently available in the literature we obtain:
a) the distribution of the ratio $\underline{r}$ = $P_{ext.}/P_{core}$ as a function of the
total radio luminosity P_{tot}; b) the RLF $n(P)$ of the cores in elliptical
galaxies; and c) the frequency of cores as a function of the optical
luminosity of the parent galaxy.

The present analysis is based on: i) a sample of 53 elliptical radio
galaxies from the B2 catalogue as studied by Colla et al. (complete sample)
ii) a sample of 34 radio galaxies brighter than m_v = 17.0, from the 3CR
catalogue (almost a complete sample) and an incomplete subset of 25 optically weaker 3CR radio galaxies. All these radio galaxies have been
observed at 5 GHz at Westerbork and Cambridge (see references above).
Several core sources have been detected (31, 20 and 9, respectively)
among them corresponding to percentages ranging from 35 to 60%. Since
many cores are near the detection limits, we expect that there may be more
which have not been detected, especially among more distant galaxies.
From the available maps it is possible to estimate upper limits on the
intensity of undetected core sources. On the other hand, for radio
galaxies which appear "core-like" (9 in the B2 sample and 1 in the 3CR
sample), we have estimated an upper limit to the intensity of an extended
component.

a) Through proper use of the upper limits on $\underline{r}$, we estimate the "true"
percentage of cores in radio galaxies, as a function of P_{tot}(408 MHz)
and of the ratio $\underline{r}$ (at 5 GHz). The result is given in Table 1. Each
box contains the ratio between cores which could have been detected if
they had equal or smaller $\underline{r}$ values. Table 1 shows: i) for each interval
of P_{tot} the fraction of radio galaxies with a core is > 70%; ii) the

core luminosity does not seem to increase with the total luminosity. However, a correlation of the type $P_{core} \propto P_{tot}^{1/2}$ is equally possible as a null correlation. This result, if subsequently confirmed, is rather important in connection with the problem of an energy supply to the extended radio components. It would seem that the core luminosity is not a good indicator of the energy supply required to maintain the current luminosity of the extended component.

b) We obtain a RLF at 5 GHz of cores of elliptical galaxies from Table 1 and from the RLF of radio galaxies. It is displayed in Fig. 1, along with that derived from the survey at 5 GHz of nearby elliptical galaxies by Ekers and Ekers. Because the selection was made at a much lower frequency, where the extended components dominate, our determination may be biased against core-like flat spectrum sources. However, agreement with the results of Ekers and Ekers shows that this selection effect is not significantly large and that most of the cores are therefore associated with extended components. In the range $21 < \log P_{5000} < 24$, about 50% of the cores are associated with extended components of a similar or stronger radio luminosity (as noted also by Ekers and Ekers). The RLF of the cores seems to have a break around $\log P_{5000} \simeq 24.5$, as does the general RLF. The slope beyond the break is not well determined, but it may be even steeper than that of the general RLF since there are no known core radio galaxies much stronger than $\log P_{5000} \simeq 25$. A determination of the cores RLF at these radio luminosities should be derived in the future from optical identifications of radio sources in the 5 GHz NRAO radio survey.

c) Colla et al. (1975), using 5 GHz Westerbork observations of the B2 sample, gave evidence that the frequency of cores in elliptical galaxies depends on the optical luminosity of the parent galaxy in a way similar to the overall radio emission. Adding the galaxies of the 3CR samples, for which accurate magnitudes are available, confirms this result. Specifically, the fraction of ellipticals in the ranges $(-22 < M_p < -21)$, $(-21 < M_p < -20)$, $(-20 < M_p < -19)$, with cores having $\log P_{5000} > 22.8$, is $(19 \pm 10)\%$, $(6 \pm 2)\%$ and $(1 \pm .5)\%$, respectively.

Table 1

Percentages of Cores of Different r as Function of P_{tot}

$\log r$	$2.5 \pm .5$	$1.5 \pm .5$	$.5 \pm .5$	< 0.0
$\log P_{tot}$				
$> 26.$	$29 \pm 14\%$	$34 \pm 12\%$	$4 \pm 4\%$	
$25.5 \pm .5$	$5 \pm 5\%$	$30 \pm 11\%$	$20 \pm 9\%$	$23 \pm 9\%$
$24.5 \pm .5$	$13 \pm 9\%$	$20 \pm 11\%$	$25 \pm 11\%$	$24 \pm 11\%$
$23.5 \pm .5$		$8 \pm 8\%$	$34 \pm 14\%$	$24 \pm 11\%$

8. QUASARS

It is not possible to determine a local RLF for quasars, since their local density (z < 0.1) is too low for a direct determination. All of our information comes from objects at larger distances (z >> 0.2), where evolutionary effects are dominant. Therefore, a determination of the RLF of quasars at the present epoch is strictly connected to the deter- mination of their evolution law. Evidence concerning the evolution of quasars has been reviewed by Schmidt (1972a) who showed that it can be accounted for by a pure density evolution described either by an expo- nential law $\rho(z) = \rho_0 \, 10^{5\tau}$ (τ being the look-back time in units of the age of the universe) or by a power law $\rho(z) = \rho_0 \, (1+z)^n$, with n = 6±1.

We assume, in that which follows, the second expression with n = 5. From the complete samples of radio quasars (QSS's) available in the literature, 3CR (see Smith et al. 1976) and 4C (Lynds and Wills, 1972), a RLF extrapolated to z = 0, by means of the assumed evolution, is derived This is shown in Fig. 1. A comparison with the overall RLF of radio sources shows that, for $\log P_{408} > 27.5$, quasars contribute to most of the space density of radio sources, while at $\log P_{408} < 27.0$, the contri- bution of the radio galaxies is dominant (see also Schmidt, 1972b). Chang of n by ±1 would change the local space density of quasars by about a factor 3 at the bright end but would not change the lower end significantl The total space density of quasars down to $\log P_{408} \simeq 26.0$ is $(10 \pm 2.5) \cdot 10^{-9}$ Mpc^{-3}. This value may be compared with the density of all quasars obtained from optically selected samples regardless of their radio emissio (QSO's) (Braccesi et al. 1970; Schmidt, 1970). The optical luminosity function of quasars, determined from these optical samples adopting the same evolution law as for the radio quasars, is a power law with slope 1.5 ± .15. Above an absolute optical luminosity, at 2500 Å, $F_{2500} = 10^{22} \cdot$ the integrated space density, at zero redshift, is $(1.3 \pm .4) \cdot 10^{-7}$ Mpc^{-3} and above $\log F_{2500} = 21$ it might be up to $\simeq 10^{-5}$ Mpc^{-3}. It has been shown (Fanti et al. 1973) that strong radio emission is a property of only those quasars with $\log F_{2500} > 22.4$, while optically weaker quasars are very seldom strong radio emitters. Thus, it appears that only about 8% of all quasars with $\log F_{2500} > 22.4$ are radio sources with radio luminosity larger than 10^{26}.

Several attempts have been made to extend the determination of the RLF at lower radio luminosities. Most of these attempts are done through the use of the so-called $\Psi(R)$ function. This represents the fraction of all quasars, per unit volume, having ratio R between the monochromatic radio and optical luminosities at 500 MHz and at 2500 Å, respectively (see Schmidt, 1970). The introduction of this function came from Schmidt' finding (1970) that the redshift distributions at the 18th magnitude for QSS's and QSO's are very similar, which seemed to imply that the radio and optical luminosities are correlated. $\Psi(R)$ may be determined on the basis of complete samples of QSS's, by knowing the number—magnitude, $N(<m)$ relationship for QSO's (see Schmidt, 1970; Fanti et al. 1973) and by the formula

$$\Psi(R) \cdot \Delta R = \frac{N(<m,r)}{N(<m)}$$ (8.1)

where $N(<m,r)$ is the number of QSS's in the considered sample which are brighter than m and possess a radio to optical flux ratio equal to $R \pm \Delta R/2$. Alternatively, it can be found by searching radio emission from optically selected quasars. $\Psi(R)$ is reasonably well determined for $\log R > 3.5$ from the 3CR and 4C samples. Attempts to determine $\Psi(R)$ below that value have been made in the past by Fanti et al. (1973, 1975) and Katgert et al. (1973) on the basis of: i) a sample of quasars from the B2 catalogue; ii) a search of radio emission down to $\simeq 5 - 10$ mJy, from the sample of QSO's studied by Braccesi et al. (1970).

From the first sample it appeared that $\simeq 30 - 100\%$ of quasars (with $\log F_{2500} > 22.4$) have radio luminosities with $\log R > 2.0$. From the second sample it was determined that at most 20% of all quasars have $\log R > 2.0$. Since the two determinations of $\Psi(R)$ were thought to be significantly different, and were also obtained from two samples at different depths in space, the discrepancy was assumed as evidence of an evolutionary effect in quasars with $2.0 < \log R < 3.0$ (Fanti et al. 1973, 1975). Recently other attempts have been made to determine $\Psi(R)$, at $\log R < 3.0$, using: i) a sample of QSS's from the Westerbork survey (de Ruiter, Willis and Arp, 1976); ii) a search for radio emission from a new sample of 51 QSO's studied spectroscopically by Schmidt (1974) plus 8 bright QSO's from the Tonantzintla catalogue (Iriarte and Chavira, 1957). This search led to 8 radio detections above 10 mJy (Fanti et al. 1976). This figure is in excellent agreement with the previous result by Katgert et al. concerning the QSO's studied by Braccesi et al. The various determinations of $\Psi(R)$ are shown in Fig. 4 (again only for those quasars with $\log F_{2500} > 22.4$). The new radio data on QSO's have been combined with the previous measures by Katgert et al. There is still a systematic difference (at $\sim 2\sigma$ level) between the $\Psi(R)$ determinations from the optically selected and the radio selected samples in the range $2.0 < \log R < 3.0$. However, the previously suggested explanation of an evolutionary effect has now been ruled out, since most of the new radio observed QSO's span the same range of apparent optical luminosity and, therefore, of redshift as the radio samples. The difference may be just a large statistical fluctuation or it may be due to an incorrect $N(<m)$ relationship at the very bright end ($16.0 < m < 17.5$), which is used to compute the $\Psi(R)$, and, therefore, to an evolutionary behaviour at small redshifts ($z < 0.5$) different from the assumed power law.

If we assume the determination from the QSO samples in the range $2.0 < \log R < 3.0$ as the best estimate of $\Psi(R)$, we see that at $\log R > 1.5$ only about 17% of all quasars are radio emitters. Presumably, the bulk of them have $\log R$ values much lower than 1.0. $\Psi(R)$ might also have a maximum near $\log R = 2.5$ and then decrease at low R's. This shows that the phenomenon of strong radio emission is not more frequent among the quasars than it is among the elliptical galaxies. We are led to think either that the radio emission is a transient phenomenon in the life of quasars or that only in a small number of cases the physical conditions favour a strong emission at radio wavelengths.

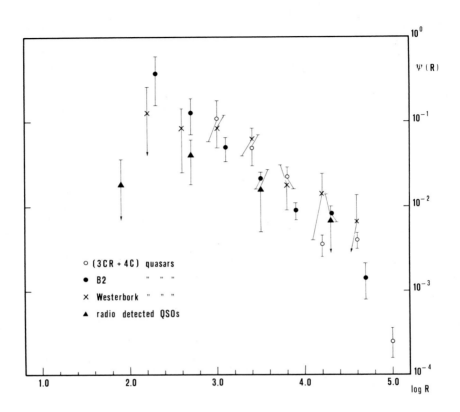

Fig. 4. The Ψ(R) function.

REFERENCES

Auriemma, C., Perola, G.C., Ekers, R.D., Fanti, R., Lari, C., Jaffe, W.J. and Ulrich, M.-H., 1976, in preparation.
Braccesi, A., Formiggini, L. and Gandolfi, E., 1970, Astron. Astrophys. 24, 247.
Cameron, M.J., 1971a, Mon. Not. Roy. Astron. Soc. 152, 403.
Cameron, M.J., 1971b, Mon. Not. Roy. Astron. Soc. 152, 429.
Caswell, J.L. and Wills, D., 1967, Mon. Not. Roy. Astron. Soc. 135, 231.
Colla, G., Fanti, C., Fanti, R., Gioia, I., Lari, C., Lequeux, J., Lucas, R. and Ulrich, M.-H., 1975, Astron. Astrophys. 38, 209.
Ekers, R.D. and Ekers, J.A., 1973, Astron. Astrophys. 24, 247.
Ekers, R.D., Ekers, J.A., Rogstad, D.R. and Smeding, A., 1976, in preparation.
Fanti, C., Fanti, R., Ficarra, A., Formiggini, L., Giovannini, G., Lari, C. and Padrielli, L., 1975, Astron. Astrophys. 42, 365.
Fanti, C., Fanti, R., Lari, C., Laan, H. van der, Padrielli, L. and de Ruiter, H.R., 1976, in preparation.
Fanti, R., Formiggini, L., Lari, C., Padrielli, L., Katgert-Merkelijn, J.K. and Katgert, P., 1973, Astron. Astrophys. 23, 161.

Felten, J.E., 1976, submitted to Astrophys. J.

Fomalont, E.B. and Miley, G.K., 1975, Nature 257, 99.

Heeschen, D.S., 1970, Astrophys. Letters 6, 49.

Iriarte, B. and Chavira, E., 1957, Bull. Obs. Ton. No. 16, 3.

Jaffe, W.J. and Perola, G.C., 1976, Astron. Astrophys. 46, 275.

Jaffe, W.J., Perola, G.C. and Valentijn, E.A., 1976, Astron. Astrophys. 49, 179.

Katgert, P., Katgert-Merkelijn, J.K., Le Poole, R.S. and Laan, H. van der, 1973, Astron. Astrophys. 23, 171.

Kellermann, K.I., Clark, B.G., Niell, A.E. and Shaffer, D.B., 1975, Astrophys. J. Letters 197, L113.

Lynds, R. and Wills, D., 1972, Astrophys. J. 172, 531.

McHardy, I.M., 1974, Mon. Not. Roy. Astron. Soc. 169, 527.

Merkelijn, J.K., 1971, Astron. Astrophys. 15, 11.

Owen, F.N., 1975, Astrophys. J. 195, 593.

Pfleiderer, J.L., 1973, Mitt. Astron. G. 32, 108.

Pooley, G.G. and Henbert, S.N., 1974, Mon. Not. Roy. Astron. Soc. 169, 477

Riley, J.M., 1975, thesis.

Riley, J.M. and Pooley, G.G., 1975, Mem. Roy. Astron. Soc. 80, 105.

Ruiter, H.R. de, Willis, A.G. and Arp, H.C., 1976, submitted to Astron. Astrophys. Suppl.

Sandage, A., 1972, Astrophys. J. 178, 25.

Schmidt, M., 1970, Astrophys. J. 162, 371.

Schmidt, M., 1972a, Astrophys. J. 176, 273.

Schmidt, M., 1972b, Astrophys. J. 176, 303.

Schmidt, M., 1974, Astrophys. J. 193, 509.

Smith, H.E., Spinrad, H. and Smith, E.O., 1976, submitted to Publ. Astr. Soc. Pacific.

DISCUSSION

H.E. Smith: If I understand correctly, your luminosity function requires that $<M_p>$ for powerful radio galaxies increase by $0^m.4$ per decade of radio power. However, if you look at 3CR galaxies with $m < 19'$ where there is some measure of completeness, then there is a difference of less than 0.3 mag over about 4 decades of radio power.

Perola: I have shown that $<M_p>$ depends on the power of the sources onl if P is less than 10^{24} WHz^{-1} and that at larger powers $<M_p>$ is independet of P. This result is in full agreement with what you say about 3CR galaxies, whose powers, at z larger than 0.03, are larger than $10^{24.5}$. In fact the result you mention also indicates that $<M_p>$ in the large power range stays about constant beyond the limit z=0.1 imposed in the determination of the bivariate n(P) in our Fig. 2b.

Conway: Data by Edwards at Jodrell Bank gives the power emitted by central components vs. the power emitted by the whole source for 26 bright 3C sources, and suggests that these are proportional to each other for radio galaxies. The ratio is higher for quasars.

R. Fanti: If one plots core luminosity against luminosity of the whole

source, one actually finds a proportionality between the two luminosities. But this is largely due to a selection effect. More powerful
radio galaxies are generally far away and therefore, owing to the limited
sensitivity of radio telescopes, one detects only stronger cores.

Rowan–Robinson: I have tried to tie together what we know about activity
in galaxies - radiogalaxies, N-galaxies, Seyferts and quasars - into a
single picture. I assume that the probability of a galaxy being active
depends on optical luminosity, declining from 100% for the most luminous
to 2% for the least luminous. I then construct probability functions
for the luminosity of radio cores, optical cores, extended radio sources
(for active ellipticals) and disc emission (for all spirals). I assume
a range of dust optical depths in front of the optical cores to explain
the infrared emission from active galaxies.

QSOs are then simply active galaxies whose optical cores outshine the
parent galaxy, while in Seyferts the optical core has to contribute at
least 1% of the total light. The distribution functions for the various
observable parameters fit most of what we know about active galaxies
(the bulk of the data reviewed by Fanti & Perola, for example). A surprising prediction is that most optically selected QSOs are in spirals,
explaining their low radio to optical ratio. Although the scheme has
yet to be evolved to earlier epochs, a prediction could be that compact,
flat spectrum radio sources do not show much evolution.

PHYSICAL CONDITIONS IN RADIO GALAXIES AND QUASARS*

Donald E. Osterbrock
Lick Observatory, Board of Studies in Astronomy and
Astrophysics, University of California, Santa Cruz

It is well known from the pioneering work of Baade and Minkowski that radio galaxies very often have strong emission lines in the spectra of their nuclei, indicating the presence of relatively large amounts of ionized gas. For instance, in the early survey of radio galaxies by Schmidt (1965), of the 35 galaxies observed, 32 had at least [O II] $\lambda3727$ in their spectra and well over half had relatively strong [O II] and other observable emission lines as well. In the recent review of optical identifications and spectroscopy of the revised 3C catalogue of radio sources by Smith et al. (1976), 137 radio galaxies are listed. Of these descriptive spectral information is given for 98, of which 49 show strong-emission line spectra, 19 intermediate-strength emission, 12 weak emission, and 18 a pure absorption-line spectrum without detectable emission lines. The fraction of objects with emission line-spectra is much higher than for normal galaxies. It is thus apparent that though the presence of emission lines is neither a necessary nor a sufficient condition that a galaxy be an observable radio source, nevertheless a large fraction of radio galaxies do contain ionized gas in their nuclei.

For the past several years, therefore, a group of us at Lick Observatory have been using the image-tube scanner (Robinson and Wampler 1972, 1973; Miller et al. 1976) with the 3 m telescope for a spectrophotometric survey of radio galaxies and related objects. The aim is to use the well-known techniques of analyzing gaseous nebulae from measurements of their emission-line spectra to try to understand the physical conditions in the ionized gas in the nuclei of radio galaxies and ultimately the energy source, energy-input mechanism and basic nature of the radio-galaxy phenomenon (or phenomena).

First let us discuss Cyg A = 3C 405, one of the brightest radio sources in the sky, known from the work of Baade and Minkowski (1954) to have very strong emission lines. Our spectrophotometric study of this galaxy has already been published (Osterbrock and Miller 1975).

*Lick Observatory Contribution 410

D. L. Jauncey (ed.), Radio Astronomy and Cosmology, 183-191. All Rights Reserved.
Copyright © 1977 by the IAU.

In form Cyg A is a not untypical cD galaxy with a double nucleus.
Possibly the double nucleus is the apparent result of the presence of
a dust lane similar to that in the much nearer radio galaxy Cen A, for
a direct photograph of Cen A taken with a short focal-length telescope
appears very similar to the large-telescope direct plate of Cyg A.

We were able to measure the relative intensities of 24 emission
lines in the spectrum of Cyg A and in addition to set upper limits to
the strengths of 4 other important lines. A wide range in ionization
is observed, from [O I], [N I] and [S II], through strong [O III],
[N II] and [Ne III] to [Ne V], [Fe VII] and [Fe X]. The emission lines
all have similar profiles, with full widths at half maximum of approximately 500 km s^{-1}. The H I Balmer lines have a very steep gradient
which can be attributed to interstellar extinction, part of which must
occur in Cyg A. The amount of extinction derived in this way corresponds to E_{B-V} = 0.7 magnitudes, and was used to correct the observed
line intensities not only of the H I lines but of the other recombination and collisionally excited lines. The wide range of observed lines
indicates a wide range of ionization in the nucleus of Cyg A. The
[O III] ratio implies a mean temperature of approximately 15000° in
the O^{++} region while the [N II] ratio implies a mean temperature approximately 10000° in the N^{+} emission region. Furthermore, these
ratios indicate that the electron densities $N_e < 10^6$ cm^{-3} in the ionized
regions. The [S II] ratio indicates $N_e \approx 10^3$ cm^{-3} in the S^{+} region.
None of these parameters are strikingly dissimilar from the physical
conditions in a typical high-surface brightness planetary nebula.
Furthermore, the relative strengths of the lines indicate that the
abundances of the light elements are similar to the abundances in
planetary nebulae or H II regions of our Galaxy.

The continuous spectrum of Cyg A shows no detectable absorption
lines. The observed continuum approximately fits the power law
$F_\nu \propto \nu^{-n}$ with n = 3.3, or if the correction for extinction derived
from the observed emission lines is applied, n = 1.6.

The mass of ionized gas in Cyg A may be estimated from the ob-
served H I line flux. The luminosity $L(H\beta) \approx 2 \times 10^{42}$ ergs s^{-1} with
the distance 3.4×10^8 pc derived from the observed redshift. If we
assume the mean electron density in the ionized gas $N_e = 10^3$ cm^{-3}, the
mass of ionized gas is approximately 2×10^7 M$_\odot$. The derived mass
is inversely proportional to the assumed density. With $N_e = 10^3$ cm^{-3},
the ionized volume corresponds to a sphere with radius 100 pc or approximately 0.1 at the distance of Cyg A, far smaller than the observed bright central region which is several seconds in diameter.
Clearly in Cyg A as in many other emission-line galaxies the ionized
gas has a highly non-uniform distribution within the observed volume
or, in other words, the filling factor is quite small.

The wide range of ionization observed in the emission-line spectrum of Cyg A must give some information on the energy-input mechanism.
One possible energy-input mechanism, the conversion of kinetic energy

to heat (shock-wave heating) can be eliminated, because of the weakness of [O III] 4363 and the resulting low calculated mean temperature in the O^{++} region mentioned above. Therefore a plausible working hypothesis is that the energy-input mechanism is photoionization
as in planetary nebulae and H II regions, though hot stars can be
ruled out as the source, because models as well as observations show
that their radiation cannot produce the wide range of ionization observed in radio galaxies. However, photoionization by a power-law
spectrum or other spectrum extending far to the ultraviolet may produce the observed emission lines. The observed Cyg A line intensities
corrected for interstellar extinction agree reasonably well with a photoionization model calculated by McAlpine (1971) for a power-law input
spectrum from an assumed central source with n = 1.2, not too different
from the index derived for the optical continuum of Cyg A, n = 1.6.
This model assumes N_e = 10^4 cm^{-3}, a filling factor ϵ = 0.01 and a corrected helium abundance N(He)/N(H) = 0.09. The observed Cyg A line
spectrum agrees even better with the observed spectrum of NGC 1952,
which is known to be photoionized by a synchrotron source, except that
in it all the He lines are stronger as the result of an abundance effect. Thus at present it is plausible to suppose that the Cyg A emission lines arise in a gas with relatively normal abundances ionized by
a central synchrotron power-law source. Other possible sources, for
instance a hot, massive superstar, are not excluded; if the ultraviolet fluxes expected from such sources can be calculated, they should
be used as the input spectra for photoionization models to be compared
with Cyg A.

Of the radio galaxies observed in our spectrophotometric survey
at Lick to date, approximately 2/3 have narrow emission lines with
widths similar to the emission lines in Cyg A, approximately
500 km s^{-1}. The narrow-line radio galaxies reduced and discussed by
our group to date include 3C 98, 3C 178, 3C 192, 3C 327 and PKS 2322-12
by Costero and myself (1976) and 3C 33, 3C 184.1, 3C 433, 3C 452 and
5C 3.100 by Koski (1976). Most of these galaxies have emission-line
spectra similar to that of Cyg A, though weaker (by different amounts)
with respect to the continuous spectrum. Two of them, 3C 178 and
PKS 2322-12, have a somewhat lower general level of ionization, with
[O III] λ5007 comparable in strength to Hβ, and [Ne V] and [Fe VII]
emission undetectable. In all these galaxies, in contrast to Cyg A,
absorption lines of an integrated stellar spectrum can be seen. These
lines are weaker than in normal galaxies without emission lines,
probably indicating that the observed continuum is a combination of
a galaxy spectrum with a featureless power-law or synchrotron continuum
of the Cyg A type. In several of the narrow-line radio galaxy spectra,
H I absorption lines can be seen in the near ultraviolet, indicating
the presence of early-type stars. These features do not occur in all
narrow-line radio galaxies, however; in some the absorption-line
spectrum is a good match to a normal elliptical gallaxy.

The spectra of the narrow-line radio galaxies may be approximately
corrected for interstellar extinction on the assumption that the H I

lines arise by recombination. The calculated amounts of extinction
are large, ranging from E_{B-V} = 0.3 for 3C 192 to about E_{B-V} = 1.1 for
3C 178. These same extinctions would give corrected continuous spectra much bluer than observed in normal elliptical galaxies, suggesting
that the gas and dust are closely associated in a small volume in the
radio galaxies, and that the stellar population occupies a larger volume and does not suffer the full extinction (Warner 1973).

The emission-line spectra of all these radio galaxies have
[O III] line ratios corresponding to N_e = 10^4 to 10^6 cm^{-3}, if T $\approx$
10000°, and thus give no evidence for shock-wave heating. It seems
likely that the objects with emission-line spectra similar to Cyg A
can also be reasonably well fitted by photoionization models with
power-law input spectra. The low-ionization objects have spectra
similar to the nuclei of the spiral galaxies M 51 and M 81 (Peimbert
1968, 1971) though the [O I] and [S II] lines are not quite as strong
in these latter objects. The emission lines are also significantly
narrower in the two spiral galaxies than in the two radio galaxies.
The input mechanism is not known in either case. Photoionization
models with a steeper power law, such as $F_\nu \propto \nu^{-2}$ or ν^{-3} should be
calculated to see if they will reproduce the combination of strong
[O I] and [S II] dependent on a large partly ionized region, with
fairly weak [O III] indicating a relatively low general level of
ionization.

In addition to radio galaxies, we have also been observing the
spectra of Seyfert galaxies, which appear in many ways to be closely
related objects. In the classification scheme of Khachikian and
Weedman (1971, 1974), the analogues to narrow-line radio galaxies are
Seyfert 2 galaxies. To date, Koski has reduced and discussed the
spectra of 19 of these objects. Most of them have emission-line
and continuous spectra indistinguishable from the spectra of narrow-
line radio galaxies, and it thus appears that the ionized gas nuclei
of these two classes of objects are very similar physically.

About 1/3 of the emission-line radio galaxies we have observed
have, in contrast to the narrow line-radio galaxies, relatively broad
(10000 km s^{-1} or more) H I emission lines and narrow forbidden lines
similar to those in narrow-line radio galaxies. These objects include
3C 227, 3C 382, 3C 390.3, 3C 445 (Osterbrock, Koski and Phillips 1975,
1976) and 3C 120 (Phillips and Osterbrock 1975). Several of them had
previously been reported to have broad or double emission lines by
Lynds and Burbidge. In all these broad line-radio galaxies except
3C 120   the H I emission lines have weak narrow components with the
same widths and redshifts as the forbidden lines, and the broad components have irregular nonsymmetric profiles. In each of these four
radio galaxies the H I emission lines have different and unusually
steep decrements. Broad weak He I and He II emission features are
detectable in some of these galaxies. Variations in the H I line
profiles of 3C 390.3 were clearly observed between 1974 and 1975. The
narrow emission lines in all these galaxies have relative intensities

approximately the same as in Cyg A, but [O II] λ3727 is weaker and
[O III] λ4363 is significantly stronger in the narrow-line spectra
of the broad-line radio galaxies. This probably indicates electron
densities $N_e \approx 10^6$ to 10^7 cm^{-3} in their O^{++} zones.

In the broad-line region on the other hand N_e must be quite high,
as shown by the complete absence of any forbidden lines, which must be
due to collisional de-excitation. So far as I know, this interpretation was first published in the context of Seyfert galaxies by Woltjer
(1968). Quantitatively it appears that $N_e \gtrsim 10^8$ cm^{-3} in the broad-
line region. The mechanism responsible for emission of the broad H I
emission is clearly of great interest. The fact that the continuous
spectra of all four galaxies show only weak stellar absorption lines,
and approximately follow power-law forms suggest that they all have
strong nonthermal contributions. Furthermore the total equivalent
widths of Hβ emission in all four galaxies are roughly the same, suggesting that photoionization followed by recombination has something
to do with the excitation. However, the measured Balmer decrements
do not follow the recombination gradient for any assumed interstellar
extinction, nor are the measured Balmer decrements the same in all four
galaxies, so the excitation is certainly not due to recombination alone.
Probably collisional excitation and Balmer-line self-absorption, both
strongly modify the H I emission. Some calculations of these effects
have been made by Shields (1974), Adams and Petrosian (1974) and
Netzer (1975), but further theoretical work using accurate collision
cross-sections and modelling the strong density fluctuations that must
occur is clearly necessary.

The total amount of gas in the broad-line region is not very
large, of order 10 M$_\odot$ and the radii of the ionized regions deduced
assuming $\epsilon = 1$ are quite small of order 0.01 pc. It is therefore easy
to understand the time variation of the H I profile in 3C 390.3. The
broad irregular profile of the H I lines must result from mass motions
of the ionized gas in a relatively small number of clouds or streams.
There is a good correlation between line width and Hβ luminosity in
these four radio galaxies.

The emission-line spectra of the broad line radio galaxies are
similar in many ways to those of Seyfert 1 galaxies, and we have therefore also been observing these objects. To date the spectra of 35
Seyfert 1 galaxies have been reduced and discussed (Osterbrock 1976).
None of these Seyfert galaxies has H I emission line widths as great
as in the two broad-line radio galaxies with the widest lines, 3C 382
and 3C 390.3, though they cover the range of the other three radio
galaxies, and one of these, 3C 120, was first discovered as a Seyfert
galaxy (Arp 1968) and is often so described. Most of the measured
line ratios of the Seyfert 1 galaxies cover the range within which the
broad-line radio galaxies fall, though there are a few differences that
appear observationally significant. One is that the Hα/Hβ/Hγ ratios
are steeper in the broad-line radio galaxies than in the Seyfert 1
galaxies. Another is that the Fe II emission features are much weaker

in the broad-line radio galaxies than in the average Seyfert 1 galaxy, though there is a wide range in strength of these features in Seyfert 1 galaxies and a few are known with Fe II weak or unobservable. Very strong variations in the H I emission-line profiles have been observed in one Seyfert 1 galaxy, NGC 7603 (Tohline and Osterbrock 1976) and probably also occurred in two others, Mk 358 and Mk 291, since we have begun observing these objects.

Though the concept of the division of Seyfert galaxies into two classes, 1 and 2 as proposed by Khachikian and Weedman (1971) is a good one, it is not completely clearcut, and there are a few inter-mediate cases in which a strong narrow H I component with the same width and redshift as the forbidden lines is superimposed on a broad component characteristic of the Seyfert 1 galaxies. We might call such objects Seyfert 1.5 galaxies (Osterbrock and Koski 1976) but actu-ally there seems to be a continuous variation in properties ranging from pure Seyfert 1 (broad H I profiles) to pure Seyfert 2 (narrow H I profiles with relative intensities $H\beta/\lambda5007 \approx 0.1$).

Though nearly all the radio galaxies we observed may be classified either as broad (H I)-line radio galaxies, or narrow (H I)-line radio galaxies, one object falls outside both these groups. It is PKS 1345 + 12, which is being investigated by Grandi (1976). All the emission lines in this galaxy, recombination and forbidden, have intermediate widths (about 1200 to 1600 km s^{-1} full width at half maximum) and all have asymmetric profiles with similar forms. Furthermore, the [O III] and [Ne III] lines have a slightly different redshift from the other emission lines such as H I, [O I], [S II], [N II] and [O II]. Differ-ences in redshift depending on level of ionization have previously been found in the Seyfert 1 galaxy I Zw 1 by Phillips (1976), but PKS 1345 + 12 is the only case known to me in which the profiles are similarly asymmetric, but the redshifts are different.

We have looked for correlations between the optical emission-line properties of radio galaxies and their radio properties. A relatively large fraction (about 2/3 in our sample) of the radio galaxies are nar-row-line radio galaxies, the analogues of Seyfert 2 galaxies, although among Seyfert galaxies as a group, about 3/4 are Seyfert 1's, and only 1/4 Seyfert 2's (Khachikian and Weedman 1974). Likewise, of the ten Seyfert galaxies that have been measured as radio sources by de Bruyn and Willis (1974) or Kojoian et al. (1976), five are Seyfert 2's, and at least two of the remaining five, NGC 4151 and Mk 6, are better clas-sified as Seyfert 1.5 than Seyfert 2 (Osterbrock and Koski 1976). In fact, four of the five broad-line radio galaxies (all except 3C 120) have this same type of composite broad + narrow H I lines profiles, strongly suggesting that the narrow emission-line spectrum is physically connected with the radio emission more closely than the broad emission-line spectrum is.

At the Minkowski Symposium, van der Laan (1976) pointed out that a larger fraction of radio galaxies with compact radio sources in their nuclei have emission lines in their spectra than of radio galaxies without compact sources. Among our sample a higher fraction of broad-line radio galaxies (3C 382 and 3C 390.3, two of five) are known to have compact central sources than of the narrow-line radio galaxies (Cyg A and 3C 452, two of eleven).

Practically all the radio galaxies have power-law spectra with indices α = 0.75 $\pm$ 0.10, while the Seyfert galaxies have a slightly larger range but essentially the same mean value. Only 3C 120 has a greatly deviant radio spectrum (Veron et al. 1974). In optical form, a large fraction of the broad-line radio galaxies are N galaxies, while more of the narrow-line radio galaxies are cD, DE or E galaxies (Mathews et al. 1964). Many of the nearby Seyfert galaxies show signs of spiral structure, but their bright nuclei relate them closely to N galaxies (Morgan et al. 1971, Morgan 1971).

Besides 3C 390.3, several Seyfert-galaxy nuclei are known to vary in light in time scales as short as a month (see e.g. Selmes et al. 1975). All of them are Seyfert 1 galaxies. It appears that the optical activity is connected with the presence of high-velocity ionized gas, but that radio emission, though sometimes present in this phase, is more often observed in the "quiescent" phase in which the ionized gas has velocities of order 500 km s^{-1}. The similarity of both radio and narrow-line optical properties suggests that narrow-and broad-line galaxies are different stages in the evolution of one and the same type of objects. Since many of the broad-line galaxies are double radio sources, it seems likely that they have had earlier outbursts, and thus that they evolve back and forth between broad-and narrow-line stages.

Quasars, of course, appear very closely related to radio and Seyfert galaxies, and indeed much of the motivation for our optical study of the latter objects is our hope that they will help us to understand quasars. An excellent spectrophotometric survey of low-redshift quasars has recently been published by Baldwin (1975a). Most of them have optical spectra similar to broad-line radio galaxies and Seyfert 1 galaxies. The Fe II emission features are weak in the quasars, as in the broad-line radio galaxies, though two quasars, 3C 273 and PKS 0736 + 01 have Fe II strong (Baldwin 1975b), as in many Seyfert 1's. Otherwise the temperatures, densities, line widths and abundances seem quite similar in quasars and Seyfert 1 galaxies.

It appears that we are beginning to get optical measurements that are capable, at least in some cases, of distinguishing between different types of radio sources and Seyfert galaxies. The problems of refining these observations, and through them understanding the nature (or natures) of radio sources remain challenging problems for the future.

I am very grateful to my colleagues and collaborators mentioned in the text for allowing me to quote their results, many still unpublished, and particularly to Dr. S.A. Grandi for providing numerous

references to radio measurements. I am also most grateful to the
National Science Foundation for partial support of my research on these
topics over the years.

REFERENCES

Adams, W.M., and Petrosian, V.: 1974, Astrophys. J., 192, 199.
Arp, H.C.: 1968, Astrophys. J., 152, 1101.
Baade, W., and Minkowski, R.: 1954, Astrophys. J., 119, 206.
Baldwin, J.A.: 1975a, Astrophys. J., 196, L91.
Baldwin, J.A.: 1975b, Astrophys. J., 201, 26.
Costero, R., and Osterbrock, D.E.: 1976, Astrophys. J., submitted.
de Bruyn, A.G., and Willis, A.G.: 1974, Astron. Astrophys., 33, 351.
Grandi, S.: 1976, Astrophys. J., in preparation.
Khachikian, E.Y., and Weedman, D.W.: 1971, Astrofizika, 7, 389.
Khachikian, E.Y., and Weedman, D.W.: 1974, Astrophys. J., 192, 581.
Kojoian, G., Sramek, R.A., Dickinson, D.F., Tovmassian, H., and
    Purton, C.R.: 1976, Astrophys. J., 203, 323.
Koski, A.T.: 1976, Ph.D. thesis, Univ. of California, Santa Cruz.
MacAlpine, G.M.: 1971, Ph.D. thesis, Univ. of Wisconsin.
Mathews, T.A., Morgan, W.W., and Schmidt, M.: 1964, Ap. J., 140,35.
Miller, J.S., Robinson, L.B., and Wampler, E.J.: 1976, Advances in
    Electronics and Electron Physics, 40.
Morgan, W.W.: 1971, Astron. J., 76, 1000.
Morgan, W.W., Walborn, N.R., and Tapscott, J.W.: 1971, Pontificiae
    Academiae Sci. Scripta Var. No. 35, 27.
Netzer, H.: 1975, M.N.R.A.S., 171, 395.
Osterbrock, D.E.: 1976, Astrophys. J., in preparation.
Osterbrock, D.E.,and Miller, J.S.: 1975, Astrophys. J., 197, 535.
Osterbrock, D.E., Koski, A.T., and Phillips, M.M.: 1975,Ap. J., 197, L41.
Osterbrock, D.E., Koski, A.T., and Phillips, M.M.: 1976, Ap.J., 206, 898.
Osterbrock, D.E., and Koski, A.T.: 1976, M.N.R.A.S., accepted.
Peimbert, M.: 1968, Astrophys. J., 154, 33.
Peimbert, M.: 1971, Bol. Obs. Tonantzintla Tacubaya, 6, 97.
Phillips, M.M., and Osterbrock, D.E.: 1975, Pub. A.S.P., 87, 949.
Phillips, M.M.: 1976 Astrophys. J., accepted.
Robinson, L.B., and Wampler, E.J.: 1972, Pub. A.S.P., 84, 161.
Robinson, L.B. and Wampler, E.J.: 1973, Astronomical Observations with
    Television-Type Sensors, p. 69.
Schmidt, M.: 1965, Astrophys. J., 141, 1.
Shields, G.A.: 1974, Astrophys. J., 191, 309.
Selmes, A.M., Tritton, K.P., and Wordsworth, R.D.: 1975, M.N.R.A.S.,
    170, 15.
Smith, H.E., Spinrad, H., and Smith, E.O.: 1976, Pub. A.S.P., submitted.
Tohline, J.E., and Osterbrock, D.E.: 1976, Astrophys. J., submitted.
van der Laan, H.: 1976, Pub. A.S.P., submitted.
Veron, M.P., Veron, P., and Witzel, A.: 1974, A.A. Sup., 13, 1.
Warner, J.W.: 1973, Astrophys. J., 186, 21.
Woltjer, L.: 1968, Astron. J., 73, 914.

DISCUSSION

Rowan-Robinson: Does your V – 6 cm index refer to the visual magnitude of the nucleus or of the whole galaxy? Are the extinctions deduced in Type 2 Seyferts consistent with their absence of broad wings being due to dust?

Osterbrock: To the whole galaxy. No, I do not think that the same amount of extinction that is observed in the central narrow components of the HI lines in Seyfert 2 galaxies could possibly suppress broad wings of the type observed in Seyfert 1, or Seyfert 1.5 galaxies if they were also present in Seyfert 2's.

Wampler: You have shown that Seyfert Galaxies can be separated from Radio Galaxies by spectral and radio criteria. When this is done one finds that there is also a segregation in form as the radio galaxies are elliptical while the Seyferts are spiral. Would you like to comment on this?

Osterbrock: The separation appears to be very strong, though more work needs to be done on obtaining large-scale direct plates of Markarian and Zwicky Seyfert galaxies and classifying them. These differences must have physical significance, but I do not know yet what it is.

de Felice: Can you comment further on the observed separation between Seyfert Galaxies and Radio Galaxies regarding the FeII emission line strength?

Osterbrock: The fact that this correlation (weakness of FeII emission in broad-line radio galaxies) exists is extremely interesting but we do not understand the reason for it. Mark Phillips is working observationally on FeII emission in Seyfert galaxies, and we hope that his results will throw some light on how the excitation occurs, and therefore on the physical differences between broad-line radio galaxies and Seyfert galaxies.

ABSORPTION SYSTEMS IN HIGH REDSHIFT QSOs

A. Boksenberg
Department of Physics and Astronomy
University College London
Great Britain.

In addition to the characteristic emission lines, absorption lines frequently are seen in the spectra of QSOs, usually those with high redshift ($z_{em} \gtrsim 1.8$). About 10 percent of all QSOs listed in the compilation of Burbidge et al. (1976a) are recorded as having at least one 'identified' absorption system, meaning that a pattern of several selected observed lines can be matched with the apparent wavelengths of transitions (generally from the ground level) in a physical plausible group of atoms or ions at the same, although arbitrary, redshift (Bahcall 1968, Aaronson et al. 1975). Identified absorption line redshifts range from being comparable with the associated emission line redshifts, to having very much smaller values with relative velocities exceeding 0.5c in the QSO frame. Added to this, there are many QSOs having absorption lines not yet recognised as belonging to identified systems, both those objects already having one or more identifications, and others with none.

It is important to realise, however, that whether or not absorption lines are detected depends very much on observational technique. As in the similar case of interstellar absorption lines in the Galaxy, only when the instrumental spectral resolution is adequately high can such lines be seen at all. The survey resolutions used when first attempting to identify features in QSOs usually are too low to bring out any but the strongest of the absorption lines that may be present and this can give a misleadingly sparse picture of the line population. This point is well illustrated by the spectra of the QSO PHL 957 ($z_{em} = 2.69$) observed at low and high resolutions by Coleman et al. (1976). The low resolution spectrum, in their Fig.1, shows only a few absorption lines, including an extremely strong line of Lyα in a system at $z_{abs} = 2.309$. In contrast, the high resolution spectra in their Fig.3 reveal a rich absorption spectrum, and they list 203 individual lines (see also: Lowrance et al. 1972, Wingert 1975).

Again, for interstellar absorption lines, only by extending the range of observation into the middle and far ultraviolet (by observing from above the atmosphere) do lines of most of the available

D. L. Jauncey (ed.), Radio Astronomy and Cosmology, 193-222. All Rights Reserved.
Copyright © 1977 by the IAU.

transitions become accessible: lines arising from ground-state transitions in a wide range of ionization stages of the more abundant
elements (HI, MgI-II, CII-IV, FeII-III, SiII-IV, NI-III, NV, OVI, etc.,
e.g. Morton and Hu 1975). The same absorption lines as are seen in
the interstellar gas also appear in QSO spectra. Furthermore, in both
cases, lines arising from excited fine-structure levels generally are
absent, but occasionally are observed. For QSOs, as has been well
described by Lynds (1972), with increasing redshift the detection of
absorption lines depends strongly first on the MgII doublet($\sim \lambda 2800$)
then on the CIV doublet ($\sim \lambda 1500$), and finally on Lyα ($\lambda 1216$) which
enters the observable range at $z \simeq 1.65$. Ionization conditions in the
absorbing regions may in general work against the support of Mg$^+$, and
this would result in an observational bias for the detection of the
higher redshift systems. On the other hand, the fact that nearly all
QSOs with $z_{em} \gtrsim 1.8$ have absorption lines in their spectra may be
attributed to the dominance of neutral hydrogen, which persists in the
absorbing regions as a relatively large residual despite the usually
unfavourable ionization conditions.

    With some few exceptions, the QSO absorption lines are very
narrow and generally are unresolved even at the highest observed
spectral resolutions. Lines which appear single at intermediate
resolution are frequently revealed as finely split at high resolution.
The observations of the QSO B2 1225+31 shown in Fig.1, which we
obtained with Sargent and Lynds, serve as an example of this. The
upper spectrum, at intermediate resolution ($\Delta\lambda \sim 2A$), shows several
deep absorption lines each without apparent structure; the lower
tracings are selected from spectra observed at high resolution
($\Delta\lambda \sim 0.7A$) and reveal fine velocity components and narrow features
not detected in the upper spectrum. Structure such as this appears
to be remarkably similar in all cases comparably observed (e.g.
Boksenberg and Sargent 1975). A small portion of a spectrum of the
BL Lac object Pks 0735+178 obtained (with Sargent) at still higher
resolution ($\Delta\lambda \sim 0.3A$) is shown in Fig.2. At the top are the observed
MgII H and K lines, having a redshift near 1.424; next, a computed
profile based on four component clouds and convolved with the
instrument profile is compared with the observed data points, for the
K line; and at the bottom is the corresponding unconvolved computer
profile for the K line. It is clear that there are (at least) four
velocity components distributed over the range 165 km s^{-1}; the
narrowest has a measured full width at half minimum of < 20 km s^{-1}.
It is suggestive that the width of each component is comparable with
the range of velocities over which cloud components are observed in
the local interstellar gas (Boksenberg et al. 1975), and that the
total width of the complex falls within the range expected for a
galaxy (Burbidge 1975).

    Referring again to Fig.1, evidently there is a dramatic increase
in the density of absorption lines shortward of the Lyα emission line
compared with longward. This is commonly seen in all high-redshift
QSOs adequately well observed (Lynds 1972). Most of these lines

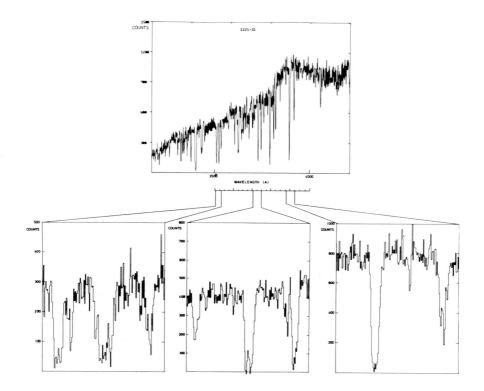

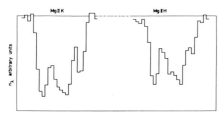

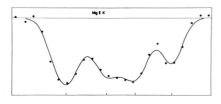

Fig.1: (above). Spectra of the QSO B2 1225+31 shown (upper) at intermediate resolution (Δλ ∿ 2A) and (Lower) selected sections at high resolution (Δλ ∿ 0.7A).

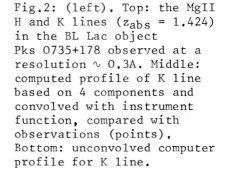

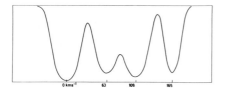

Fig.2: (left). Top: the MgII H and K lines (z_{abs} = 1.424) in the BL Lac object Pks 0735+178 observed at a resolution ∿ 0.3A. Middle: computed profile of K line based on 4 components and convolved with instrument function, compared with observations (points). Bottom: unconvolved computer profile for K line.

remain unidentified, in the sense that no corresponding patterns of other lines are observed with which to group them and enable redshift systems to be recognised. The most plausible explanation for these lines is that they represent Lyα absorption in clouds at lower redshift than the QSO against which they are observed and for which lines of other elements (and other Lyman lines) are too weak to be detected (Lynds 1971, 1972).

There are a few QSOs, already alluded to, which exhibit absorption features having a strikingly different appearance from the sharp lines seen in all other cases. Such an object is PHL 5200 (Lynds 1967, Burbidge 1968), whose spectrum (observed at $\Delta\lambda \sim 15A$, with Sargent) is shown in Fig.3a. The emission redshift of PHL 5200 is 1.985 and it has very broad absorption bands shortward of Lyα, NV λ1240 and SiIV λ1393,λ1402 in the wavelength range shown here, and also CIV λ1548,λ1551. The bands extend from $z_{abs} \sim 1.86$ right up to the emission redshift, a range of ~ 12500 km s^{-1}. At slightly higher resolution, sharp absorption lines also are seen in this object (Burbidge 1968, Lynds 1972). The broad absorption features almost certainly arise in material flowing out from the QSO itself (Scargle et al. 1970). The range in velocities is comparable with the turbulent velocities commonly observed in the broad emission line regions in QSOs and Seyfert nuclei (e.g. Osterbrock et al. 1976, Boksenberg et al. 1975). Similar features, of several thousand km s^{-1} width, are seen in the equivalent ultraviolet spectra of stars, as in the case of ζ Pup (Fig.3b), where there is no doubt that material is being ejected (Morton 1975). Other QSOs of this class probably having intrinsic absorption are RS 23 (Burbidge 1970) and the recently discovered object Q1246-057 (Malcolm Smith, private communication). The latter is distinct in that it shows a clear separation of ~ 15000 km s^{-1} between a broad-lined absorption system (width ~ 4000 km s^{-1}) and the emission redshift.

Narrower absorption lines somewhat reminiscent of those in QSO spectra also appear in the spectra of some Seyfert galaxies. Such a case is Markarian 231, which is placed among the QSOs in optical luminosity (Boksenberg et al. 1976). A low resolution spectrum ($\Delta\lambda \sim 7A$) of the nucleus of this object is shown in Fig.4. Apart from the interesting broad-lined emission spectrum at a redshift of 12600 ± 200 km s^{-1} (the numbered multiplets refer to permitted transitions in FeII) there are several absorption features belonging to three separate systems indicated by bracketted roman numericals. System I contains resonance lines of CaII and NaI, and λ3889 of HeI, at a redshift of ~ 8000 km s^{-1}. System II represented by the D-lines and λ3889, is at a redshift of 6250 km s^{-1}. System III, featuring the K-line and members of the Balmer series, is at 12900 km s^{-1}, close to the emission redshift; this may represent the intrinsic absorption spectrum of stars in the galaxy. Inserted in Fig.4 are high resolution ($\Delta\lambda \sim 0.5A$) profiles of the H, K and λ3889 lines, shown (with arbitrary vertical displacements) on a velocity scale with the positions of several possible velocity components in System I marked. The mean

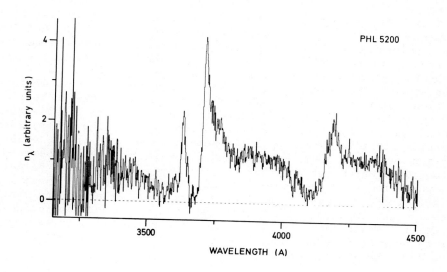

Fig. 3a: Spectrum of the QSO PHL 5200 showing broad absorption
bands.

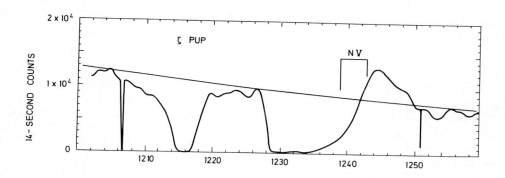

Fig. 3b: Spectrum of ζ Pup (Morton 1975)

outward velocity is ~ 4700 km s^{-1} for System I and ~ 6350 km s^{-1} for System II. Although at first sight these lines appear similar to the narrower absorption lines seen in QSOs, the presence of the HeI line at once points to a clear distinction: in QSOs absorption lines have been seen to arise only from ground or excited fine structure states, whereas the HeI line arises from a metastable level 23 eV above the ground state. The presence of the HeI line argues strongly in favour of a close association of the absorbing regions with the nuclear activity in Markarian 231, and suggests that these are a geometrically favoured sample of the broad emission line region. The lower redshift absorption lines in Markarian 231 resemble those seen in NGC 4151 (Anderson and Kraft 1969, Weymann and Cromwell 1972, Boksenberg and Penston 1976) which not only has the HeI line, but also the Balmer lines which too are never seen in QSOs. Furthermore, all these lines vary greatly in strength on a timescale as short as about one month, making another important distinction between these and absorption lines in QSOs, for which there is no evidence for variability in strength or velocity over periods up to several years (with the possible exception of PHL 5200: Burbidge 1968).

Having dealt with the types of absorption lines which probably are intrinsic to the objects in whose spectra they appear, and clearly are characteristically different from the majority of absorption lines seen in QSO spectra, we now consider the origin of the latter. If the QSOs are at cosmological distances (which we shall assume here) the narrow absorption lines may, in principle, be intrinsic to the QSOs, arising in material ejected from them at velocities up to the very high values observed, or be produced in material residing at intervening cosmological distances in line to the QSOs, either in intergalactic clouds or directly associated with galaxies. In the remainder of this paper we review some of the arguments and evidence for and against these possibilities.

One of the first QSOs found to have a rich absorption line spectrum is 3C 191. In this object it was noticed that z_{abs} is only slightly lower than z_{em} and it was natural to think that this implied a physical relationship between the emitting and absorbing material. The evidence for such a relationship apparently was strengthened by the presence of absorption lines corresponding to the SiII $\lambda 1264$ and $\lambda 1533$ transitions from the $J = 3/2$ excited fine-structure level of the ground state (Stockton and Lynds 1966), suggesting that radiation from the QSO was providing the necessary excitation. As has been shown by Bahcall and Wolf (1968), the relative population of excited fine-structure levels is a powerful indicator of the physical conditions in the absorbing material. Bahcall et al. (1967) further analysed this object and deduced from the relative SiII line strengths that: (a) the absorption arose in a region of density $n \lesssim 10^3 cm^{-3}$, with equality holding if the fine-structure excitation is induced only by electron collisions and (b) the distance of the absorbing

material from the QSO continuum source is $R \gtrsim 10^{2\pm1}$ pc, with equality
holding if excitation is due only to ultraviolet fluorescence. Higher
resolution studies by Williams et al. (1975) revealed that the absorption system is split into two components (z_A = 1.945 and z_B = 1.949)
with a separation characteristic of the CIV resonance doublet ('line-
locking': see below). A doubling of the absorption spectrum previously
had been inferred by Scargle (1973) on the basis of a model invoking
radiation pressure driven outflow for the absorbing material. This
seemed to strengthen the case for the absorbing material being
intrinsic to the QSO. However, on closer study Williams et al. (1975)
found that the indications are not quite so clear. Assuming that the
ionization of the absorbing material is due to photoionization by the
QSO continuum source, and that the elemental abundances are solar,
they found that for reasonable values of spectral index the computed
ionization is much too high to account for the observed absorption
spectrum of 3C 191 at densities $n < 10^3 cm^{-3}$ required by the fine-
structure population unless R is greater than $\sim 10^3$ pc, and then excitation of the fine-structure levels cannot occur by ultraviolet
radiation (excitation due to infrared radiation was discounted as
implausible). This led to the conclusion that the fine-structure
excitation is due to collisions and hence $n \simeq 10^3 cm^{-3}$; the observed
level of ionization then requires $R \approx 10^4$ pc. The photoionization model
gave a satisfactory fit to the observed absorption equivalent widths
for a single cloud, but ran into difficulties when both clouds were
considered, because essentially all of the ionizing radiation along
the line of sight must be absorbed in the inner cloud, leaving none to
produce NV probably but not definitely observed in the other. It seems
to us that the deduction that the fine-structure excitation occurs by
collisions not ultraviolet fluorescence, coupled with the apparent
difficulty suffered by the two-cloud photoionization model, severely
undermines the claim that the absorbing material is clearly intrinsic
to 3C 191. Furthermore, it has since been found that excited fine-
structure levels can be populated in absorption systems with redshift
greatly different from the emission redshift, as in the cases of
Pks 0237-23 (Boksenberg and Sargent 1975) and OQ 172 (Baldwin et al.
1974), which now makes the original case qualitatively less compelling;
and such transitions also are observed in the interstellar gas
(Morton 1976), a fact which sustains the cosmological hypothesis.

As was mooted above, the intrinsic hypothesis gains some support
from the apparent phenomenon of 'line-locking' observed in a number of
cases (Strittmatter et al. 1973, Williams et al. 1975, Boksenberg and
Sargent 1975). Such a process may occur naturally if radiation
pressure were in some way instrumental in affecting the ejection of
material from the QSO (Mushotsky et al. 1972, Scargle 1973). Qualitatively, the physical picture invoked for this effect is as follows:
gas driven out from a QSO in some way (not necessarily by radiation
pressure alone) will absorb radiation emitted from the QSO line and
continuum regions, but which has been filtered through absorption by
gas closer to the QSO and moving at different velocities. In their
local frame, ions of any particular species in the outflowing material,

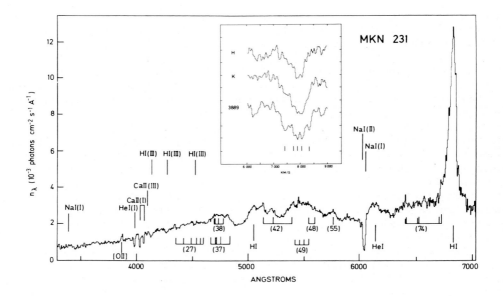

Fig. 4:   Low resolution ($\Delta\lambda \sim 7A$) spectrum of the Seyfert galaxy
          Markarian 231.   The insert shows high resolution ($\Delta\lambda \sim 0.5A$)
          profiles of the CaII and HeI absorption lines.

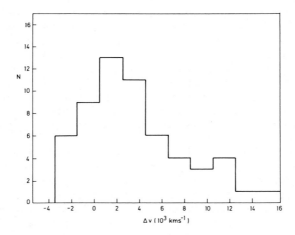

Fig. 5:   Histogram of the distribution in relative velocities of
          emission and absorption systems for 39 QSOs.

if at the appropriate velocity, will be exposed to redshifted radiation
from the central source that may have the blue wing of an emission
line, an ionization edge, or a redshifted absorptuon line from intervening gas, falling at the frequency of one or more of their particular resonance lines. Such ions will experience less radiation pressure
than those at neighbouring velocities and a balance with the inward
gravitational force may be set up to provide an accumulation of ions
in velocity space at certain velocities at which they experience no
net inward or outward acceleration. This would be manifest as 'line-
locking', in which the long wavelength component of a resonance
doublet such as CIV λ1548,λ1551 in one absorbing cloud coincides with
the short wavelength component in another moving out at a higher
velocity (as is apparent in the case of 3C 191), or line-edge locking
when an absorption line in a cloud coincides with a strong wavelength
gradient in the radiation flux such as may occur at the HI or HeII
ionization edges. Thus, in individual objects, certain preferred values
may occur in $(1 + z_i)/(1 + z_j)$, i.e. in λ_m/λ_n where these are the rest
wavelengths of features appearing at the same observed (redshifted)
wavelength in two systems of different redshifts z_i and z_j and where
z_i includes z_{em}. Quantitatively, radiative acceleration of QSO clouds
has been studied by Kippenhahn et al. (1974), Kippenhahn et al. (1975)
and Mestel et al. (1976) who showed, using simple models, that near-
relativistic velocities could be achieved and that instabilities could
occur to break up uniform flow. However, no quantitative explanation
of line-locking or line-edge locking yet exists and, indeed, has
serious theoretical difficulties (J.J. Perry, private communication).
Furthermore, no theoretical treatment yet reproduced the observed
character of the absorption lines; in particular, their extraordinarily
narrow widths introduces special difficulties, as discussed by Williams
(1972), Weymann (1973), McKee and Tarter (1975), and Weymann (1976).
On the other hand, it is important to note that the radiative driving
mechanism has no difficulty in explaining the broad emission lines
in QSOs and Seyfert galaxies (Blumenthal and Mathews 1974), which
clearly are intrinsic to these objects.

    The occurrence of line-locking and line-edge locking recently has
been reviewed and empirically studied by Burbidge and Burbidge (1975).
Although the evidence for these phenomena seems at first sight to be
suggestive, it is far from being conclusive and indeed may not be
apparent at all when considered from a broader statistical base
(Aaronsen et al. 1975, Sargent and Boroson 1976, Carswell et al. 1976,
Carswell, Boksenberg and Sargent (1976,in preparation). Furthermore,
Carswell et al. (1976) note that there is evidence for possible NV
and SiIV line-locking in the QSOs 0736-06 and 4C 24.61, but that in
both cases the resonance doublet concerned does not appear prominently
in the spectrum; they suggest that the splittings probably are due to
chance and comment that this may be true for other claimed line-
lockings where the corresponding lines are indeed observed. Related
comments in a similar context have been made by Bahcall (1975).

A strong argument against the intrinsic hypothesis is made in the case of the QSO 3C 286 by the recent radio observations of Wolfe et al. (1976). This object has a radio absorption line at 839.4 MHz, which is identified with the 21 cm HI line at a redshift of z_{abs} = 0.692 (Brown and Roberts 1973), and an optical emission redshift of z_{em} = 0.849. From very-long-baseline observations Wolfe et al. were able to observe two closely spaced regions in the absorbing medium, one in front of each of the two strongest components of continuum emission, with full widths at half-minimum of 3.7 km s^{-1} and 7 km s^{-1} separated in radial velocity by 3 km s^{-1}. They considered a simple model in which the gas is confined to a thin spherical shell of radius R centred on one of the emission components and enclosing both, and derived the value R $\geq$ 20 kpc and a separation of 260 pc between the observed absorbing regions (assuming z_{em} is cosmological, H_0 = 50 km s^{-1}Mpc^{-1}, and q_0 = 0). On comparing the momentum of such a shell with that supplied by the centrally located source of optical and ultraviolet continuum radiation they find it to be greater by a very large factor, and conclude that radiation emitted by 3C 286 cannot drive the absorbing gas to the relative velocity observed ($\sim$ 0.1c). Other mechanisms, including the action of relativistic particles and magnetic fields stored in the radio source, are found to be implausible, and models involving large gravitational redshifts can be ruled out. However, the observations are consistent with absorption by gas in an intervening galaxy.

Redshifted 21 cm absorption also has been observed in the BL Lac object AO 0235+164 at z_{abs} = 0.52385 (Roberts et al. 1976). This redshift is identical, within the errors, to the optically derived redshift (Burbidge et al. 1976b) and is the first instance in which both radio and optical high-redshift absorption lines are detected; it provides a link, if one were needed, between objects of both types. The 21 cm line shape shows structure on a scale of $\sim$ 6 km s^{-1} and an overall range of $\sim$ 50 km s^{-1}. For T_s = 100 K the neutral hydrogen total column density in the direction of the source is $\simeq$ 2.3 x 10^{21}cm^{-2}.

The suggestive evidence afforded by 21 cm observations that at least some absorption lines in QSO spectra are due to material in or associated with intervening galaxies (Wolfe et al. 1976), is strongly supported by the more direct observations of Haschick and Burke (1975) and Grewing and Mebold (1975). Both groups took radio spectra near 21 cm wavelength of several close QSO-galaxy pairs whose projected separations at the galaxies are sufficiently small to expect a chance of detecting absorption of the QSO radiation by galactic neutral hydrogen. Hydrogen emission from the galaxies was detected in all cases. A narrow absorption feature (half-width $\leq$ 5 km s^{-1}) was found in the spectrum of one of the QSOs, 4C 32.33, close to the systemic velocity of the neighbouring galaxy NGC 3067 (of type Sb), giving a neutral hydrogen column density N_{HI} $\leq$ 2.7 x 10^{17}$T_s$cm^{-2} where T_s is the spin temperature of the gas (this is the result of Haschick and Burke, who give the slightly smaller upper limit). Assuming T_s = 100 K, N_{HI} $\leq$ 2.7 x 10^{19}cm^{-2}. No absorption features were detected in the other cases, but the observed line is not much stronger than the instrumental detection limit, so it cannot be inferred that no comparable absorption occurs in these, merely that it may be present undetected in the noise.

For the absorption near NGC 3067, Haschick and Burke derive a 'disc distance' (the radial distance from the centre of the galaxy at which the line of sight to the QSO intersects a disc lying in the galactic plane) of about 60 kpc ($H_O \simeq$ 50 km $s^{-1}Mpc^{-1}$) and point out that this implies that the neutral hydrogen responsible for the absorption feature lies at a distance of more than 4 Holmberg radii from NGC 3067.

Although for the great majority of QSOs the absorption redshifts are smaller than the emission redshifts, there are some which have $z_{abs} > z_{em}$ (e.g. Lynds 1972, Weymann et al. 1976). The relative velocities of these absorption systems with respect to the emission systems has never been observed to exceed a few thousand km s^{-1}. No spectroscopic differences between the $z_{abs} > z_{em}$ systems and the more common blueward absorption systems (both those with $z_{abs} \lesssim z_{em}$ and $z_{abs} << z_{em}$) are apparent, either in the types of neutral or ionized species observed or in their characteristic line widths and fine splittings. A histogram of the distribution in relative velocities of emission and absorption systems for $z_{abs} \approx z_{em}$ for 39 QSOs drawn from Burbidge and Burbidge (1975), Carswell et al. (1976) and a recently observed sample of QSOs with $z_{em} \sim 2$ by Carswell, Boksenberg and Sargent (1976) is shown in Fig. 5. In some objects there is more than one system, and systems with $z_{abs} > z_{em}$ and $z_{abs} < z_{em}$ co-exist. The fact that $z_{abs} \approx z_{em}$ strongly implies that the absorbing material is located in the vicinity of the QSOs; but it need not be directly associated with them, as is suggested by the case of PHL 1222 (Weymann and Williams 1976). This object has z_{em} = 1.903 and an absorption system with z_{abs} = 1.934, among others; from the absence of absorption lines arising from the excited fine-structure levels of CII and SiII and a consideration of the ionization equilibrium assuming photoionization by the QSO Weymann and Williams tentatively deduced that the gas must have a density $n_e \lesssim$ 15 cm^{-3} and a distance > 4 x 10^5 pc from the QSO. If gravity is responsible for the observed infall velocity of 3190 km s^{-1} the simplest model of a cloud falling from rest at infinity requires a QSO mass > $10^{14}M_\odot$, much greater than the value $\sim 10^8M_\odot$ sometimes cited (Burbidge and Perry 1976). On the other hand, Weymann and Williams point out that the inferred mass of > $10^{14}M_\odot$ and the velocity and distance of the cloud to the QSO are similar to the masses, velocity dispersions and sizes of large clusters of galaxies, and is suggestive evidence that systems with $z_{abs} > z_{em}$ are clouds of gas moving in the gravitational field of such clusters. The fact that carbon and silicon are observed in the PHL 1222 cloud argues for the actual association of the gas with a galaxy in the cluster, in which case the radiation field producing the observed ionization could have a strong contribution from the galactic stars; this makes the derived cloud distance very much a lower limit. Alternatively, the gas may

have been tidally disrupted from a galaxy and is more accurately
identified as intracluster gas, and then the stellar radiation field
is relatively weak. If the above interpretation of $z_{abs} > z_{em}$ systems
is true it is a small step to include all systems with $z_{abs} \approx z_{em}$ in
the same category. The shape of the histogram in Fig.5 lends support
to this suggestion: it can be decomposed into a sharply peaked component with equal distribution in relative velocity about z_{em}
($\pm$ 3500 km s^{-1}) and a shallower one with $z_{abs} > z_{em}$ extending to large
relative velocities, merging with the systems of $z_{abs} \ll z_{em}$. The
sharply peaked component can be interpreted as being due to absorption
by intervening material contained in the supposed QSO clusters themselves, either in or out of galaxies, whereas the extended component
could arise from material in unrelated clusters, and field galaxies
(Lynds 1972). For the latter we would expect their dispersion in
peculiar velocities about the mean Hubble flow to increase as (1 + z)
and, say, would be about 1000 km s^{-1} at z = 2, taking 300 km s^{-1} as the
present value of dispersion (Burbidge 1975). An explanation not
including cluster galaxies, but only unbound field galaxies, is not
consistent with the observations (Weymann et al. 1976).

We now turn to a direct consideration of the hypothesis that the
QSO absorption lines with $z_{abs} \ll z_{em}$ are due to cosmologically
distributed intervening material. Bahcall and Peebles (1969) pointed
out that this question is open to statistical analysis, and they
proposed two specific tests. Both of these are based on the expectation that the absorbing material is randomly distributed along the
line of sight to the QSO: test 1 examines the distribution in total
number of absorption redshifts found in each QSO of a uniformly
observed sample; test 2 examines the relative frequency of occurrence
of different values of absorption redshifts either in one object, or
in several not necessarily uniformly observed. In test 2 the possibility that the absorbers may also have a z-dependent evolution must
not be forgotten (Bahcall 1971). Such tests have been attempted in a
direct or implied fashion by, for example, Roeder (1972), Aaronsen et
al. (1975), and Carswell et al. (1976), but the observational material
they used has not been sufficiently consistent or uniform to give
valid results. The importance of strict observational uniformity (for
test 1) was stressed by Bahcall and Peebles (1969) and reiterated by
Bahcall (1971). The first uniformly observed sample of absorption
lines in QSOs on which a proper test can be based has been obtained
recently by Carswell, Boksenberg and Sargent (1976) (hereinafter
referred to as CBS). A total of 17 QSOs contained in a relatively
narrow redshift range (1.9 $\leq z_{em} \leq$ 2.2) were observed with the same
instrument at uniform resolution (about 5A), wavelength coverage and
(approximately) signal-to-noise. Great pains were taken to define
the detection of absorption lines in a consistent way by use of
numerical limits on a probability criterion based on the signal-to-
noise in the data. Spectra of two of the objects observed are shown
in Fig.6. A typical noise curve is shown with the spectrum of
3C 9 (z_{em} = 2.012): the smooth line is the 1σ noise level in the sky-
subtracted spectrum of the object; the small spikes indicate the

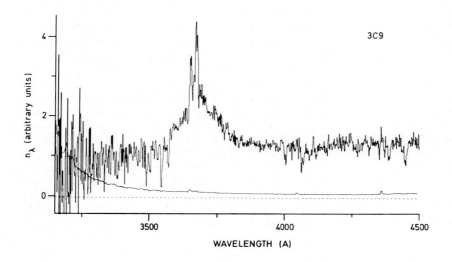

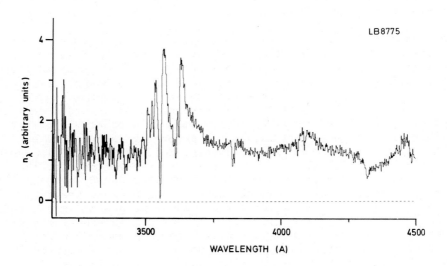

Fig. 6: Spectra of two of the QSOs from the z ∿ 2 survey of Carswell, Boksenberg and Sargent. A noise curve is shown for 3C 9 as explained in the text.

greater noise at the positions of the sky emission lines, and the
rapidly increasing noise towards shorter wavelengths is due to the
loss of signal at the encroachment of atmospheric extinction (the mean
level of the QSO spectrum is everywhere approximately corrected by
use of an observed standard). The strong emission line is Lyα (with
greater or lesser contributions from NV λ1240 and SiIII λ1206) and
clearly is cut by absorption lines in these examples. Considerable
absorption structure also is present shortward of the Lyα emission,
merging with the noise near the atmospheric cutoff. As mentioned
earlier, for most well-observed high redshift QSOs it is striking that
the density of absorption lines is much greater in the wavelength
region below the Lyα emission line than above and that most of the
former remain unidentified. The only plausible explanation is that
the unidentified lines actually are Lyα absorption lines in clouds at
lower redshift than the QSO, and for which lines of other elements are
not detectable (Lynds 1971). Further implications of this will be
discussed later. Accepting this explanation, it is easy to see that
a non-random observational bias could act in favour of the higher
redshift objects in the range of the sample, since the detection of
lines shortward of the Lyα emission line becomes rapidly easier as the
position of this line moves to longer wavelengths, away from the noisy
region approaching atmospheric cutoff. The highest redshift object
in the sample, Pks 0424-13 (z_{em} = 2.165), is particularly favoured
in this respect and indeed it has by far the largest number of
unidentified absorption lines.

The absorption line spectra were subjected to a computer search
program in an attempt to identify redshift systems. As is usual in
such analyses, a number of absorption lines, both shortward and longward of emission Lyα remain unidentified. Since the Bahcall-Peebles
test is valid for any uniform selection criterion, several ways of
examining the data were tried: for example, one case included only the
most probable identified redshift system, and in another all unidentified lines were included and counted as belonging each to its own
redshift system. Furthermore, in examining the distribution in number
of absorption redshifts per object to test the assumption that the
material of lower redshift is cosmologically distributed, systems with
$z_{abs} \approx z_{em}$, assumed to be in the vicinity of the QSOs as previously
discussed, should be omitted. This was done by supposing that a redshift within 4000 km s^{-1} (in the QSO frame) of z_{em} is associated with
a QSO while one below this limit is not. Then, to avoid the non-
random observational bias just pointed out, only those absorption
systems with z_{abs} < 1.88 were included, where this value was chosen to
be clear of any possible cluster material associated with the object
of lowest redshift in the sample, PHL 1222. In all cases examined
(with z_{abs} < 1.88) no significant difference between the observed and
Poisson distributions was found. Although more similar work must be
done (and is already planned) to increase the statistical significance
of the results, it is clear that the observations are not inconsistent
with the hypothesis of intervening cosmological material. It is
interesting to note that Carswell et al. (1976), commenting on their

observations but not entering into a statistical analysis, state that
the richness of the absorption spectrum of 3C 9, contrasting with the
complete absence of lines in LB 8755 and BSO 6, suggests that the
absorption line regions cannot be randomly distributed at lower cosmological redshifts and therefore are intrinsic to the QSOs. CBS did
not observe BSO 6 in their survey, but did include 3C 9 and LB 8755:
absorption lines were detected in LB 8755, and both this object and
3C 9 were included in the statistical analysis just described, which
yielded random distributions. For studies of this kind, this points
once more to the strict need for uniform observational material and
consistent treatment in its reduction and analysis.

   As an additional outcome of the survey by CBS, there is an
interesting peak near z_{abs} = 1.6 in the number of identified redshift
systems plotted against redshift. This seems to occur for objects
clustered around R.A. 23 - 1h and 0 - 40° Dec. Although the result is
only of marginal statistical significance, if the absorption arises
in intervening cosmological material, this may be evidence for large
scale clustering in the Universe, which perhaps is not surprising as
galaxy clusters and possibly superclusters are a common observational
feature locally. If this enhancement is real then not only does it
provide confirmation that intervening cosmological material is
responsible for at least some of the absorption lines, but paradoxi-
cally also makes the basic form of the Bahcall-Peebles tests invalid:
thus, departures from the expected random distributions in numbers of
redshift systems (possibly with a mean varying smoothly with z_{abs} for
test 2) do not necessarily imply that the absorbing material is
intrinsic to the QSOs. Further evidence of this nature comes from a
high resolution study of the QSO Pks 0237-23 by Boroson et al. (1976):
on examining the redshift distribution of CIV lines, they find a clear
excess in number of systems near z_{abs} = 1.65. This already is apparent
in Fig.1 of Boksenberg and Sargent (1975). It is suggestive that the
velocity spread in observed redshift grouping is a few thousand km s^{-1},
typical of a large cluster of galaxies.

   We may go further than simply studying the distribution in redshift. Wagoner (1967) has computed the probability that light received
from a distant QSO be intercepted by an intervening galaxy and show
absorption lines (see also Roeder and Verreault 1969). Wagoner used a
conventional galactic luminosity function, including only spiral
galaxies without evolution, and found the probability P(z) that photons
emitted by a source of cosmological redshift z have passed within the
Holmberg dimensions of an intervening galaxy (i.e. the locally
measured 25 mag (arcsec)$^{-2}$ isophote) is a strong function of z, and
in the range 0.6 < z < 1.88 (appropriate for the systems in the survey
data of CBS) his results give P $\simeq$ 0.08 - 0.06 for q_0 = 0 - $\frac{1}{2}$. Corresponding values of the mean number of redshift systems observed by CBS
are 0.55 ± 0.15 for the systems defined by several lines, i.e. the
'identified' systems, and 2.1 ± 0.8 when all unidentified lines are
counted in as single systems (the limits indicate the range from
'certain' systems to 'certain + probable'). However, for the latter,

the unidentified lines assumed to be Lyα span only the far more restricted redshift range from z = 1.88 to a value near z = 1.65 defined by the atmospheric cutoff, and not the whole range appropriate for the identified systems in which lines of other elements at longer wavelengths still are observable at much lower redshift. Wagoner's value for the restricted range is P ≃ 0.015 − 0.01 for $q_0 = 0 - \frac{1}{2}$ (within the Holmberg dimensions). CBS's corresponding value for 'Lyα only' systems is 0.7 ± 0.3. A point to mention is that the spectral resolution used by CBS (Δλ ∿ 5A) corresponds to a range in velocity ∿ 400 km s^{-1}; this masks the fine splittings apparent at higher resol-- ution and ensures that in general any intervening galaxy is counted only once. Comparing the observations with Wagoner's predictions we note that the computed results fall short by about one order of magni- tude for the 'identified' systems and nearer two orders for the 'Lyα only' systems. All other things being equal, this requires that the effective dimensions of a galaxy seen by its absorption lines be larger by factors about three and ten respectively than its Holmberg dimensions.

However, as pointed out by Bahcall and Spitzer (1969), the gas density required to produce absorption lines is so low that it would not be surprising if the maximum radius for measurable absorption in a line were appreciably greater than that detected in other ways, and they suggested that the observed lines are produced in extended low density haloes of normal galaxies (see also Bahcall 1975 and Röser 1975). In support of this, Kormendy and Bahcall (1974) have shown that many spiral galaxies and small groups of galaxies indeed have very large optical haloes, several times the Holmberg dimensions at a surface brightness ∿ 27 mag (arcsec)$^{-2}$. This alone may be enough to explain the observations for the 'identified' systems. We distinguish here between 'identified' systems and 'Lyα only' systems. Because the former include lines of heavy elements, we associate these systems with gas in the inner region of a galaxy where stars have formed; this region clearly extends well beyond the arbitrary 25 mag (arcsec)$^{-2}$ isophote used as a representative, but not strict, boundary by Wagoner (Wagoner recognised this by introducing a factor Σ by which to multiply the cross-section of a galaxy defined by the 25 mag (arcsec)$^{-2}$ isophote and obtain the effective cross-section for a given interaction). The 'Lyα only' systems, on the other hand, probably are due to very extended outer regions where the density is too low to allow star formation and the gas still has primordial composition.

In clear support of this contention, there is much direct evidence from 21 cm emission measures for the presence of neutral hydrogen at large distances from spiral galaxies (e.g. Roberts 1972, Davies 1974, Mathewson et al. 1975, Rogstad et al. 1974), in addition to the absorb- tion measurements of Haschick and Burke and Grewing and Mebold described before. The measures of Davies (1974) reveal appreciable quantities of hydrogen at distances out to 2 to 5 Holmberg radii in the M81/M82/NGC 3077 system, M31, M33, IC 342 and M51. In all cases the gas is at velocities consistent with being bound to the parent

galaxy, and in several galaxies there is as much neutral hydrogen outside the Holmberg dimensions as inside. Davies' (1974) Fig.1 showing
the enormously extensive hydrogen region associated with the
M81/M82/NGC 3077 group is particularly striking. Mathewson et al.
(1975) found HI clouds near the galaxies NGC 55 and NGC 300 in the
Sculptor group with velocities similar to the systemic velocities of
the two galaxies and at linear projected distances up to 80 kpc from
NGC 55 and 180 kpc from NGC 300 (taking 3 Mpc as the distance to the
Sculptor group). They also found a long tail of HI extending along
the major axis of NGC 300 to a projected distance of 140 kpc from the
centre. Such tails also are seen in M83 (Rogstad et al. 1974) and
IC 10 (Shostak 1974). Again, it is striking to note the enormous
extent of the HI regions compared with the relatively small optical .
dimensions of these galaxies (e.g. Mathewson et al. 1975, Fig .1 and 2;
Rogstad et al. 1974, Fig.6). The nearby 'high velocity' HI clouds and
the Magellanic Stream probably are related phenomena (de Vaucouleurs
and Corwin 1975, Mathewson et al. 1974). Evidence for large extensions
of HI beyond the observed optical dimensions of galaxies comes also
from a study of galactic rotation curves (Roberts and Rots 1973,
Rogstad and Shostak 1972).

We point out that although the extended regions of HI emission
near galaxies appear to be broken up into separate 'clouds', actually
their bounding contours simply represent the minimum signal that can
be detected by the radio technique. Typically this corresponds to a
column density of $\sim 10^{19} cm^{-2}$, comparable with the 21 cm absorption
measures (assuming T_s = 100 K). Neutral hydrogen at lower column
densities probably fills in the space between the higher density
'clouds',and also extends still further out from the associated
galaxies than do the already extensive regions mapped in 21 cm emission.
Such regions of lower density would be readily detected spectroscopically at Lyα wavelength if seen as an absorption line against a continuum source, and may appear like the Lyα absorption lines observed
in the spectra of high-redshift QSOs. The limiting detectable column
density of HI when observed in this way typically may be $\sim 10^{13} cm^{-2}$
(assuming absorption here is occurring near the linear section of the
curve of growth and $z_{abs} \sim 2$): this is 10^6 times lower than has been
measured by radio techniques.

If a considerable fraction of all spiral galaxies possess such
enormous hydrogen haloes, even if several orders of magnitude less
dense than those mapped locally at 21 cm, then the observed frequency
of 'Lyα only' systems in QSOs of z $\sim$ 2 is entirely explained by the
large effective cross-section they present. (Incidentally, the
assumption that the'Lyα only' systems are to be identified with neutral
hydrogen is well borne out by the appearance of corresponding Lβ and
other Lyman lines in QSOs of high enough redshift to bring these into
the observable spectral range). That such material probably is primordial is indicated by the similarity in appearance between the Lyman
lines having corresponding lines of heavy elements and those which do
not. The spectrum of OQ 172 is a good illustration of this (Baldwin
et al. 1974). It is unlikely that the distinction between such systems

is explained by different degrees of ionization, with the 'hydrogen'
systems being essentially neutral, because the general QSO radiation
field at, say, $z \gtrsim 2$ is expected to keep all unshielded tenuous matter
moderately to highly ionized (Arons 1972).

The spectrum of the high redshift (z_{em} = 3.4) QSO OH 471(Carswell
et al. 1975) in Fig.7 (obtained with Sargent) clearly shows the multitude of hydrogen absorption lines shortward of emission Lyα, extending
in this figure to Lyβ/OVI. We estimate the mean difference in velocity
between adjacent detected absorbing regions in the range 2.78 < z < 2.95
as $\sim$ 1200 km s^{-1}. Assuming that the material is cosmologically
distributed in a Friedmann universe (Λ = 0, pressure negligible) the
mean interception distance is in the range 3 - 6 Mpc for
H_o = 50 km s^{-1} Mpc^{-1} and q_o = $\frac{1}{2}$ - 0. It is interesting that we obtain
a similar result in this redshift range for the material in line to
the QSOs OQ 172 (z_{em} = 3.53) and 0830+115 (z_{em} = 2.97), the latter
from our observations with Sargent and Lynds. There is also a suggestion of periodic clustering in the density of the absorption lines in
OH 471 with a velocity separation $\sim$ 10000 km s^{-1}, and a corresponding
interception distance 26 - 52 Mpc when making the same assumptions as
before.

We point out in passing that the presence of both primordial and
enriched material argues against the intrinsic interprepation of QSO
absorption lines, because for this we expect ejected material all to
be enriched, as is observed for the line emitting regions.

Going further in the above vein, it is our impression from the
comparatively little evidence yet available that there are propor-
tionately few absorption systems definitely containing heavy elements
in the highest range of redshifts observed, say $z \gtrsim 3$. If so, we may
be seeing evidence that most galaxies have not yet fully formed before
$z \sim 3$. Interpretation of the observations in terms of absorption in
protogalactic material previously has been made by Arons (1972) and
Röser (1975). If, as it seems, the space density of QSOs dramatically
decreases at some high redshift (say, $z \sim 3.5$) then correspondingly
the protogalactic haloes may be largely neutral.

Returning to Wagoner's estimates for the number of galaxies in
line to a distant source, there are two more factors which could
considerably increase the chance of encounter, apart from the points
already made. The first is the possibility that the number and size
of galaxies in general may be grossly underestimated because of
observational selection, as has been pointed out by Disney (1976).
Thus, an apparently insignificant dwarf actually may be the core of a
large galactic system, most of it not seen above the sky background
on conventional plates. The observations of Kormendy and Bahcall
(1974), Arp and Bertola (1969, 1971) and de Vaucouleurs (1969) support
this. The second concerns elliptical galaxies, which are not included
in Wagoner's estimates. The contention that there is little gas in

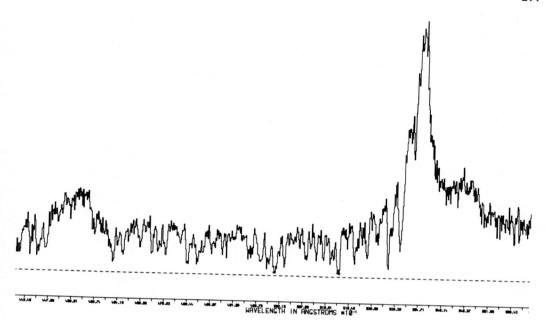

Fig. 7:   Spectrum of OH 471 (z_{em} = 3.4). The strong emission line
is Lyα, the weaker is Lyβ/OIV.

these objects is based on the general lack of 21 cm emission, although
now some marginal detections may exist (Huchtmeir et al. 1975). It is
consistent with the previous discussion that although the radio technique may (just) not be sufficiently sensitive to detect neutral
hydrogen in most ellipticals it does not follow that absorption lines
of hydrogen (and heavy elements) will not be detected optically.
Further weight is given to this by the likelihood that all gas in
elliptical galaxies is ionized by ultraviolet stars (Rose and Tinsley
1973); such gas may be readily detectable optically (at high redshift)
but not be evident when observed at 21 cm. Added to this is the
possibility that no major difference exists between ellipticals and
spirals at the protogalaxy stage (Gott and Thuan 1976), which would
increase the expected number of absorption lines at high redshift.
Evolutionary effects in general were not included by Wagoner.

      In conclusion, we contend that apart from the very broad absorption
lines seen in a few QSOs, such as those in PHL 5200, which most
likely are due to intrinsic mass outflow from these objects, the great
majority of narrow absorption lines in QSO spectra, ranging in relative

velocity from a large fraction of c outward to a few thousand km s^{-1} inward, all of which are characteristically similar and resemble interstellar lines, can be most naturally explained as being produced in cosmologically distributed intervening material, and there is no need to invoke any intrinsic mechanism for these.

With grateful thanks I acknowledge many illuminating discussions and communications with J.N. Bahcall, J.G. Bolton, E.M. Burbidge, G.R. Burbidge, R.F. Carswell, R.D. Davies, M.J. Disney, R. Kippenhahn, L. Mestel, S.W. Moore, J.J. Perry, B.A. Peterson, R.C. Roeder, W.L.W. Sargent, P.A. Strittmatter, R.J. Weymann, R.E. Williams and many others. I am especially grateful to W.L.W. Sargent who introduced me to this subject in our long standing and, due to him, inspired observing collaboration.

References

Aaronson, M., McKee, C.F., and Weisheit, J.C. 1975, Ap.J., **198**, 13.

Anderson, K.S., and Kraft, R.P. 1969, Ap.J., **158**, 859.

Arons, J. 1972, Ap.J., **172**, 553.

Arp, H., and Bertola, F. 1969, Ap.Letters, **4**, 23.
――――――――― 1971, Ap.J., **163**, 195.

Bahcall, J.N., Sargent, W.L.W., and Schmidt, M. 1967, Ap.J.(Letters), **149**, L11.

Bahcall, J.N. 1968, Ap.J., **153**, 679.

Bahcall, J.N., and Wolfe, R.A. 1968, Ap.J., **152**, 701.

Bahcall, J.N., and Peebles, P.J.E. 1969, Ap.J.(Letters), **156**, L7.

Bahcall, J.N., and Spitzer, L. 1969, Ap.J.(Letters), **156**, L63.

Bahcall, J.N. 1971, A.J., **76**, 283.

Bahcall, J.N. 1975, Ap.J.(Letters), **200**, L1.

Baldwin, J.A., Burbidge, E.M., Burbidge, G.R., Hazard C., Robinson, L.B., and Wampler, E.J. 1974, A.P., **193**, 513.

Blumenthal, G.R., and Mathews, W.G. 1974, Ap.J., **198**, 517.

Boksenberg, A., Kirkham, B., Michelson, E., Pettini, M., Bates, B., Carson, P.P.D., Courts, G.R., Dufton, P.L., and McKeith, C.D. 1975, Phil.Trans.R.Soc.Lond.A., **279**, 303.

Boksenberg, A., and Sargent, W.L.W. 1975, Ap.J., **198**, 31.

Boksenberg, A., Shortridge, K., Allen, D.A., Fosbury, R.A.E., Penston, M.V., and Savage, A. 1975, Mon.Not.R.Astr.Soc., **173**, 381.

Boksenberg, A., Carswell, R.F., Allen, D.A., Fosbury, R.A.E., Penston, M.V., and Sargent, W.L.W. 1976, Mon.Not.R.Astr.Soc., (in press).

Boksenberg, A., and Penston, M.V. 1976, Non.Not.R.Astr.Soc., (in press).

Boroson, T., Sargent, W.L.W., Boksenberg, A., and Carswell, R.F. 1976 (in preparation).

Brown, R.L., and Roberts, M.S. 1973, Ap.J.(Letters), $\underline{184}$, L7.

Burbidge, E.M., 1968, Ap.J.(Letters), $\underline{152}$, L111.

Burbidge, E.M. 1970, Ap.J.(Letters), $\underline{160}$, L33.

Burbidge, E.M., and Burbidge, G.R. 1975, Ap.J., $\underline{202}$, 287.

Burbidge, E.M., Caldwell, R.D., Smith, H.E., Liebert, J. and Spinrad,H. 1976b, Ap.J.(Letters), $\underline{205}$, L117.

Burbidge, G.R. 1975, Ap.J.(Letters), $\underline{196}$, L7.

Burbidge, G.R., Crowne, A.H., and Smith, H.E. 1976a, (submitted to Ap.J.Suppl.,).

Burbidge, G.R., and Perry, J.J. 1976 (preprint).

Carswell, R.F., Strittmatter, P.A., Williams, R.E., Beaver, E A., and Harms, R. 1975, Ap.J., $\underline{195}$, 269.

Carswell, R.F., Boksenberg, A., and Sargent, W.L.W. 1976 (in preparation).

Carswell, R.F., Coleman, G. Strittmatter, P.A., and Williams, R.E. 1976 (preprint).

Coleman, G., Carswell, R.F., Strittmatter, P.A., Williams, R.E., Baldwin, J., Robinson, L.B., and Wampler, E.J. 1976, Ap.J., $\underline{207}$, 1.

Davies, R.D. 1974, in J.R. Shakeshaft (ed), 'The Formation and Dynamics of Galaxies', IAU Symposium No. 58, p.119.

de Vaucouleurs, G. 1969, Ap.Letters, $\underline{4}$, 17.

de Vaucouleurs, G., and Corwin, H.G. 1975, Ap.J., $\underline{202}$, 327.

Disney, M.J. 1976, Nature, $\underline{263}$, 573.

Gott, J.R., and Thuan, T.X. 1976, Ap.J., $\underline{204}$, 649.

Grewing, M., and Mebold, U. 1975, Astr. & Ap., $\underline{42}$, 119.

Haschick, A.D., and Burke, B.F., 1975, Ap.J.(Letters), $\underline{200}$, L137.

Huchtmeier, W.K., Tanmann, G.A., and Wendker, H.J. 1975, Astr. & Ap., $\underline{42}$, 205.

Kippenhahn, R., Perry, J.J., and Röser, H.-J. 1974, Astr. & Ap., $\underline{34}$, 211.

Kippenhahn, R., Mestel, L., and Perry, J.J. 1975, Astr. & Ap., $\underline{44}$, 123.

Lowrance, J.O., Morton, D.C., Zucchino, P., Oke, J.B., and Schmidt, M. 1972, Ap.J., $\underline{171}$, 233.

Lynds, C.R. 1967, Ap.J., $\underline{147}$, 396.

Lynds, C.R. 1971, Ap.J.(Letters), $\underline{164}$, L73.

Lynds, C.R. 1972, in IAU Symposium No. 44: External Galaxies and
    Quasi-Stellar Objects, ed. D.S. Evans. Dordrecht: Reidel Publ.Co.,
    p.127.

Mathewson, D.S., Cleary, M.N., and Murray, J.D. 1974, Ap.J., <u>190</u>, 291.
    ——————————— 1975, Ap.J.(Letters), <u>195</u>, L97.

McKee, C.F., and Tarter, C.B. 1975, Ap.J., <u>202</u>, 306.

Mestel, L., Moore, D.W., and Perry, J.J. 1976, Astr. & Ap., <u>52,</u> 203.

Morton, D.C., and Hu, E.M. 1975, Ap.J., <u>202</u>, 638.

Morton, D.C. 1976, Ap.J., <u>203</u>, 386.

Mushotzky, R.F., Solomon, P.M., and Strittmatter, P.A. 1972, Ap.J.,
    <u>174</u>, 7.

Osterbrock, D.E., Koski, A.T., and Phillips, M.M. 1976, Ap.J.(Letters),
    <u>197</u>, L41.

Roberts, M.S., and Rots, A.H. 1973, Astr. & Ap., <u>26</u>, 483.

Roberts, M.S., Brown, R.L., Brundage, W.D., Rots, A.H., Haynes, M.P.,
    and Wolfe, A.M. 1976, A.J., <u>81</u>, 293.

Roeder, R.C., and Verreault, R.T. 1969, Ap.J., <u>155</u>, 1047.

Roeder, R.C. 1972, Ap.J., <u>171</u>, 451.

Rogstad, D.H., and Shostak, G.S. 1972, Ap.J., <u>176</u>, 315.

Rogstad, D.H., Lockhart, I.A., and Wright, M.C.H. 1974, Ap.J., <u>193</u>, 309.

Rose, W.K., and Tinsley, B.M. 1974, Ap.J., <u>190</u>, 243.

Röser, H.-J. 1975, Astr. & Ap., <u>45</u>, 329.

Sargent, W.L.W., and Boroson, T.A. 1976 (submitted to Ap.J.,).

Scargle, J.D., Caroff, L.J., and Noerdlinger, P.D. 1970, Ap.J.(Letters),
    <u>161</u>, L115.

Scargle, J.D. 1973, Ap.J., <u>179</u>, 705.

Shostak, G.S. 1974, Astr. & Ap., <u>31</u>, 97.

Stockton, A.N., and Lynds, C.R. 1966, Ap.J., <u>144</u>, 451.

Strittmatter, P.A., Carswell, R.F., Burbidge, E.M., Hazard, C.,
    Baldwin, J.A., Robinson, L., and Wampler, E.J. 1973, Ap.J., <u>183</u>, 767.

Wagoner, R.V. 1967, Ap.J., <u>149</u>, 465.

Weymann, R.J. 1973, Comm.Ap. and Space Phys., <u>5</u>, 139.

Weymann, R.J. 1976, Ap.J., <u>208</u>, 286.

Weymann, R.J., and Cromwell, R. 1972, in IAU Symposium No. 44:
    External Galaxies and Quasi-Stellar Objects, ed. D.S. Evans,
    Dordrecht: Reidel Publ.Co., p.155.

Weymann, R.J., Williams, R.E., Beaver, E.A., and Miller, J.S. 1976 (preprint).

Williams, R.E. 1972, Ap.J., 178, 105.

Williams, R.E., Strittmatter, P.A., Carswell, R.F., and Craine, E.R. 1975, Ap.J., 202, 296.

Williams, R.E., and Weymann, R.J. 1976, Ap.J.(Letters), 207, L143.

Wolfe, A.M., Broderick, J.J., Condon, J.J., and Johnston, K.J. 1976, Ap.J.(Letters), 208, L47.

DISCUSSION

H.E. Smith: I have several comments.

1)   In the high redshift QSO's, Margaret Burbidge and I have uniform spectroscopic data on several, but I speak in particular of M 0830+115 (z_e = 2.97).   One sees a sharp discontinuity in the absorption line density at Ly α emission, of a factor of five or so.   In general one cannot identify most of the lines shortward of Ly α emission and it is most reasonable to assume that the majority are Ly α absorption from clouds with insufficient column density to produce absorption from less abundant species.   In the case of 0830+115 one then requires on the order of 100 separate intervening systems.   If, as is common, the lines break up at higher dispersion then even more systems are required. This object is not extreme, rather it seems to be the rub for high z systems.

2)   For systems with absorption lines from heavy elements (CIV, SiIV, etc.) one must restrict the cross section to less than a Holmberg radius.   The observations clearly show a strong decrease in heavy element abundance with radius (on the order of a factor of ten or more) for spiral galaxies. If our understanding of galaxies is correct the abundances of heavy elements must be exceedingly low outside the optical radius since these regions simply could not have processed material through stars in any amount.

3)   Some objects, PHL 5200 and RS23 in particular, are clearly ejecting material with velocities on the order of 10,000 km s^{-1} which can be seen from the broad P-Cygni type profiles on the lines 10,000 km s^{-1}.   Ray Weymann has shown us observations of galactic novae which show similar profiles at some stages which then break up into large numbers of very narrow absorption lines.   The analogy to absorption line systems in QSO's may be quite apt.

4)   If line-locking coincidences can firmly be established, then one must accept that these systems have been ejected from the QSO.   There are a number of suggestive cases, but I'm not sure if one can say for certain that it is operating in any given case.

5)   Finally I am impressed by the case of the highly active BL Lac object AO 0235+164.   It is peculiar in that it has not only shown two very strong outbursts ($\Delta m \simeq 5$ mag) in the past 30 years, but that it is the <u>only</u> object known with two absorption line systems with z < 1. This suggests that the absorbing material may be material associated with previous outbursts.   With respect to the distribution of absorption line redshift systems, I'd like to emphasize the difficulty of ident- ifying systems in high redshift QSO's.   In many cases one sees a sharp discontinuity in the density of lines at Lyman α emission.   In these cases we are probably seeing absorption line systems containing only Lyman α, distributed uniformly in redshift from just below z_{em} to a redshift with respect to the QSO corresponding to a relation velocity ~ 0.5 c.

Boksenberg: This is clearly for the highest redshift objects (say, $z \gtrsim 3.0$) which have been adequately well observed, for example OH471, OQ172 and 0830+115.   Furthermore the line density in a given redshift range (but between emission Lα and Lβ to avoid confusion) is about the same for all these objects.   For intermediate redshift objects (z = 2) the sharp discontinuity in the density of lines at emission Lα is not as pronounced in some cases as in others (indeed some objects have a distinctly sparce appearance below emission Lα).   From a brief con- sideration of such data I very much favour the explanation that the absorption is occurring in intervening material and is not intrinsic to the QSO's.

Burke: I see no reason at our present state of knowledge to assume that the HI clouds in the vicinity of galaxies are primeval, lacking heavy elements.   They could be well-enriched material ejected from the parent galaxy, and such ejection could occur at a very early stage in the life of the galaxy.   The present sensitivity of HI detection in absorption is approaching $10^{18}/cm^2$.

Carswell: I do not believe that the Palomar $z_{em} \simeq 2$ survey data is in conflict with either of the "intrinsic to the QSO" or "intervening cosmological material" hypotheses.   There were some problems in analysing that material because the signal-to-noise was poor below the Lα emission, and there were a large number of unidentified lines even above that feature.   It appeared at first sight, taking all possible systems, that the observed distribution was significantly different (< 0.1% by chance) from the poisson distribution the cosmological cloud model predicts.   However there are a large number of choices available, depending on how you believe you have to select objects from our sample to make a genuinely homogeneous one, and it seems possible to make a case for any answer you want on this basis.   The data we have is really not good enough to distinguish between the two models.

On the basis of less homogeneous material obtained at Steward Observatory by E. Coleman, P.A. Strittmatter and R.E. Williams and my- self, I believe that there is a serious conflict with the cosmological cloud hypothesis, but further observational work remains to be done before the statistical test can usefully be applied. However, cases we have with no absorption lines and one with about 18 shortward of emission Lα suggest that a poisson distribution does not fit the data well.

QSO-GALAXY PAIRS AND OPTICAL OBSERVATIONS

J.N. Bahcall and B.F. Burke

We would like to call to the attention of optical observers some potentially crucial observations. As has been discussed in this symposium, 21 cm absorption-lines caused by the neutral hydrogen halos of two nearby galaxies (NGC 3067 and NGC 6053) have been recently detected in the spectra of two quasars. It has also been suggested (see Bahcall, Ap. J. (Letters) 200, L1, (1975) and references quoted therein) that the extended halos of galaxies may cause many of the ultraviolet absorption lines that are seen in the spectra of large-redshift quasars. A problem with this suggestion is that it implies that the halos of galaxies (at large distances from where most of the stars are presently observed) contain appreciable quantities of heavy elements. Many astronomers believe that this is impossible (cf., however, Bahcall 1975), but direct evidence is lacking on the heavy element content of large galactic halos (radii $\sim 10^2$ Kpc).

We urge optical astronomers to undertake sensitive searches for possible heavy-element absorption lines at the known redshifts of the 21-cm-absorbing galactic halos (e.g., 1494 km/s for NGC 3067, 4C 32.33). In the visible this may be done by looking for traditional interstellar lines (e.g., H and K) that could occur at the same redshift as the 21-cm absorption lines. With a Large Space Telescope, one would look for the resonant absorption lines of such abundant heavy elements as C, N, and Si. It is very difficult to estimate accurately the expected equivalent widths since the physical conditions in the halo clouds are not well known (and may be different from the conditions that existed when the quasar absorption lines were produced). Nevertheless, the search for heavy-element absorption lines in the spectra of quasars (that are part of quasar-galaxy pairs) at the redshifts of the galaxy halo offers a unique opportunity to possibly observe heavy elements in extended galactic halos. The astrophysical and cosmological implications of such a discovery could be great.

E.M. Burbidge: Gene Smith covered very well the arguments for the multiple redshift systems being associated with the QSO itself, rather than being produced in intervening galaxies, or intergalactic clouds, and I have only a few remarks to add.

First: I believe that people get worried by the use of the adjective "intrinsic" for these redshift systems - people seem to link this instinctively with "intrinsic" redshifts - i.e. non-doppler redshifts. So I would like to emphasize that we are indeed talking about doppler displacements, due to pressure-driven gas outflow from a central continuum source.

That radiation pressure must produce outflow of gas has been clear from theoretical work, for example by Bill Mathews and colleagues at U.C. Santa Cruz, and by Kippenhahn, Mestel and Perry at Max Planck Institute, Munich, and by others. This must occur even in the emission-line producing gas. Outflow, even if it starts with a very optically thick supernova-like shell moving out at around 10^4 km/sec as in PHL 5200, must accelerate as does all gas outflow going towards lower density, and instabilities will set in and will lead to break-up as in expanding nova shells where very narrow absorption lines can be seen.

Now a word about the distribution of absorption-line QSO's. There are some objects (in particular PKS 0237-23) where there is a large number of redshift systems running all the way from 2.20 to 1.36. Yet there are still 2 objects at redshifts $\gtrsim$ 2 which do not have even one absorption-line redshift system (from recent high-resolution work by the U. of Arizona astronomers). This is very difficult to explain on the intervening galaxy hypothesis. Third, I should reiterate that apart from Ly α the most characteristic lines seen are the CIV doublet. In our galaxy, rocket and Copernicus observations show this is produced in circum-stellar shells associated with hot stars, not in the cool inter-stellar gas. Thus it is difficult to associate such features with cloudlets of cold HI several Holmberg radii outside galaxies.

Lastly, a word about line-locking. Radiation - pressure driven outflow can be affected by a strong gradient in flux with wavelength - i.e. by coincidences between absorption lines and ionization edges and other strong absorption lines from different ions at different redshifts. This will produce preferred ratios in (1 + z). This effect is well seen in a number of QSO's, and Jeff Scargle has also discussed its occurrence in certain hot stars where UV lines seem to produce the same kind of line-locking.

Scargle: One of the general objections often made to the picture in which the narrow high-velocity QSO absorptions are "intrinsic", or due to outflowing gas, is the difficulty of understanding the large ratio of $V_{total}/\Delta V$ (ΔV = velocity dispersion within the system). This number is a few x 10^4 for the most extreme cases in QSO's. It does not seem to be generally realized that there exist shell stars for which the ratio is nearly as great, namely a few x 10^3. Thus whatever difficulties exist for QSO's also exist for shell stars, for which the intrinsic nature of the absorptions is well-established.

Bahcall: Which QSO's at high redshift do not have absorption redshifts?

E.M. Burbidge: They are BSO6 and LB8755; they are not particularly faint and very detailed studies have been made of them.

Van der Laan: How many cases of $z_{abs} > z_{em}$ are there now and do they contain multiple z_{abs} systems?

E.M. Burbidge: There are about half a dozen, and about half contain multiple absorption redshift systems.

NEUTRAL HYDROGEN ABSORPTION FEATURES
IN THE SPECTRA OF QUASAR – GALAXY PAIRS

B.F. Burke

We have continued the search for 21 cm neutral hydrogen in absorption in quasar – galaxy pairs to a total of 8 pairs. In addition to the feature found in the pair 4C32.33-NGC3067 (Haschich and Burke, Ap. J. (Lett) <u>200</u> L137, 1975) we have found a further feature in the pair 1749+701-NGC6503. That 2 out of 8 pairs show absorption features strengthens the evidence that neutral hydrogen exists far away from spiral galaxies, and suggests that the average cross section is approximately 6 times larger than the visual image. Thus, galaxies may be considered as candidates for the discrepant absorption lines seen in QSO's.

Secondly, we note the similarity of the observed 21 cm absorption lines: their column densities and velocity widths lie within one order of magnitude. Furthermore the apparent velocities of the identified clouds range from 70 to 192 km sec^{-1} with respect to the galaxies. These properties are close to the characteristics of high velocity clouds and we suggest that our clouds are of the same nature. This may supply the needed absorption line structure and may also act in favour of an association of high velocity clouds with galaxies.

3C 286 : A COSMOLOGICAL QSO?

A.M. Wolfe, J.J. Broderick, J.J. Condon and K.J. Johnston

Ap. J. (**Letters**) <u>208</u>, L47 (1976)

Wolfe: To compare the narrow 21 cm absorption features you find in the QSO-galaxy pairs with the line in 3C 286 it is important to compare the linear dimension subtended by the radio source at the absorbing galaxy. In 3C 286 the VLB sources subtend ~ 200 pc at the "intervening galaxy"; but in 4C 32.33 the region subtended is < 10 pc. Thus it is easier to understand the narrow widths of the absorption feature in the latter object than in the former. Could you say something about the linear distances subtended in the new objects you have discovered?

Shaffer: 1749 + 701 is unresolved at about 0.01 arc seconds.

G.R. Burbidge: The simplest interpretation of the galaxy-QSO pairs is that they are physically associated. In this case presumably the QSO has been ejected from the galaxy. If this were the case it would not be surprising if much of the gas seen in this galaxy was ejected also; i.e. it may not be a normal galaxy as far as a hydrogen halo is concerned.

Burke: The absorption lines are very narrow, suggesting undisturbed gas. No quasars have been located close to the M81-M82 system, which has associated hydrogen clouds well outside the Holmberg diameters, and we suggest that the observed absorption lines in our work come from similar outlying gas.

SOUTHERN HIGH REDSHIFT QSO'S

J.G. Bolton, R.D. Cannon, D.L. Jauncey, B.A. Peterson,
Ann Savage, M.G. Smith, K.P. Tritton, and A.E. Wright.

A considerable number of southern QSO's with high redshift have
been recently discovered from both optical and radio investigations.
The optical searches have made use of objective prisms on southern
Schmidt telescopes and favour the discovery of QSO's with a strong
Lyman α emission line within the wave length range of the plate response.
Detections from radio investigations have resulted from identification of
neutral or non UV excess stellar objects on the basis of accurate, $\leqslant$ 2",
radio positions.   Some of the objects found in the latter search could
not have been found either by using colour criteria or objective prism
spectra.   One such example is PKS 0528 - 250 (Jauncey, Wright, Peterson
and Condon, in preparation) whose spectrum is shown in the earlier paper
by Bolton.   This object has a prominent absorption spectrum with a
redshift of 2.813 but has no emission lines.   Another object, PKS
2126-15, has an emission line redshift of 3.275 with a very rich absorption line spectrum below Ly α emission.

Examination of the first plates taken with the 0.73 degree objective
prism on the U.K. 1.2 metre Schmidt telescope has shown that it is
possible to detect quasars with remarkably low line to continuum ratios.
Objects previously selected as possible quasars on the basis of colour
can be separated into quasars or galactic stars with a high degree of
certainty and this results in a considerable saving of large telescope
time in spectroscopic examination.   Two examples of U.K. Schmidt spectra
are shown in Figure 1.

The Tololo Schmidt objective prism plates have provided a considerable number of very high redshift quasars where the line to continuum
ratio for Lyman α is very strong.   Quantitative data on the emission-
line spectra of the Tololo QSO's are summarised by Osmer and Smith
(Ap. J. in press).   Comparison of the spectral characteristics of
emission-lines and continuum in the newly-discovered Tololo QSO's and a
set of 8 quasars previously identified with radio sources shows the two
sets to be indistinguishable, with the possible exception of the NV/Lα
ratio (Osmer, Ap. J. in press).   The absorption-line spectra of these
QSO's are now being studied.   Q 1246-057 (z_{em} = 2.212) has already been
mentioned earlier by Boksenberg.   Absorption lines in the broad-lined
system (z_{abs} = 2.05) with velocity dispersion ~4000 km^{-1} were even
visible (with hindsight) on the Schmidt discovery plate.   High velocity
dispersion has been found in PHL 5200 for example (Lynds, C.R., Ap.J.
(Lett) 164, L73, 1967; Burbidge, E.M., Ap. J. (Lett) 152, L111, 1968),
but this new object provides the first example where a high-velocity-
dispersion cloud is clearly separated from the emission region by speeds
of thousands of kilometres per second.   It seems unlikely that such
broad lines can arise in an intervening galaxy.   A preliminary analysis
by Carswell, Smith and Whelan of the spectrum of Q0002 - 422, z_{em} =
2.758 shows a definite absorption system with a redshift of only z_{abs} =
0.835; all the expected lines are present (FeII, MgII, MgI) with
reasonable equivalent widths.

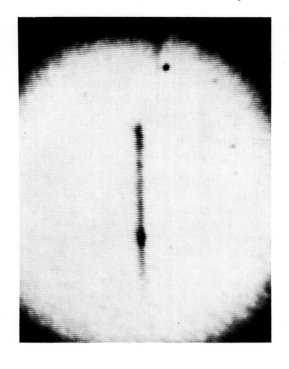
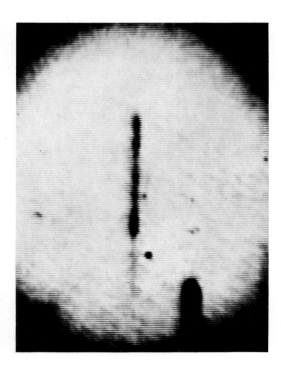

(a) Q 2225-404

(b) PKS 2227-399

Fig. 1

Fig. 1.

Spectra of quasars obtained with the U.K. Schmidt.   The right hand end
of each spectrum corresponds to the green response cut off of the IIIaJ
emulsion.   The spectra were photographed from the T.V. screen of the
modified Zeiss blink microscope at the Royal Observatory, Edinburgh.
The "scans" were obtained by photographing the trace of a single line of
the T.V. monitor on an oscilloscope.   Spectrum (a) is of the optically-
selected QSO, Q 2225-404, of redshift 2.02.   Note (i) the strong Lα
emission (the left-most blob) at an apparent wave length of 3363 Å
(ii) the sharp drop in continuum intensity shortwards of Lα, (iii) the
other emission lines, presumably including C IV and Si IV.   The right-
most blob is an artifact of the reduced prismatic dispersion in the
green combined with the IIIaJ emulsion response.   Spectrum (b) is of
the radio selected QSO, PKS 2227-399 of redshift 0.319 which exhibits
considerable optical variability.   Note the very strong MgII emission
feature towards the left end of the spectrum.   This is the first time
that MgII has been seen on objective-prism plates.

RELATIONS BETWEEN THE OPTICAL AND RADIO PROPERTIES OF EXTRAGALACTIC RADIO SOURCES

E. M. Burbidge
University of California, San Diego

I.   INTRODUCTION

Correlations between the radio and optical properties of radio sources have proved elusive and the main conclusion to be drawn from this is that there is a great variety of objects in the universe that emit nonthermal radiation, so that attempts to use these objects for cosmological purposes can be frustrated unless one can find some way of selecting objects that do have common intrinsic properties. Despite this, the search for relations and correlations is interesting quite apart from cosmology, because such correlations should provide a groundwork for a physical theory or theories of what is really happening in sources of nonthermal radiation.

I want first briefly to discuss radio galaxies, because the problems here are better definable than for the QSOs, for which the basics, even distances and redshifts, and the very obvious heterogeneity of such objects, pose very great problems in their understanding. Then I shall turn to the QSOs and present some new results.

II.   RADIO GALAXIES

A recent catalogue of 3CR sources (Smith, Spinrad, and Smith 1976) lists 137 radio galaxies, and of those for which optical spectroscopic characteristics are given 45 have strong emission lines, 23 have emission lines of moderate strength, 12 have weak emission and 23 have only absorption lines. Thus there appears to be some correlation between a radio source being strong enough to be listed in the 3CR catalogue and the presence in its optical spectrum of emission lines other than just weak [O II] or Hα. On the other hand, when one considers galaxies that are weaker radio sources, the situation changes. Data compiled a year ago (Burbidge 1975) for 222 galaxies

D. L. Jauncey (ed.), Radio Astronomy and Cosmology, 223-235. All Rights Reserved.
Copyright © 1977 by the IAU.

showed that 152 had no strong emission lines in their spectra, while
70 did have strong emission lines. Weaker radio galaxies include
ellipticals and spirals with nuclear emission. Nuclear emission from
spirals becomes quite common at a low level. The E galaxies contribute heavily to the objects that have only absorption lines in their
optical spectra while spirals may have emission lines of weak or
average strength, as well as the fewer cases that have strong emission
lines.

The range of morphological types of radio galaxies have changed
little since the original categorization by Matthews, Morgan, and
Schmidt (1964). There are 15 N systems in the 3CR set of 137
galaxies. It is worth remembering that the original definition by
Morgan (1958) for the N-type galaxies is "systems having small,
brilliant nuclei, superposed on a considerably fainter background."
The "db" or dumbbell galaxies, i. e. double ellipticals in a common
stellar envelope, have however been found to be much rarer than
appeared in the Matthews, Morgan, and Schmidt list. The giant E or
D galaxies, very often the brightest members of clusters of galaxies,
still predominate.

Several papers in this symposium have discussed correlations
between radio spectral index and luminosity of radio galaxies, and
the paper by Osterbrock describes in detail optical spectrophotometric
results for a number of radio galaxies. I think one can say that there
is a tendency for a compact radio or optical object to possess strong
emission lines in its optical spectrum. There is especially a tendency
for objects with complex radio spectra to be associated with compact
optical objects; N systems certainly fall into this category. However,
there is one notable exception to the statement that compactness goes
with the presence of strong emission lines, and that is the BL Lac
objects. I will discuss these in Section V of this paper.

III.   QSOs -- GENERAL CONSIDERATIONS

One of the important questions which one would like to be able to
answer is "What is the relation between radio-emitting and radio-
quiet QSOs?" QSOs were first discovered as radio sources, but
apparently they are far outnumbered in the sky by the radio-quiet
objects. We have estimates from Sandage and Luyten (1967) of the
number of radio-quiet QSOs per square degree down to a magnitude
limit of 19. 7; the number usually adopted is 5 per square degree.
This number may, however, need revision; it may not be simply
extendable to fainter objects; it may even not be constant across the
sky. These are all questions which need to be looked into and new

programs for finding radio-quiet QSOs from objective prism spectra, both by M. Smith and P. Osmer in the southern hemisphere and by A. A. Hoag in the northern hemisphere, are expected to throw some light on this question soon. Harding E. Smith, A. A. Hoag and I have been observing some of the objects in the region of sky covered by the Hercules cluster of galaxies; these objects were discovered by A. A. Hoag in his objective "grism" study using the 4-meter telescope at Kitt Peak. The preliminary results are very interesting, since there appear to be a large number of objects -- 16 candidate QSOs in the first field of approximately one square degree which was searched, of which 8 have been confirmed, and something like 11 candidates in the second field nearby but off the Hercules cluster. This search method is subject to observational selection in that objects of high redshift will predominantly be found, since they show $Ly\alpha$ emission which is the strongest line. We need to extend such research to objects of lower redshift. The old way of finding radio-quiet QSOs, by looking for objects with UV excess, may still be the best way of finding low-redshift objects, and, since this method may discriminate against high-redshift objects, a combination of the two may cover the whole redshift range.

Let me now turn to some comments on the models for QSOs in which photoionization is the input that controls the region giving rise to the emission lines. One important question is what happens at the Lyman limit; this can be observed in objects at high redshift. Among such high-z objects there is a range from those in which there is no discernible discontinuity at the Lyman limit to those in which there is a very heavy and almost complete absorption there so that very little radiation is emitted immediately shortward of 912 Å. This emphasizes once again the gross differences that exist from one object to another. The emission-line spectrum in any one object also clearly indicates that there are inhomogeneities of temperature and density in the region producing the emission lines. In addition, the line-emitting material must be concentrated in blobs which do not fully cover the continuum source, for the following reason. The ratio of $Ly\alpha$ to $Ly\beta$ emission in those objects of sufficiently high redshift to show $Ly\beta$ is such that $Ly\alpha$ is very strong and $Ly\beta$ is weak or almost absent. This can be interpreted as being due to an optically thick line-emitting region. Every $Ly\alpha$ photon which is produced throughout the emission region escapes after many scatterings. But $Ly\beta$ photons are degraded into $Ly\alpha + H\alpha$ and thus the $Ly\beta$ that one sees comes only from a thin skin of the emitting region. But optically thick material must not fully cover the continuum source because it would then cause a very large absorption at the Lyman limit, and as already noted there are objects in which this is not the case. Thus the line-emitting

material may be distributed in blobs which partially cover the continuum source. Another possible model is one in which the emitting
gas does not have spherical symmetry about the central object.
Indeed, this might well be the case if the central object is a strongly
gravitating collapsed mass in which there was initial angular momentum.

In connection with models, before describing some specific
results that I would like to present, I want to mention the absorption-
line spectra that appear at many redshifts in objects of high z. It
appears promising to explain these by means of radiation-pressure-
driven outflow of gas, which can be accelerated to high velocity and
is then subject to instabilities so that it breaks up into small filaments.
A new consideration is that the outflow might also be driven by cosmic
ray pressure. Some work along these lines is being done at the Max
Planck Institute for Astrophysics in Munich, in collaboration with
G. Burbidge.

Another question to consider is whether or not the apparent
abrupt drop in the redshift distribution at $z \gtrsim 2.3$ is real. It is not
clear to what extent this may be due to observational selection. If
objects are identified with radio sources on the basis of their having
a UV-excess or appearing as blue stellar objects on the Palomar
Atlas, there might very well be such a selection against high-z QSOs,
because some of those objects have heavy absorption beyond their
Lyman limit and some have a multitude of absorption lines shortward
of Lyα, so they will not appear strong in the ultraviolet and therefore
may well be missed. The new searches designed to pick up high-
redshift quasars should provide new statistics here.

IV.   QSOs -- OBSERVATIONAL RESULTS

Before describing these, let me outline ways in which optical
astronomers can work together with radio astronomers:

1.   One can work with a well-defined radio sample and get
     optical observations of as many objects as possible at low
     signal to noise.

2.   One can work on particular, limited, radio samples, for
     example, those selected by means of their radio spectral
     properties.

3.   One can work on individual objects which have been selected
     either from their radio or optical properties as being of
     particular interest.

Programs underway include examples of all three.

1. Optical Survey of Molonglo Radio Sources

Identifications and optical observations of sources in the MC2 and MC3 catalogues have been made by a group including C. Hazard, H. Murdoch, J. Baldwin, H. E. Smith, E. J. Wampler, and myself. The optical observations were made at Lick Observatory with the 3-m telescope and the Wampler-Robinson image dissector scanning spectrograph. The $+11°$ and $+16°$ of the MC2/3 from R.A. 11 1/2 to 17 h have been fairly well covered. There are 65 candidate QSO identifications, of which 42 have had reliable spectroscopic observations. Of the BSOs, 38 proved to be QSOs and 4 were galactic stars. This indicates a high degree of reliability, around 90%, for the identifications. Work in other regions of the sky covered by the MC2 and MC3 catalogue is less advanced, but a similar ratio of reliability of identification has been found in the R.A. 20 to 24 h region. For the whole two strips, we have so far 57 confirmed QSOs and 2 BL Lac objects. In addition to BSOs, several red stellar objects which might be identifications of the radio sources were observed, in an attempt to be sure not to discriminate against high-z QSOs. Of these, 9 proved to be galactic stars and 1 was one of the 2 BL Lac objects found. This strongly suggests that in this sample there do not exist a large number of very high redshift red or neutral color objects. There are 8 objects which have continuous spectra, i.e. our observations have sufficiently high signal to noise to indicate that these are not galactic stars and may well be candidate BL Lac objects.

Of the 600 known QSOs, this Molonglo sample comprises 10%. It is of interest to consider properties such as the mean redshift and distribution of redshift. We find $\langle z \rangle = 1.162$, which is very similar to $\langle z \rangle$ found for the 4C QSOs. The distribution in redshift is also similar to that found from previous samples and again indicates a paucity of objects with redshift $z > 2.2$. It is interesting that there is a suggestion of a peak at a redshift of about 0.8 to 0.9, which does not appear in the distribution for the total 600 known QSOs. There are of course not enough observations so far to say whether this is a real peak.

We have looked into the question whether there is a systematic variation of redshift around the sky at a level corresponding to 10% in velocity or variation in numbers of QSOs around the sky. There may be some indications of patchiness, but it is too early to say anything definite one way or the other.

With this good sample of objects, it is interesting to look for a correlation of the continuum spectral index with important emission lines such as Lyα, C IV, Mg II, and Hβ. In a preliminary search we found no correlation, although the simple photoionization model would predict a dependence of equivalent width of emission lines upon the shape of the optical continuum. I believe the only conclusion to be drawn from this is again that the QSOs comprise objects which have a very wide range of physical conditions.

I want to turn now to very preliminary results by the same group of observers on sources in the Molonglo faint survey, MC5, using identifications by D. Jauncey, C. Hazard and H. Murdoch. H. E. Smith and I obtained spectra of some half dozen objects and found the first four to be high-z QSOs:

Object	Redshift	Object	Redshift
0758+120	2.66	0830+115	2.97
0824+110	2.29	0938+119	3.19

Again, a wide range of physical conditions must be present in these four objects. 0938+119 and 0830+115 are like the Ohio cm-excess objects OQ 172 and OH 471 which have flat or peaked radio spectra. In 0938+119 there is a clear discontinuity near the Lyman limit but as in OH 471 this does not occur at the wavelength corresponding to z_{em} but at a shorter wavelength, and it is probably produced by a multitude of absorption lines. 0830+115, like OQ 172, does not have a Lyman discontinuity but has hundreds of narrow absorption lines shortward of Lyα emission. It is interesting to note that none of the 5 QSOs with highest z (OQ 172, OH 471, 0938+119, 0830+115, and 4C 5.34) have complete absorption in the Lyman continuum continuum at a wavelength corresponding to that given by their emission-line redshift system. Yet the absence of Lyβ emission again indicates that the regions producing the emission lines is optically thick as discussed above.

Concerning the emission-line strength in these high-z objects, the ratio C IV to Lyα is larger than the "normal" value of 1/4. O VI is present in 0830+115 and 0938+119.

2. Red QSOs and BL Lac Objects

I want now to turn to consideration of two groups of objects, those which I shall call the red QSOs, and the BL Lac objects. There are six 3C objects which have been identified as QSOs, but which are distinctly red in color:

| 3C 68.1 | 3C 154 | 3C 418 |
| 3C 108 | 3C 212 | 3C 422 |

Boksenberg showed that 3C 68.1 has a very steep slope in its optical continuum with $\alpha \approx 6$. These objects are quite difficult to observe because they tend to be faint. Harding E. Smith has studied 3C 418. It has a $m_V \approx 21.0$, and a continuum slope in the optical region $\alpha = 4.3$. This value of 4.3 may possibly be reduced to about 4 if one corrects for an average extinction in our own Galaxy. The redshift $z = 1.687$. It has Mg II, C III], and C IV emission lines of normal strength. That these objects are fairly rare is indicated by the fact that only these six have emerged in the approximately 50 QSOs in the 3C catalogue. However, it must be remembered that due to their red color and their faintness they may have been discriminated against in identification and their actual relative numbers may be higher than indicated by these results.

The BL Lac objects are extremely compact in both their radio and optical radiating regions, as is indicated by the characteristic features by which one defines BL Lac objects -- that they show very strong variability on very short time scales, and a high degree of polarization which itself is variable in both extent and angle. What distinguishes the BL Lac objects from the optically violent variable QSOs is the lack of reasonably strong emission lines in them. They were first categorized as having no emission lines at all, but some, when high S/N observations were obtained, have been shown to have weak emission lines. The lack of strong emission lines can be due to two causes -- either a lack of ionizing radiation, or a lack of gas to be ionized. The former is suggested by the fact that the BL Lac objects resemble the red QSOs in having steep spectral indices and they appear neutral or reddish in color on the Sky Survey plates. One may note that there is plenty of gas seen in the spectrum of one very interesting BL Lac object, viz. AO 0235+164. This absorbing gas was shown by Burbidge et al. (1976) and Rieke et al. (1976) to possess two redshifts, $z = 0.524$ and 0.851; it is the first object found to have two absorption-line redshifts less than $z = 1$.

If one attempts to explain the weakness of the emission lines in BL Lac objects as being due to steepness of the optical continuum, however, it is interesting to note that the C IV line is quite strong in the red QSO 3C 418, and it requires very short-λ radiation to produce C^{+++}. This argues for there being a considerable flux of either high-energy particles or shock waves to ionize the gas, or else that the continuum is complex and increases again at shorter wavelengths than can be observed from the ground.

V.   THE RADIO SOURCE 3C 303

3C 303 has a very interesting history and is an example of my third category of cooperation between radio and optical observers. Wyndham (1966) identified the source as an N-type galaxy, but noted that there was a blue stellar object about 20'' away which might also be the source.   Sandage (1973) measured z = 0.141 for the galaxy, from strong emission lines.   Later observations with higher resolution indicated that the radio source was complex and encompassed both the galaxy and the BSO.   Optical observations using an electronographic camera were made by Wlérick and Lelièvre (1975), who noted that there was actually a group of three objects within a few seconds of arc which all had a strong UV excess.   These lay within the complex radio source and appeared to be connected with it.   The brightest of the group of UV excess objects was quite faint, being around 20th magnitude, and at the time my attention was drawn to 3C 303, by P. Véron, it was not possible to observe with a conventional spectrograph because of sky background.   It was necessary to wait until the Wampler-Robinson image dissector scanner was available on the Lick 3-m telescope, so that sky subtraction could be performed.

Kronberg has made detailed radio observations at NRAO and Westbork and has mapped the structure of the radio source and has noted that the optical object, the N galaxy and the group of UV-excess objects lie, as described by the French workers earlier, within the radio source and seem to be associated with its strongest isophotes. One of the fainter objects lies right on the compact center of the western component of the radio source.   Thus, before optical observations were made, it appeared that all the optical objects were likely to be physically associated.   The brightest of the UV excess objects, according to Wlérick and Lelièvre, has B = 20.05, B - V = 0.08, U - B = -0.59.   After two seasons of observation, H. E. Smith and I have found that this object is indeed a QSO, with a redshift z = 1.57 from lines of Mg II, C III], and C IV, whereas the galaxy has a redshift less than 10% of this.   Accepting the evidence from the radio structure as indicating that the radio source is one single source with complex parts, then we have indeed what I would call the best case I know of a QSO physically associated, it appears, with a galaxy of much lower redshift.   Now there is of course always the possibility that the QSO is a radio-quiet object which happens by chance to fall within the radio contours and only appears to be connected with them. And this points up again the necessity to find out just what is the surface density of radio-quiet QSOs across the sky.

Nevertheless, the coincidence of the close grouping of 3 UV-excess objects within a few seconds of arc of each other, and one of these being a QSO, is very striking. All three lie less than 20" from the N system. It is, of course, very desirable that the other two UV-excess objects be observed spectroscopically. Unfortunately, they are 1 1/2 - 2 mag. fainter than the QSO. However, we shall attempt to observe the object that coincides with the compact radio core.

REFERENCES

Burbidge, E. M.: 1975, Lectures in Radio Galaxies (Urbino Summer School, ed. G. Setti, D. Reidel Publ. Co., Dordrecht-Holland).

Burbidge, E. M., Caldwell, R. D., Smith, H. E., Liebert, J., and Spinrad, H.: 1976, Astrophys. J. Letters 205, L117.

Lelièvre, G. and Wlérick, G.: 1975, Astron. and Astrophys. 42, 293.

Matthews, T. A., Morgan, W. W., and Schmidt, M.: 1964, Astrophys. J. 140, 35.

Morgan, W. W.: 1958, Pub. Astron. Soc. Pacific 70, 364.

Rieke, G. H., Grasdalen, G. L., Kinman, T. D., Hintzen, P., Wills, B. J., and Wills, D.: 1976, Nature 260, 754.

Sandage, A.: 1973, Astrophys. J. 180, 687.

Sandage, A. and Luyten, W. J.: 1967, Astrophys. J. 148, 767.

Smith, H. E., Spinrad, H., and Smith, E. O.: 1976, Pub. Astron. Soc. Pacific, in press.

Wyndham, J.: 1966, Astrophys. J. 144, 459.

DISCUSSION

Bahcall: Is there a concentration of the red QSOs towards the galactic plane?

E.M. Burbidge: There is no obvious concentration near the plane, and so I do not believe that we can account for these objects by saying that they are reddened by galactic absorption.

Walsh: Does the proportion of QSOs that are red increase near the Sky Survey print limit? Identifications down to about 19.5 magnitude based on accurate radio positions show very few such objects.

E.M. Burbidge: The objects that I have listed are fairly faint and it may be that there are more of them than has been suspected. Observations of neutral or red identifications for compact radio sources will, I hope, lead to a better idea of their frequency and apparent magnitude distribution.

Kronberg: I have mapped 3C 303 at 3.7 and 11.1 cm (resolution ~ 2.2 and ~ 7" respectively) with the NRAO interferometer, and more recently at $\lambda 6$ cm (~ 7" resolution) at Westerbork together with R.G. Strom. An accurate radio-optical comparison using two deep Palomar 48-inch plates (taken by S. van den Bergh) enabled us to show that the cores of both main radio components lie within ~ 1" of two objects which are 16" arc apart (Kronberg, Ap.J.(Lett) 203, L47, 1976). One of these is a 17th magnitude N-galaxy (Z = 0.14) and the other a 22nd magnitude member of a trio of faint objects. Lelièvre and Wlérick (Astron. and Astrophys. 42, 293, 1975) who had independently identified the field have found that all three faint objects (M_v = 21.70, 19197 and 21.75) have UV excess. The radio maps at 11 cm and 6 cm show further extended structure on both sides of the N-galaxy, and confirm the presence of faint radio structure between the two main radio components. A radio extension to the southwest of the western radio component includes the entire trio of UV excess objects. It is the brightest whose redshift (z = 1.57) has recently been determined.

The fact of the double radio-optical coincidence is curious. Our radio maps show with virtual certainty that 3C 303 is a single radio system, and not an accidental super-position of unrelated radio sources. The question of whether the two systems of different redshift are physically related or not relies on whether the 20^m quasar and its two fainter companions coincide only by chance with the western radio component of 3C 303. In any case the optical field is interesting, and the result underlines the need to do such radio-optical comparisons for more sources. We need more deep optical plates and spectra, now that we have detailed multi-frequency radio maps at optical resolution.

Webster: At first glance this looks to be a fairly standard type of radio source with 2 outer steep-spectrum lobes and one central flat spectrum component centred on the bright galaxy. Then this seems to me to be a good case for associating the whole radio source with the galaxy, but what is there to link the quasar with the preceding radio lobe?

Kronberg: The quasar with measured redshift lies approximately in the centre of the southwest extension of the main western radio component, although not on a compact (< 1") radio core. One of its fainter (also UV excess) companions *does* coincide within ~ 1" with the bright radio core of the tadpole-like western component.

P. Véron: Lelièvre and Wlérick (Astron. and Astrophys. <u>42</u>, 293, 1975) have measured the colors of the faint object which coincides with the western radio component of 3C 303, (a few seconds NE from the QSO whose redshift was measured by M. Burbidge). This very faint object (V = 21.7) has a strong UV excess with U-B = -0.90 ± 0.25, so it is a good quasar candidate. The positional agreement is also very close, since it lies within 1 arc second of the mean Cambridge radio position as determined by Branson *et al.* and by Pooley and Henbest.

Ryle: Concerning the discussion about 3C 303; I looked back at the map published by Pooley & Henbest in 1974. This was made with the 5 km telescope at λ = 6 cm with an angular resolution of 2" arc, and showed an unresolved E component < 1".2 and a W component 4".4 x 1".3, with a strong polarisation in the northerly extension. These early observations also suggested a faint and extensive envelope and showed no feature associated with the large redshift blue object.

In order to examine the low T_s features with better sensitivity we observed the source last night with the 5 km telescope at λ = 11 cm, where the sensitivity for a resolved source is some 10 times better, and the resolution is still 3".7 arc. The map shows a jet running from the compact E source towards the W component, and a faint diffuse extension to the E of the compact source, whose length is comparable with the separation of the main components. The distribution is reminiscent of those of 3C 66 and 3C 219, and suggests that in all these cases the compact source represents the nucleus of a double structure whose components have developed in different ways.

The large redshift blue object does not appear to be associated with any significant radio emission, although the northern blue object lies within the extended W component.

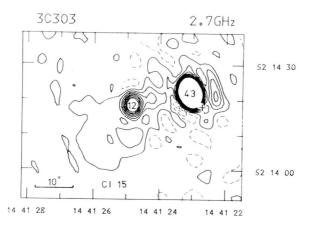

Map of 3C303 made with the 5-km telescope at 2.7 GHz. For the two intense components the higher level contours have been omitted and the peak level is indicated. The cross marks the position of the brightest of the three UV excess objects.

Miller: Is the blue object very near the centre of the western radio component extended, and is it possible to get a spectrogram of it?

E.M. Burbidge: It can be done; it's faint and it's going to take time, but it would be well worth the effort.

Rowan-Robinson: I would like to direct a question to those who identify radio sources with quasars: If there had been only the quasar for which the redshift has been measured and the right hand radio source, would it have been accepted as an identification?

E.M. Burbidge: We would have looked at it as a possible candidate.

Ryle: No, the positional disagreement is far too great and the optical object is not associated with any significant radio emission.

Swarup: For Ooty occultation sources we do not consider any object identified if it is not along the axis of the source or coincident with one component within the error bars.

Hazard: It is doubtful if the QSO near 3C303 would have been identified in the Molonglo Survey. The galaxy would have been considered the more probable identification although the QSO would have been noted and examined spectroscopically. It is my opinion that the claimed associations between QSO's and galaxies etc. is a reflection of a very high background QSO density per sq degree.

Arp: As someone who has actually observed and searched for ultra violet excess objects on photographic plates - I can testify that such a close configuration of strong ultra violet excess objects as the three near 3C303 is extraordinary.

DEEP IDENTIFICATIONS OF 3CR RADIO SOURCES

M.S. Longair

Gunn, Kibblewhite, Riley and myself are continuing deep optical surveys of the fields of unidentified and poorly identified 3CR radio sources. Our criterion is $m_{pg} > 19$. We have been fortunate to obtain observing time on the Palomar 5-m telescope and the Kitt Peak 4-m telescope.

A major procedural advance has been to use Kibblewhite's laser scanning automatic measuring machine to produce scans and contour maps of the optical fields. We have developed a three-stage reduction procedure by which we tie together the radio and optical frames of reference on the deep plates to 0.3 arc sec. A further advance is the possibility of obtaining magnitudes at very faint magnitudes by integrating over the digitised image of the optical counterpart. So far we have studied the first 4 fields of our sample and obtained 100% success in our identification programme. These sources are 3C133, 184, 190, 217. 3C190 is a relatively bright galaxy and was previously identified but the other identifications are believed to be new. In the case of 3C184 (see figure), the estimated magnitude of the galaxy is 23.2 ± 0.5. We have plates of a further 25 fields.

3C 184

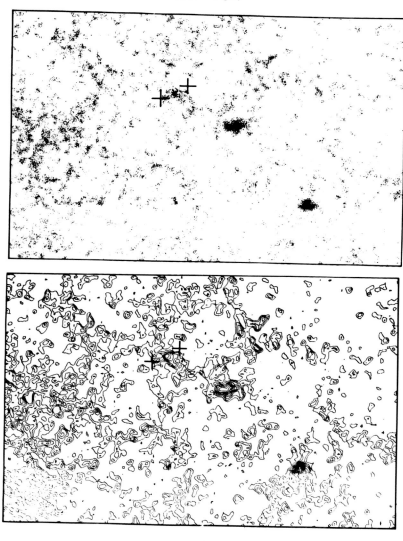

10″

Scans and contour map of the field around 3C 184. The identified galaxy lies between the two crosses.

RELATIONS BETWEEN RADIO AND OPTICAL PROPERTIES OF RADIO SOURCES –
RADIO ASTRONOMER'S POINT OF VIEW

J.M. Riley and C.J. Jenkins
Mullard Radio Astronomy Observatory, Cavendish Laboratory,
Cambridge, England.

One particular aspect of the relations between the radio and optical
properties of radio sources has been examined for a complete sample of
3CR sources, namely the relation between the radio structure of a source
and its optical identification. Possible differences between the radio
structures of quasars and radio galaxies are investigated, and the data
provide clues as to the optical nature of the unidentified sources.

1. THE DATA

A complete sample of 166 3CR sources has been observed in total
intensity with the Cambridge 5-km telescope at 5 GHz with a resolution
of 2" arc in RA and 2" $\cosec \delta$ in Dec.(Jenkins et al 1977, and references
therein). This sample was selected as follows : (a) the flux density
of each individual source at 178 MHz is greater than or equal to 10 Jy
according to the compilation given by Kellermann et al (1969), (b) $\delta \geqslant$
10° and (c) $|b| \geqslant 10^{\circ}$. Spinrad's most recent compilation of the identification status of each 3CR source was used. The sources in the 166
sample have been divided into 3 classes according to their identification (i) galaxies, including N galaxies, (ii) quasars and (iii) unidentified, including those for which the identification is uncertain.

2. THE RESULTS

2.1. LAS Distribution

The distributions of the largest angular size of the radio structures (LAS) for the 3 optical classes are shown in Fig.1. The distribution for the unidentified sources is similar to that for the galaxies
shifted to a slightly smaller angular size. The distribution for quasars differs from both in having more very compact sources (< 1" arc)
but even for the resolved sources the LAS for the quasars are smaller
on average than for the unidentified sources. These results indicate
the likelihood that the majority of the unidentified sources are asso-

D. L. Jauncey (ed.), Radio Astronomy and Cosmology, 237-243. All Rights Reserved.
Copyright © 1977 by the IAU.

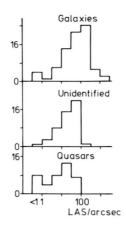

Fig.1.   The distributions of the LAS of the radio structures of sources
in the 166 sample.

ciated with galaxies and are slightly more distant on average than the
sources already identified with galaxies.

2.2.   Radio Structural Types

The sources in the sample have been divided into 4 classes according
to their radio structure;   compact components coincident with the optical
objects are not taken account of in this scheme except where there is
no other radio structure.   The classification is as follows : (i) compact, unresolved by the 5-km telescope e.g. 3C 147, (ii) double, with
components on either side of the associated optical object e.g. Cygnus
A, (iii) asymmetric double, with one compact component coincident with
the associated object and only one component displaced from it, e.g. 3C
273 and (iv) other, which includes any source which does not fall into
one of the other 3 classes e.g. 3C 83.1B, and any source for which there
is uncertainty as to which of the three classes it belongs.   The distribution of these radio structural types within the 3 optical classes

Table 1

	Compact	Double	Asymmetric Double	Others
All	4%	70%	1%	25%
Galaxies $\log(P_{178}) > 26$	7%	78%	0%	15%
Quasars	25%	50%	13%	12%
Unidentified	4%	45%	–	51%

is shown in Table 1 as the percentage of those sources with a given optical class which have a given structure. Any unidentified source with two components has been classified as <u>double</u> when both components are resolved, and as <u>other</u> when at least one component is unresolved as in this case the source could be a <u>double</u> or an <u>asymmetric double</u>; nearly all the unidentified sources classified as <u>other</u> are of the latter type. It is difficult to compare the distributions for radio galaxies and quasars meaningfully as the quasars are nearly all much more powerful than the radio galaxies. However comparison of the quasar distribution with that for radio galaxies with luminosities $P_{178} > 10^{26}$ W Hz^{-1} sr^{-1} indicates a very much higher proportion of <u>compact</u> and <u>asymmetric double</u> sources amongst the quasars.

2.3. Symmetries in Double Sources

For the sources classified as <u>double</u> it is possible to investigate the symmetry of the two components with regard to their flux densities and separation from the associated object. The distributions of the ratio of the total flux density of the brighter component to the total flux density of the fainter are shown in Fig.2(a) for all double sources

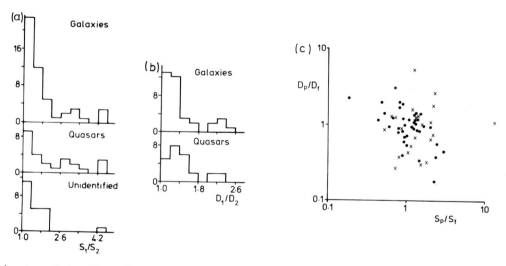

Fig.2. (a) The distributions of the ratio of the total flux density of the brighter component to that of the fainter component for all double sources in the sample. (b) The distributions of the ratio of the distance of the peak of the component further from the optical object to that of the component near to it for identified double sources in the sample. (c) The ratio of the distance of the peak of the preceding component from the optical object to that of the following one plotted against the ratio of the total flux density of the preceding component to that of the following one for double sources identified with galaxies (●) and quasars (×).

in the sample.   The galaxies and quasars both show tails towards large
flux density ratios, though the galaxies are more peaked towards 1;   the
distribution for the unidentified sources is similar to that for the galaxies.   The distributions of the ratio of the distance of the peak of the
component further from the optical object to that of the component nearer
to it are shown in Fig.2(b) for all identified sources.   Again the galaxies show a tendency to be more peaked towards 1.   A plot of the ratio
of the distance of the peak of the preceding component from the optical
object to the distance of the peak of the following one against the ratio
of the total flux density of the preceding component to that of the following one for double sources identified with galaxies and quasars is
shown in Fig.2(c).   It can be seen that there is a correlation for the
galaxies in the sense that the nearer component is brighter, whilst no
such effect exists for the quasars which also show a much broader spread.
It is difficult to say whether any of these effects reflect a genuine
difference between the structures of the double sources associated with
galaxies and quasars or whether it is just a question of the power of
the radio source, as the quasars are nearly all much more powerful than
the galaxies.

2.4.   Compact Central Radio Components

Compact central radio components (ccrc), taken as being unresolved
radio components associated with the optical objects in extended sources,
have been detected in many of the sources in the sample.   The distributions in total luminosity at 5 GHz for all identified sources and those
with ccrc are shown in Fig.3(a);   redshifts estimated from the apparent
magnitudes have been used where no measured redshifts exist.   It can
be seen from Fig.3(a) that the majority of the galaxies with $P_{5000} <$
10^{25} W Hz^{-1} sr^{-1} possess ccrc whilst there is an apparent lack of ccrc
in the galaxies of higher power.   This absence of ccrc in the higher
power sources is not a result of inadequate resolution as it was shown
in section 2.1 that the majority of the sources associated with galaxies
are well resolved with the 2" arc beam;   it therefore seems likely that
the sensitivity is insufficient to detect ccrc in the more distant powerful sources.   That this is indeed the case is indicated in Fig.3(b)
which is a plot of the ratio of the flux in the ccrc to the total flux
at 5 GHz against total power at 5 GHz for radio galaxies with total flux
densities at 5 GHz greater than 1 Jy;   as only the brightest sources have
been used any ccrc with more than 1 percent of the total flux could in
principal be detected and there is no observational selection against
populating any part of Fig.3(b).   From this diagram it can be seen that
the fraction of flux in the ccrc increases with decreasing power, in
agreement with the results of Perola & Fanti (this volume), and there
are no ccrc with luminosities at 5 GHz greater than 10^{24} W Hz^{-1} sr^{-1};
these results are suggestive of a narrow luminosity function for ccrc.
The luminosity distribution for the ccrc associated with galaxies in the
sample is shown in Fig.3(c) from which the sharp cut-off at 10^{24} W Hz^{-1}
sr^{-1} is obvious.   It is possible to construct the luminosity function
for central components in galaxies and it is found that this fits well
to an   extrapolation of the luminosity function for nearby bright

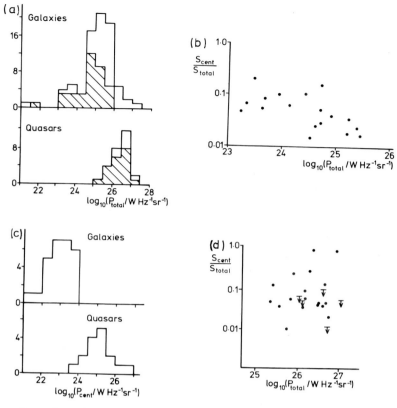

Fig.3.   (a)   The distributions in total luminosity at 5 GHz for all sources identified with galaxies and quasars;   the hatched boxes are sources with ccrc.   (b)   The ratio of the flux density of the ccrc to the total flux density at 5 GHz plotted against total power at 5 GHz for galaxies with total flux densities at 5 GHz greater than 1 Jy. (c)   The distributions in the luminosity of the ccrc at 5 GHz for galaxies and quasars.   (d)   As in (b) for all the quasars in the sample.

ellipticals.

       From Fig.3(a) it is apparent that nearly all the quasars in the sample have ccrc indicating a difference from the powerful radio galaxies. A plot of the ratio of the flux in the central component to the total flux at 5 GHz against total luminosity at 5 GHz for the quasars is shown in Fig.3(d) from which it is clear that there are no trends similar to those seen for the radio galaxies and there is a very broad spread in the luminosity of the ccrc, shown in the luminosity distribution in Fig.3(c).   The luminosities of the quasar ccrc are all much higher than those of the galaxy ccrc.   These differences reflect a genuine differ- ence between galaxies and quasars.

It is interesting that none of the unidentified sources in the sample are triples in the sense of being double sources with ccrc; this again is consistent with them being associated with more distant galaxies so that the sensitivity is not adequate to detect ccrc in them.

This work will be discussed in more detail by Jenkins (in preparation).

3. CONCLUSIONS

The main difference between galaxies and quasars from the point of view of their radio structures is the much greater luminosity of the ccrc of quasars; there are also more purely compact radio sources associated with quasars.   There may be a tendency for the double sources associated with quasars to be more asymmetric both as regards the fluxes of the components and their separation from the associated object.   The present observations of the unidentified sources are consistent with them being more distant galaxies as (i) they possess no ccrc, (ii) the distribution of LAS is similar to that for galaxies only somewhat smaller and (iii) the distribution of radio structural type is probably similar to that for the more powerful galaxies.

REFERENCES

Jenkins,C.J., Pooley,G.G. & Riley,J.M., 1977.   Mem.R.astr.Soc., in
      preparation.
Kellermann,K.I., Pauliny-Toth,I.I.K. & Williams,P.J.S., 1969.   Astrophys.
      J.,157,1.

DISCUSSION

Ekers: Would you clarify your statement about the consistency between your luminosity function for cores in radio galaxies with that for the cores of bright elliptical galaxies?   This is quite important since we need to know whether the central object is a good indicator for the correct identification or whether all E galaxies have them.   I think our data indicates that the probability of core emission does approach unity but at power levels 10 - 100 times lower than that for the 3C radio galaxies.

Riley: The luminosity function for the core in radio galaxies fits well to an extrapolation of the luminosity function for bright ellipticals of higher power.

Condon: If there is a correlation between the excess optical luminosity that makes a galaxy into a QSO, and the radio flux density in a compact core, then it might be possible to identify sources as follows: if $P_{5GHz} < 10^{24}$ W/HZ/Sr (in the core), the source is a galaxy, if $P_{5GHz} > 10^{24}$, it is a QSO. How well would this reproduce the observed identifications?

Riley: I think it would reproduce the identifications almost perfectly (except that you need the redshifts first!)

Schilizzi: The conclusion that most radio galaxies have compact radio components in their centers has been arrived at independently by VLBI measurements at 8.1 GHz between Owens Valley and Goldstone in California, (fringe spacing ~ 0".08). 39 galaxies associated with large (minutes of arc) radio sources were observed. For only a few of the sources was there any previous evidence of a central component. 17 showed fringes, 12 of the detections are amongst the 17 nearest sources (cosmologically) in the sample and are of fairly low luminosity. This strongly suggests that, with sufficient sensitivity, one could detect compact central radio components in most, if not all, radio galaxies.

LONG TERM OPTICAL VARIABILITY OF QUASARS

N. Sanitt

The magnitudes of 45 radio and 32 optically selected quasars from the samples of Schmidt, M., Astrophys. J. <u>151</u>, 393 (1968), Lynds, R. and Wills, D., Astrophys. J. <u>172</u>, 531 (1972), and Braccesi, A., Formiggini, L. and Gandolfi, E., Astron. and Astrophys. <u>5</u>, 264 (1970) are estimated by a method of measuring image diameters on the Palomar Sky Survey prints. These are combined with known photographic or photoelectric magnitudes measured between ten and fifteen years after the epoch of the Palomar Sky Survey, and the distribution of the differences in the two magnitudes should have a standard deviation of around 0.4 mag. The 45 radio quasars have a much larger standard deviation with 9 out of 45 (20%) lying outside two standard deviations from the mean. This result is consistent with known variability of these quasars. However, the optical sample shows no evidence for long term variability with all 32 quasars showing magnitude differences within two standard deviations. This implies a correlation between radio power and long term optical variability.

H.E. Smith: Do you have any color information on the variable radio QSO's?

Sanitt: The radio quasars are from the 3CR and 4C samples and exhibit an ultraviolet excess in common with the optically selected quasars, though not quite to the same degree. The more extreme ultraviolet excess of the optically selected quasars however, is due to strong emission lines and also results in a selection at certain preferred redshift values.

D. Wills: To what extent can your result be interpreted as a correlation between large redshift and lower incidence of variability, since the radio-quiet QSO's include many near z = 2?

Sanitt: The radio quiet quasars also include a number at low redshift around z = 0.4, and these also show less variability.

V

INTERPRETATION OF COSMOLOGICAL INFORMATION ON RADIO SOURCES

INTERPRETATION OF COSMOLOGICAL INFORMATION ON RADIO SOURCES

G. Burbidge
University of California, San Diego, and
Max-Planck-Institut für Astrophysik, München

The topic that I have to introduce today is concerned with the question as to whether or not we can obtain any cosmological information from radio astronomy. Alternatively, we may ask "Where does radio astronomy have an impact on cosmology?" There are several areas that must be discussed. They are:

1) The discovery and interpretation of the microwave background radiation.
2) The identification of powerful radio sources and the discovery that many of them have large redshifts. If we can prove that the large redshifts mean that the objects are at great distances, then we can use these radio sources as follows:

   (a) We can attempt to obtain a Hubble relation for the optical objects which are identified with radio galaxies;
   (b) We can look for a relation between the angular diameters of the radio sources and the redshifts of the optically identified objects and we can also look at relations between the angular diameter and the radio flux;
   (c) We can construct log N ‑ log S curves and we can carry out luminosity volume tests.

Let us introduce each of these investigations in turn.

First we briefly discuss the microwave background radiation. There is very little doubt in anyone's mind at this time but that this radiation did arise early in the history of the universe, and it is the one piece of cosmological evidence which shows unambiguously that the universe has evolved. The recent observations which show fairly clearly that the radiation is of blackbody form are extremely important in this connection. There is little controversy about this

D. L. Jauncey (ed.), Radio Astronomy and Cosmology, 247-257. All Rights Reserved.
Copyright © 1977 by the IAU.

situation, and thus this is all that I shall say about cosmological information which can be derived from the microwave background radiation.

I now turn to the many problems which are involved in category
2. Let us first consider the identification of radio sources and objects
with large redshifts. We know with some confidence that redshifts
are measures of distance for galaxies of stars. More precisely, if
z_c is the cosmological redshift, z_i and z_r the intrinsic redshift and
redshift due to random motion, and z_{obs} is the observed redshift,
then

$$(1 + z_{obs}) = (1 + z_c)(1 + z_i)(1 + z_r)$$

For normal galaxies we know from the form of the Hubble relation
that z_c is $\gg z_i$, and that z_c is $\gg z_r$. However, it has not been proved
that this is true for any other class of object.

What are the classes of objects which are identified with radio
sources? They fall into four categories, viz, elliptical galaxies, N
systems, QSOs, and BL Lac objects.

What has really been proved about the redshifts of these different
classes of objects? For the normal, genuine elliptical galaxies the
work that has been carried out over the past thirty or forty years by
Hubble, Humason, Sandage, and their colleagues leads us to suppose,
with a large degree of confidence, that these redshifts are measures
of distance, though I would add that this has never been proved.
However, the fact that we have a Hubble relation is strong evidence
in this direction.

When we come to the N systems, already there are problems.
For the majority of these, the morphological classification is strongly
correlated with the spectrum. Thus these very compact objects with
small diffuse halos around them nearly always show, when they are
identified with strong radio sources, strong, broad emission lines in
their spectra, together with a continuum which is not of stellar origin.
This continuum presumably arises both from hot gas and is in part
non-thermal in origin. An exception to this is the case of 3C 371
which certainly does show a strong stellar component. However, it
is the anomalous case among a very large number of N systems identified as powerful radio sources. Where stars can be found and the
redshift can be obtained from the stellar absorption lines and it is
found to be the same as that obtained from the emission lines, it is
reasonable to suppose that the redshift is of cosmological origin.
The difficulty (to some people) and the intriguing possibility (to others)

is that for the majority of these objects with large redshifts no stellar component has yet been found. This being the case, we can argue in one of two directions:

(a)   We can suppose that there is a stellar galaxy underlying the object that we can see, and that it has a redshift which is the same as the redshift obtained from the strong emission lines. We can then attempt to show that the energy distribution observed is consistent with the sum of the two or three components and attempt to obtain a Hubble relation. This is the method that was originally used by Sandage in 1971. Most investigators like this idea and use it. However, it is ambiguous and it can do no more than establish consistency with the cosmological hypothesis.

(b)   The alternative possibility is that such systems are not galaxies at all, but objects of quite a different type, with redshifts which are not of cosmological origin. Evidence in favor of this hypothesis is the possible periodicity in the redshifts of the N systems, and the fact that it is in some of these objects, e.g., 3C 120, where problems such as the apparent relativistic expansion speeds are encountered if the distances are obtained from the redshifts. Of course, this latter problem disappears if the objects are much closer than the distances obtained from their redshifts.

The third class of objects identified with powerful radio sources are the quasi-stellar objects. In Figs. 1, 2, and 3 you see plots made from a new catalogue of QSOs of the redshift-magnitude diagram, the apparent magnitude distribution, and the U-B against B-V plot. You all know that the nature and the distances of these objects have been under discussion for many years. All the arguments are well known. Almost every year a claim is made that proof of cosmological redshifts has been obtained, but in my view these arguments rarely, if ever, hold up. At the same time evidence suggesting that some redshifts are of non-cosmological origin also appears very frequently but is disregarded by most astronomers. There is evidence of some statistical weight that a few of the QSOs are associated with galaxies at the same redshift, but there is evidence of higher statistical weight that some are associated with galaxies with much smaller redshifts. There is a good correlation between the angular separations of pairs of QSOs and galaxies with the distances of the galaxies which Bolton has mentioned here. There is a new case of 3C 303, where in one of the radio lobes of the radio source which is identified with an N system with $z = 0.14$, there are three compact objects, one of which is a QSO with $z = 1.57$. There are close pairs of QSOs with different redshifts, etc. Most of this evidence is in favor of non-cosmological

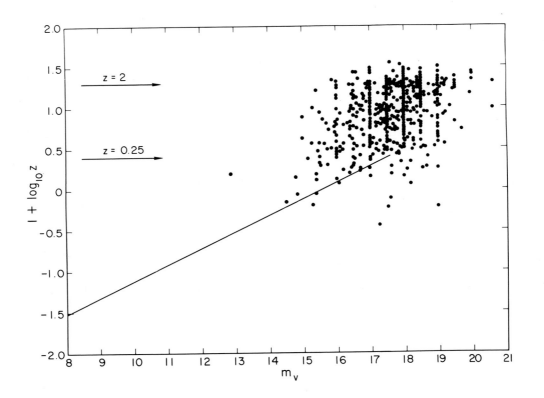

Fig. 1.   Redshift-apparent magnitude diagram for 570 QSOs taken
           from the catalogue of Burbidge, Crowne, and Smith (Ap. J.
           Supplement, February 1977).

redshifts, but it is either ignored, treated as accidental, or if people
begin to take it seriously they consider it "worrying."

Kristian attempted to argue that the fuzz around QSOs was consistent with the idea that they were embedded in galaxies at different
redshifts. However, attempts to investigate the fuzz directly have
shown so far that it is not due to galaxies in the cases of 3C 48 and
4C 37.42. It appears to me that the burden of proof that galaxies are
present still rests on those who would like to make that assumption.

Various continuity arguments have been put forward in favor of
the cosmological redshift hypothesis. For example, at this meeting
much has been made of the angular diameter - z relation, and it has
been tacitly assumed that z is a measure of distance. In general,
continuity arguments can be taken both ways.

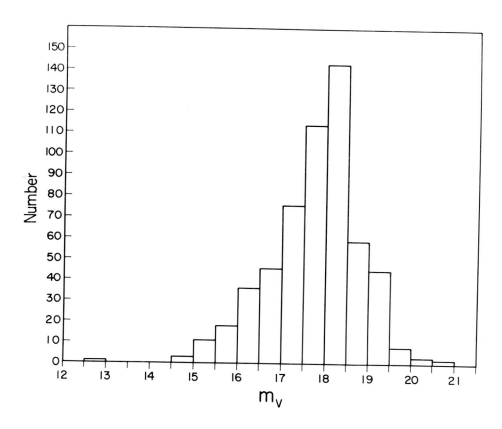

Fig. 2.   Distribution of apparent magnitudes of QSOs listed in the
catalogue of Burbidge et al. (1977).

Finally, let us turn to a brief discussion of the BL Lac objects.
Since very few emission redshifts have been discovered in these
objects, there has not been a lot to argue about. Some of them clearly
are in galaxies at modest cosmological redshifts. One, CL 4, is
thought to have its origin in our own Galaxy. One, BL Lac itself, has
been a subject of considerable debate between Lick and Palomar, and
the present situation suggests that it is certainly not a normal external
galaxy.

These then are the classes of objects with which we want to do
cosmology. Let us briefly discuss the different approaches to cosmology which are being attempted.

First we consider the bright elliptical galaxies. Here we know
that we can continue to use the Hubble relation and using the galaxies

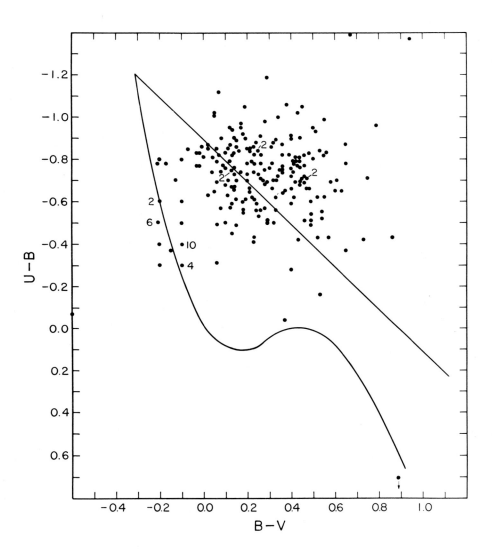

Fig. 3.   Two color diagram for all QSOs with measured colors (230)
           in the catalogue of Burbidge et al. (1977).

which have been identified as radio sources it is possible to push out
to quite large redshifts. The present situation will be discussed by
Dr. Smith later in this session. Considerable success is being
achieved and a number of redshifts greater than z = 0.5 have been
measured. The major problem will be in making the corrections for
evolution, etc.

We turn now to the redshift-apparent magnitude relation for QSOs. In Fig. 1 I have shown the redshift-apparent magnitude diagram for all of the nearly six hundred QSOs for which redshifts and apparent magnitudes are now available. Attempts have been made in the past to refine a diagram of this type in order to see if one could get a good Hubble relation for QSOs. The approach is that originally proposed by McCrea, viz., on the assumption that the redshifts are of cosmological origin, and assuming a value for the deceleration parameter q_0 one attempts to find the intrinsically brightest QSO in each redshift range, and determine the slope of the redshift-apparent magnitude relation. This was attempted in the past by Bahcall and Hills, by Burbidge and O'Dell, and by Petrosian. It will be discussed again in this session. In my view you cannot refine it enough to produce strong evidence in favor of cosmology, though you can interpret it in this way if you wish.

We turn now to the luminosity-volume test. As Schmidt originally, and Lynds and Wills after him have shown, if the redshifts of the QSOs are cosmological, then large scale evolution in the population of the 3C and 4C sources is unquestionably present. Later in this session Dr. Schmidt will update this work and discuss new samples.

Now let us consider the log N - log S studies. It is an article of faith in Cambridge that these demonstrate extensive evolution. However, if we restrict ourselves to the radio galaxies in the 3C revised catalogue, and also restrict ourselves to those with known redshifts, studies using the luminosity-volume test by Schmidt (published in the Astrophysical Journal in 1972) and more recent studies of the log N - log S relations by Narlikar and myself in 1975 (also in Ap. J.) show that there is no strong evidence, if any, for evolution. The steep slope for the log N - log S relation of -1.8 for radio galaxies in this catalogue must arise entirely from the unidentified sources. If the evolution takes place at redshifts of $z \approx 2$ or greater as is frequently claimed, it will be impossible to prove this directly, using ground-based telescopes because the galaxies are much too faint to be detected. Thus, the argument that we are seeing evolution cannot be directly proved, at least until we are able to detect objects with much larger redshifts directly using the Large Space Telescope (LST).

The assumption of evolution and the attempts to make models which have been described here and have been worked on particularly by the Cambridge radio astronomers all seem to me to be a type of parameter fitting which is probably premature. In my view we first have to prove that the objects we are looking at are really far away and only then can we argue that using the counts we can discuss the details of the evolution.

Earlier in this meeting when the counts at many frequencies were being described, I got the impression from several groups and particularly from Mills and his colleagues in Australia, that there are significant differences both in the slopes of the counts and in the numbers of sources measured at different frequencies in different parts of the sky. It was particularly striking to see the difference between the counts in the north and the counts in the south. To me the numbers in some cases appear to be highly significant. If there are real anisotropies, then at least one possibility is that many of the brighter sources are not at the great distances which had previously been assigned to them. Perhaps we should look for correlations between the distribution of radio sources in different parts of the sky and the distribution of comparatively bright galaxies in the local supercluster.

I conclude by outlining for you briefly the situation that might prevail if a significant population of the radio sources are not at great distances and are associated with QSOs which have been ejected from galaxies, as would be expected if the QSOs are comparatively local, i.e., they are at distances not greater than about 200 Mpc. We suppose that they are ejected from galaxies of various types, including spirals and the radio ellipticals. Under these conditions, what should we expect to see? The very close objects which have been ejected from comparatively nearby galaxies will be picked up as individual objects and will not be seen to be associated with their parent galaxies. This is because they will be far away from the objects from which they originated as far as the angular distances in the sky are concerned. As we go to radio sources which are further away, these will have been ejected from a population of galaxies which are also further away, and we shall see a significant number of them comparatively close to their parent galaxies. However, they will be far enough away from the parent galaxies so that they can be identified as individual objects. This would explain the correlation or the association between bright galaxies and some QSOs in the 3CR catalogue and the Parkes catalogue. As we go to even greater distances and beyond about 200 Mpc, the QSOs will be so faint that they will not appear on the sky survey plates and will no longer be detected as individual objects. Instead, their parent galaxies will be identified as the radio sources. This means that, although the identifications are incorrect, the distances for these sources will be correct. Thus, the radio luminosity function as derived from the galaxies will still be correct, and it may eventually be possible to establish that evolution is taking place through the population of parent galaxies.

In conclusion it appears to me that some types of cosmological investigations using radio sources have been premature. Much of the discussion still depends on the distances of the QSOs, and it is not proven that they are at great distances. There is no conclusive proof that the population of radio sources is changing with epoch, though it may turn out that this can be established if enough detailed work is done. But it is impossible to discuss this problem without first establishing the nature and the distances of the objects which make up the sources of radio emission. In a sense this was clear from the beginning, but so much of the discussion and the debate has been based on what to many of us is really very flimsy evidence.

Research in extragalactic astronomy at UCSD is supported in part by the National Science Foundation and in part by NASA under grant no. NGL 05-005-004.

DISCUSSION

van der Kruit: When you discussed the BL Lac objects, you did not mention the large amount of observations on AP Librae.   Does this object carry any weight in the arguments for or against the presence of a stellar component in BL Lac objects?

G.R. Burbidge: The observations reported in the literature by Disney and his colleagues are not entirely convincing.   Emission lines which were identified in the earlier work and were used in part to determine the redshift, are not present in the later observations.   Whether this is due to changes in the object (as claimed by the authors) or not, is not easy to determine.   In my view the absorption features identified in the second paper are not very convincing.

Miller: I observed the nuclear region of AP Librae with the Lick 3m image-tube scanner.   The spectra showed very clearly emission lines and the characteristic spectrum of stars in an E galaxy at a redshift near 0.048.   The visibility of the emission and galaxy features was nearly identical to that observed in 3C 371, another BL Lac object. The emission-line spectrum was very similar to that in gE galaxies with emission such as NGC 1052.   Since the emission is likely to be concentrated to the nucleus, which is a variable object, the visibility of the emission features will depend on the brightness of the non-stellar component, the size of the entrance aperture, and the dispersion of the spectrograph.   Nothing can be reliably concluded about variability of the emission lines until absolute spectrophotometry with essentially identical spectrograph set-ups is carried out over a period of time.

Ryle: I think it is important to remind optical astronomers that the general conclusions concerning the distances of radio sources and their consequent value in distinguishing between different cosmological models, do not depend on measurements of redshift, nor indeed on optical observations at all.   Twenty years ago it was shown that by relating the numbers of sources, their isotropy and the upper limit set to their contribution to the volume emissivity by measurements of the background radiation, not more than about 10% of the sources in a given flux density range could lie within the Galaxy.   Similar arguments applied to the extragalactic case showed a serious Olber's paradox unless the median value of P was at least 10^{25-26} watts $ster^{-1}Hz^{-1}$.   Independently of the identification or redshift questions, most radio sources therefore lie at cosmological distances, and local interpretations of the source counts are untenable.

G.R. Burbidge: I disagree.   As was shown several years ago by Rowan-Robinson and others, the limit set by this background does not rule out current local QSO models.   Further in the case I have just discussed, the brightest QSO's are from comparatively nearby galaxies.   Fainter ones are associated with galaxies at redshifts between about .003 and 0.02, and QSO's beyond this redshift are too faint to be identified as independent radio sources.   The sources are then identified with ten percent galaxies.   Thus the distance scale for radio sources with z (galaxies) > 0.02 is correct.   In other words, we are looking at the fine structure of the radio universe at small redshifts and identifying local QSO's, but the gross structure is at greater redshifts.

I also disagree with your view that optical astronomy is largely irrelevant in this and related problems.   Radio astronomers, except in special situations, cannot measure distances.   In this sense, optical astronomy is still all important.

Osterbrock: I have no "belief" about N galaxies, nor can I pretend to have observational data on all known N galaxies, but I should like to emphasise that in the slide I showed yesterday of the 4 broad-line radio galaxies, 3 of them are N galaxies and they all show stellar absorption features (Ca II, G, Mg I) at approximately the same z as the forbidden emission lines.   The equivalent widths of the absorption lines in all of these galaxies is only about 15 percent of the E.W. in typical elliptical galaxies, indicating strong dilution of the galaxy component by the non thermal component.

G.R. Burbidge: I certainly accept these observations; the existence of N-systems in which you can see stellar absorption features at the same redshift certainly weakens part of the argument I have made.

Wittels: Have you any comments on the lack of blue shifted objects if you try to explain QSO's, BL Lac type objects and possible N systems as ejecta from galaxies?

G.R. Burbidge: It has been known for many years and was described in our book published in 1967, that if these objects are ejected from galaxies beyond our own and if the shifts are local kinematic Doppler shifts, blue shifted objects should predominate.   The fact that they don't can be interpreted within the framework of the local hypothesis in one of several ways:

(1) it can be used as an argument against the local hypothesis,
(2) it means that the redshifts are not Doppler shifts,
(3) very contrived models can be considered in which it is
     argued that the emission takes place in a trail trailing
     behind the QSO. Then only those·objects moving away
     would be detected.

CROSS CORRELATION FUNCTIONS FOR QUASARS AND GALAXIES

M. Seldner

The cross-correlation function for galaxies and QSO's is defined
as the probability, as a function of angle, in excess of random for
finding a galaxy around a QSO. Thus, the mean projected density of
galaxies, $n_{gq}(\theta)$, around QSO's can be written

$$n_{gq}(\theta) = \bar{n}_g [1 + \omega(\theta)],$$

where $\bar{n}_g$ is the average sky density of galaxies and ω is the cross-
correlation function. The function $\omega(\theta)$ is determined for the Shane-
Wirtanen Catalogue of Galaxies and a sample of 484 published QSO's.
The result is a function similar in shape to the correlation functions
for other sets of objects such as Abell Clusters and 3CR radio galaxies,
i.e. $\omega(\theta) = A/\theta^\gamma$, $\gamma \sim 1$. The amplitude A is about 10 times larger
than would be expected if it is assumed that QSO's are at the distances
calculated from their redshifts and are correlated with galaxies in the
same manner that galaxies are correlated with each other. Division
into redshift bins shows that contributions to the positive signal come
from various redshifts and not just $z < 0.2$ as might be expected.

Webster: Can't you test the significance by looking at a number of
random positions and looking at the variance, the scatter, on the
resulting $\omega(\theta)$ curves?

Seldner: The correlation function for 500 random points with the Shane-
Wirtanen Catalogue yields a $\omega(\theta)$ which is zero at all angles, that is
it has no peak near $\theta=0$. The error bars on the QSO-galaxy graph show
the standard deviation of the mean for the first two angular bins, which
are statistically independent, the errors at larger angle should fall
off roughly as $\sqrt{\theta}$.

COSMOLOGICAL INTERPRETATION OF REDSHIFT DATA ON QUASARS THROUGH THE
V/V_{max} TEST

M. Schmidt
Hale Observatories
California Institute of Technology
Carnegie Institution of Washington

INTRODUCTION

The ultimate aim of statistical studies of redshift, magnitudes
and flux densities of quasars is to derive the general luminosity
function $\Phi(z, F_{opt}, F_{rad}, \alpha, \ldots)$ which describes the space density
as a function of redshift, intrinsic optical luminosity, intrinsic
radio luminosity, spectral index, etc. We assume throughout that the
emission-line redshifts z of quasars are cosmological. The function
Φ contains information that will be pertinent for any theory of the
formation and evolution of quasars.

We use samples of quasars with known redshift z that are complete
above given limits of magnitude m_{lim} and flux density S_{lim} over a
certain area of the sky. The derivation of the function Φ from the
observed distribution function n(z, m, S) for a complete sample is the
subject of the present paper.

USEFUL OBSERVATIONAL STATISTICS

Before proceeding, it is of interest to inquire whether there are
other distribution functions of the observables z, m and S for quasars
that carry quantitative information about the function Φ. We list in
Table 1 the six possible distribution functions of the observables.
Consider the distribution function n(z, S) which represents the
observed distribution of z and S regardless of magnitude m (i.e., summed
over all magnitudes m). At this stage it is important to remember that
quasar redshifts can only be detected at optical wavelengths, hence
require $m < m_{lim}$. Also confirmation of a suspected quasar candidate
is achieved at optical wavelengths only. As a consequence any practical
distribution function should include m if it contains z. Hence
n(z, S) is not a practical distribution function of the observables.

D. L. Jauncey (ed.), Radio Astronomy and Cosmology, 259-268. All Rights Reserved.

TABLE 1

Distribution Function	Comments
$n(z, m, S)$	quasi-stellar radio sources
$n(z, S)$	impractical (z implies m-selection)
$n(z, m)$	quasi-stellar objects
$n(S)$	impractical (no confirmation, no distance scale)
$n(m)$	impractical (no distance scale)
$n(z)$	impractical

The observed distributions $n(S)$ and $n(m)$, usually called counts, do not contain z and hence cannot yield a distance scale. In the case of a uniform space distribution in Euclidean space, the slope of the source counts $d \log n(S)/d \log S = -2.5$, independent of the luminosity function. However, quasars are observed at redshifts where deviations from Euclidean space are large, hence this special case is of limited interest. No quantitative statements concerning Φ are possible solely with counts $n(S)$ or $n(m)$.

Only the observational statistics of quasi-stellar radio sources $n(z, m, S)$, and those of optically selected quasars $n(z, m)$ appear to be practical and useful in deriving the function Φ. We shall discuss the application of the V/V_{max} method on the observed distribution $n(z, m, S)$ in the next section.

THE V/V_{MAX} METHOD

The V/V_{max} method (Schmidt 1968) tests whether the objects in the complete sample $n(z, m, S)$ can have been drawn from a population with a uniform distribution in space. We can derive for each object in the sample the maximum redshift z_{max} at which the object would just still belong to the sample (i.e., $m \leq m_{lim}$ and $S \geq S_{lim}$). It is worth noting that for a complete sample the relevant distribution function is $n(z, m, S, z_{max})$, containing four independent variables. If $V(z)$ is the volume in the Universe out to redshift z, and $V(z_{max}) = V_{max}$, then V/V_{max} is a measure of the position of the object within the volume V_{max} that is available to it within the sample limits. If the objects are drawn from a uniformly distributed population, then V/V_{max} should have a uniform distribution between 0 and 1 and

$\langle V/V_{max}\rangle$ should be 0.5, except for statistical errors.

The V/V_{max} test is the only one available that can be applied to objects limited in two or more observables (i.e., m, S, etc). It allows accurate allowance for the geometry of any assumed cosmological model of the Universe. If the density distribution is found to be non-uniform, a given density law $\rho(z)$ can be tested by checking whether V'/V'_{max} has a uniform distribution between 0 and 1, where $V' = \int \rho \, d \, V$. A method to determine $\rho(z)$ from the V/V_{max} values of the objects in the complete sample has been discussed by Lynden-Bell (1971). Once $\rho(z)$ has been satisfactorily estimated, or determined by Lynden-Bell's method, the multi-variate luminosity function is determined as $\Psi(F_{opt}, F_{rad}, \alpha, \ldots) = \Sigma(V'_{max})^{-1}$, where the summation is over all objects in the complete sample.

The reader is referred to a critique of the V/V_{max} method by Longair and Scheuer (1970) and a retort by Rees and Schmidt (1971) which highlights the main features of the method. Individual values of V/V_{max} are not particularly sensitive to the adopted shape of the optical energy distribution, contrary to a claim by Setti and Woltjer (1973).

RESULTS OF THE V/V_{MAX} METHOD

Table 2 lists $\langle V/V_{max}\rangle$ values derived for six samples of quasi-stellar radio sources. The second sample is that of Lynds and Wills (1972) based on a part of the 4C catalogue, while the third sample is mostly based on identifications by Olsen (1970) of 4C sources with redshifts (Schmidt 1975) in the declination range $20^{\circ} - 40^{\circ}$. In both cases I used the entire 3CR sample with appropriate weight, rather than the 3CR sources in the area sampled. The last sample of Table 2 is based on parts of the 5 GHz NRAO survey (Schmidt 1976).

TABLE 2

COMPLETE SAMPLES OF QUASARS IN RADIO SURVEYS

Sample	n	$\langle V/V_{max}\rangle$
3CR	44	0.64
4C(LW1)	24+3CR	0.66
4C($20^{\circ}-40^{\circ}$)	51+3CR	0.64
4C(LW2) (Wills 1974)	80	0.67
PKS($\pm 4^{\circ}$) (Wills 1974)	65	0.65
6cm(S1, S2, I)	51	0.61

The average V/V_{max} values are in the range $0.64 - 0.67$, except
for the low value of 0.61 for the 6-cm sample. Since this high-
frequency sample contains the largest fraction of sources with flat
radio spectra, it is of interest to see whether or not $<V/V_{max}>$ appears
to depend on the radio spectral index $\alpha = d \log S/d \log \nu$. On the
basis of available samples I find (see Table 3) that steep-spectrum
quasars have $<V/V_{max}> = 0.67 \pm 0.02$, and flat-spectrum quasars
$<V/V_{max}> = 0.52 \pm 0.05$ (Schmidt 1976). The transition from steep
spectrum to flat spectrum was taken to be at $\alpha = -0.2$.

TABLE 3

QUASARS WITH DIFFERENT RADIO SPECTRA

	Steep Spectrum	Flat Spectrum
α	≤ -0.20	> -0.20
n	158	28
$<V/V_{max}>$	0.67 ± 0.02	0.52 ± 0.05

The effect on V/V_{max} of errors and variability in magnitude and
flux density of quasars is of interest. A preliminary discussion
suggests that for 3CR and 4C quasars the corresponding error in
$<V/V_{max}>$ might be around $+0.02$. Quasars selected at 6 cm exhibit more
variability and the error in $<V/V_{max}>$ is estimated to be -0.01 or -0.02.
It seems most unlikely, then, that the difference in V/V_{max} of steep-
spectrum and flat-spectrum quasars is caused by these errors.

The first evidence for a low V/V_{max} of flat-spectrum quasars was
noticed in the 3CR sample by Schmidt (1968, page 406) and discussed by
Kinman (private communication), Rowan-Robinson (1973), Van der Kruit
(1973) and Setti and Woltjer (1973). Quasars from the 4C sample of
Lynds and Wills (1972) did not show the effect (Van der Kruit 1973)
but those from the Olsen $20^{\circ} - 40^{\circ}$ declination zone did (Schmidt 1974a).
The addition of substantial numbers of flat-spectrum sources from the
high-frequency NRAO survey has now resulted in a V/V_{max} of sufficient
precision to establish the effect.

Qualitative confirmation of the low V/V_{max} for flat-spectrum
quasars can be obtained from high-frequency source counts, since a
very large fraction of the flat-spectrum objects among these are
identified with quasars according to Brandie and Bridle (1974) and
Peterson and Bolton (Wall 1975). Condon and Jauncey (1974) derived a
slope of -0.85 for the integral counts of flat-spectrum sources in the
Parkes 2700 MHz survey. I find from a list of 8 GHz sources by Brandie
and Bridle (1974) a slope of -1.1 for the flat-spectrum sources. Such

a low slope is consistent with a uniform space distribution ($V/V_{max} = 0.5$) if the redshift distribution of these quasars is similar to that of the NRAO 6-cm quasars.

Finally, we consider in Table 4 the optically selected quasars, regardless of their radio flux density. These are usually selected on the basis of their U-B color. The completeness of the Braccesi sample has been discussed by Schmidt (1974b). The average V/V_{max} of 0.70 ± 0.065 deviates significantly from that expected for a uniform distribution. The Sandage-Luyten blue objects for which I have determined redshifts (Schmidt 1974c) are probably less suitable since they were mostly selected according to U-V (Setti and Woltjer 1973). Nonetheless, their $\langle V/V_{max} \rangle$ is close to that for the Braccesi sample. Our tentative conclusion is that optically selected quasars have a $\langle V/V_{max} \rangle$ close to or identical to that of radio selected steep-spectrum quasars.

TABLE 4

COMPLETE SAMPLES OF OPTICALLY SELECTED QUASARS

Sample	n	$\langle V/V_m \rangle$
Braccesi (13^h)	20	0.70
Sandage-Luyten ($0^h, 1^h, 8^h$)	34	0.66
Low-luminosity quasars	6	0.40

Samples of optically selected quasars contain a small number of quasars with optical absolute luminosities lower than those of any quasi-stellar radio sources (Schmidt 1972). Their average V/V_{max} of 0.40 ± 0.12 is lower than those for other quasars, but the difference is hardly significant.

In conclusion the V/V_{max} method has yielded the following results. First, both optically selected quasars and steep-spectrum radio quasars exhibit a $\langle V/V_{max} \rangle$ of around 0.67. Trial density laws in a $q_o = 0$ cosmology that correspond to such a $\langle V/V_{max} \rangle$ value are $\rho = (1+z)^5$ and $\rho = \exp(-10(t-t_o)/t_o)$ where t is cosmic epoch and t_o the present epoch. Flat spectrum radio quasars may have a uniform distribution in space. However, it is not clear at all why the number of presumably short-lived phenomena such as quasars should be the same at all times. It is quite possible that the evolution of flat-spectrum quasars is much more complex and that their $\langle V/V_{max} \rangle$ is accidentally close to 0.5. Perhaps the main significance of the V/V_{max} value of flat-spectrum quasars is, then, that it is significantly different from that of other quasars.

FURTHER EVIDENCE ON THE GENERAL LUMINOSITY FUNCTION

The V/V_{max} test extracts information about the distribution of sources in depth, i.e., about the z-dependent part $\rho(z)$ of the general luminosity function $\Phi(z, F_{opt}, F_{rad}, \alpha, \ldots)$. We have already seen that ρ depends on the spectral index α, hence

$$\Phi(z, F_{opt}, F_{rad}, \alpha, \ldots) = \rho(z, \alpha)\, \Psi(F_{opt}, F_{rad}, \alpha, \ldots)$$

The density law ρ may, in fact, also depend on the optical absolute luminosity F_{opt}. The observed $n(z, m)$ distribution of optically selected quasars should, in principle, allow a derivation of the dependence of ρ on F_{opt} (Schmidt 1972). However, the available samples contain mostly objects near $\simeq 18$. A survey with the Palomar 18-inch Schmidt telescope undertaken by R. Green over an area of 10,000 square degrees should yield a sufficient number of brighter quasars, near $m \simeq 15$. This large project will take several years for completion.

Quasars with redshifts larger than 2.5 are difficult to find optically, because Lyman - α emission enters the B-band, so the ultra-violet excess disappears. Osmer and Smith have succeeded in detecting quasars with $z = 2.5 - 3.1$ by means of an objective-prism Schmidt survey, a technique that requires strong Lyman - α emission but no ultraviolet excess. It appears that the numbers found (Osmer 1976, Smith 1976) may well agree with the prediction based on the density law $\rho \sim \exp(-10\, t/t_o)$, but further systematic work is needed.

An attempt to factorise the multi-variate luminosity function $\Psi(F_{opt}, F_{rad}, \alpha)$ has been made by Schmidt (1970). The fact that the observed redshift distribution of (steep-spectrum) radio quasars and optically selected quasars at optical magnitude 18 appears identical, led to the hypothesis that F_{rad} appears in the luminosity function only in its ratio to F_{opt}, i.e.,

$$\Psi(F_{opt}, F_{rad}, \alpha) = \phi(F_{opt})\, \psi(R)\, f(\alpha)$$

where $R = F_{rad}/F_{opt}$. It is easy to show that, in this case,

$$n(>f_{rad}^{lim}, f_{opt}) = G(>R^{lim})\, n(f_{opt}),$$

where the left hand side represents the optical magnitude distribution of radio quasars stronger than f_{rad}^{lim}, while $n(f_{opt})$ is the optical magnitude distribution of optically selected quasars. For derivations and details about the meaning of f_{rad} and f_{opt}, see Schmidt (1970). The function $G(>R)$ has been determined from this relation. As an example, optical counts of optically selected quasars near 17-th magnitude increase by a factor of 6 per magnitude, while those of 3CR quasars increase by a factor of only 2 per magnitude. Hence $G(>R)$ decreases by a factor of 3 per optical magnitude at the corresponding value of R.

Another advantage of this introduction of $\psi(R)$ and $G(>R)$ seemed to be that the peak in the optical magnitude distribution of 3CR quasars at 18.5 magnitude was easily represented by the proper choice of $G(>R)$. Since it decreases with optical magnitude while $n(f_{opt})$ increases, a maximum for the product is reasonable. Since $R^{lim} = f_{rad}^{lim}/f_{opt}$, we would then expect the peak in the magnitude distribution to be at an optical flux level proportional to that of the limiting radio flux density of the radio catalog. However, most reports at this Symposium on the magnitude distribution of identified radio quasars mention a peak around 19-th or 20-th magnitude. Since the limiting flux density for, say, the Westerbork identifications reported here by Dr. J. Katgert-Merkelijn is more than 100 times lower than the 3CR limit, we would have expected a peak beyond magnitude 23. The reason for this discrepancy is unclear at present.

As mentioned earlier, the flat-spectrum radio quasars appear to have an approximately uniform space distribution, in contrast to both steep-spectrum radio quasars and optically selected quasars. The relationship of their luminosity function to that of the other quasars has not been investigated yet.

REFERENCES

Brandie, G. W. and Bridle, A. H.: 1974, Astronom. J. 79, 903.
Condon, J. J. and Jauncey, D. L.: 1974, Astronom. J. 79, 437.
Longair, M. S. and Scheuer, P. A. G.: 1970, Monthly Notices Roy. Astron. Soc. 151, 45.
Lynden-Bell, D.: 1971, Monthly Notices Roy. Astron. Soc. 155, 95.
Lynds, R. and Wills, D.: 1972, Astrophys. J. 172, 531.
Olsen, E. T.: 1970, Astronom. J. 75, 764.
Osmer, P. S.: 1976, private communication.
Rees, M. J. and Schmidt, M.: 1971, Monthly Notices Roy Astron. Soc. 154, 1.
Rowan-Robinson, M.: 1973, Astronom. and Astrophys. 23, 331.
Schmidt, M.: 1968, Astrophys. J. 151, 393.
Schmidt, M.: 1970, Astrophys. J. 162, 371.
Schmidt, M.: 1972, Astrophys. J. 176, 273.
Schmidt, M.: 1974a, ESO/SRC/CERN Conference on Research Programmes for the New Large Telescopes, p. 253.
Schmidt, M.: 1974b, Proceedings of the Sixteenth Solvay Conference, p. 463.
Schmidt, M.: 1974c, Astrophys. J. 193, 509.
Schmidt, M.: 1975, Astrophys. J. 195, 253.
Schmidt, M.: 1976, Astrophys. J. (Letters) 209 (in press)
Setti, G. and Woltjer, L.: 1973, Annals New York Ac. Sc. 224, 8.
Smith, M. G.: 1976, Astrophys. J. (Letters) 206, L125.
Van der Kruit, P. C.: 1973, Astrophys. Letters 15, 27.
Wall, J. V.: 1975, The Observatory 95, 196.
Wills, D.: 1974, ESO/SRC/CERN Conference on Research Programmes for the New Large Telescopes, p. 275.

DISCUSSION

Jauncey: The n(S) and uniform space distribution of the flat, short-lived QSO's looks just like the "birth" of matter in a steady-state universe!

Longair: Is the redshift distribution for steep-spectrum radio quasars the same as that for flat-spectrum quasars?

Schmidt: At 18^m both seem to have a similar redshift distribution. A larger sample of flat-spectrum quasars is required to check this.

Rowan-Robinson: It is a prediction of the new scheme I outlined in an earlier session that there should only be weak evolution for compact, flat-spectrum sources, reflecting the normal evolution of gas and stars in a galaxy. The steep-spectrum, extended sources can arise from the interaction of a beam with an intergalactic medium whose density changes with time. I feel that we should get away from the use of arbitrary mathematical forms for evolution functions, and think about how a galactic nucleus knows what the epoch is.

R. Fanti: Could you comment on the fact that the average V/V_m for optically selected quasars is the same as for radio quasars with steep spectra?

Schmidt: I can only say that it apparently indicates that the density distribution of the two types is the same. It is too early to claim any understanding of why this is the case.

Tinsley: I would like to know what cosmological model is used in your exponential density law, since the relation between time and redshift depends strongly on q_o at large redshifts.

Schmidt: I used $q_o = 0$, so $t/t_o = (1+z)^{-1}$.

Menon: I think that the sharp distinction between the flat-spectrum quasars and the steep-spectrum quasars is due to selection effects. I am sure that if we observe most low-frequency sources and study their high-frequency spectra, most of them turn out to have flat high-frequency indices. There is already some evidence to that effect from my observations of a sample of weak Ooty occultation sources with the NRAO interferometer at 8 GHz. A high percentage of them which have normal spectra until about 2 GHz turn out to have strong central components at 8 GHz and would appear as flat spectra if observed only at high frequencies. Since the flat-spectrum sources have presumably shorter lifetimes, in considering their distribution we probably should also introduce a birth function.

Schmidt: The spectral index I used applies to frequencies below 5 GHz, and should not be dominated by the central component. Hence I do not agree at all that the effects found are due to selection effects.

Sanitt: Since there are redshift-dependent selection effects which affect the optically selected quasars, how can one be sure that the redshift distributions of optically and radio selected quasars are identical?

Schmidt: I believe that the selection for ultraviolet excess has a negligible effect on the redshift distribution for $z < 2.5$. It is the approximate identity of the redshift distribution below $z=2.5$ for optically selected and radio selected quasars that leads to the introduction of the radio luminosity function in the form $\psi(R)$.

Rees: You said that V/V_m was anomalously low for "low luminosity" quasars. Surely you would expect to obtain ~ 0.5 for nearby objects unless the evolution is extraordinarily steep at epochs close to the present?

Schmidt: It does depend on the density law, indeed. For the exponential density law $\exp(-10\ t/t_o)$ the density decreases rapidly, and even at relatively small redshifts one might still expect an average V/V_m well above 0.5.

Peterson: The mean V/V_m of flat-spectrum Parkes QSO's observed by Peterson, Jauncey, Wright and Condon (Astrophys. J. (Letters), 207, L5 (1976)) gives $\langle V/V_m \rangle = 0.52$ for ten QSO's brighter than 18^m with $S_{11\ cm} > 0.3$ Jy. This value of $\langle V/V_m \rangle$ for flat spectrum QSO's in the Parkes 11cm Survey is expected because Wall and Condon and Jauncey have shown that the slope of log N - log S for all the flat-spectrum Parkes sources is less than 1.5 and that nearly 80% of all these sources have QSO identifications. Thus both $\langle V/V_m \rangle$ and log N - log S indicate a uniform distribution in flat space-time for the flat-spectrum QSO's.

Because the colour balance of U,B plates taken for QSO identifications is magnitude-dependent, in the sense that faint QSO's appear too red and may therefore not be identified, it is important to have photometric calibrations in your survey for optical QSO's.

Schmidt: The Bright Quasar Survey films are scanned by a PDS microphotometer and photometrically reduced on the basis of photometric standard stars observed in each field. We should be able to handle any colour imbalance in the U,B exposures without difficulty.

Longair: Exponential models of the cosmological evolution of radio sources provide a good fit to the data from the V/V_m test for quasars because of the observed fact that in the 3CR sample $\langle V/V_m \rangle$ is about 0.67 independent of redshift; i.e. you get just as much evolution from $z = 0$ to 0.5 as you do from $z = 1$ to 2. Exponential evolution functions of cosmic time mimic this behaviour rather closely and are much more satisfactory than power-law models, say $(1+z)^\beta$, where all the strong evolution is pushed to large redshifts.

When you look at a much larger sample of quasars, is it still true that V/V_m is more-or-less independent of redshift?

Schmidt: I have not looked into that, but I might generalize your
remark by saying that there are methods (such as Lynden-Bell's C^- method)
that attempt to derive the density law rather than just assuming it and
seeing whether it fits.   In the past I have naively split the whole
sample into the nearer half and the farther half, to see whether one
law fits better than the other;   that has always remained rather
inconclusive and I cannot answer your question.

Bolton: I think the interesting thing to look at next is the log n –
log s or V/V_{max} for the optically selected quasars as a function of
their continuum spectral index.   If the result is the same for the
radio quasars, its interpretation is going to present some difficulties.

INTERPRETATION OF SOURCE COUNTS AND REDSHIFT DATA IN EVOLUTIONARY UNIVERSES

J. V. Wall, T. J. Pearson and M. S. Longair
Mullard Radio Astronomy Observatory,
Cavendish Laboratory, Cambridge, England

Conventional interpretation of the N(S) relation requires cosmic evolution of the radio source population. Investigators agree on the general features of this evolution: it must be confined to the most luminous sources, and must be strong, the numbers of such sources at redshifts of 1 to 4 exceeding the present numbers by a factor $\gtrsim 10^3$. There is no consensus as to whether density or luminosity evolution prevails (or both), whether a cutoff in redshift is necessary, or whether the source populations found in high-frequency surveys follow even the general evolutionary picture deduced for the low-frequency survey population. It is therefore hardly surprising that the physical basis of the evolution, the ultimate goal of N(S) interpretation, remains largely "in the realm of imaginative speculation" (P. A. G. Scheuer).

Recently the observational basis for source count interpretation has been greatly strengthened. There now exist N(S) relations at frequencies of 408, 1410, 2700 and 5000 MHz which all reach to $\sim 10^5$ sources sr^{-1}, and for which the statistical definition at S < 1 Jy is greatly improved over that available for previous investigations. Furthermore, at high flux densities at all frequencies, identifications and redshifts approach completeness and therefore luminosity distributions may be defined with greater certainty. We have undertaken a numerical investigation of the implications of this much-improved data base for cosmic histories of the different radio source populations. The early results are reported here.

1. A SCHEME FOR SOURCE COUNT ANALYSIS

We wish to determine $\rho(P,z,type)$, the generalized luminosity function, which describes the spatial density of radio sources of luminosity P and given type at all redshifts z. The function may be factored as $\rho = F(P,z,type) \cdot \rho_0(P,type)$; F is the evolution function and ρ_0 is the local luminosity function.

D. L. Jauncey (ed.), Radio Astronomy and Cosmology, 269-277. All Rights Reserved.
Copyright © 1977 by the IAU.

We have developed a new numerical scheme which guarantees that
these models will at least be consistent with luminosity distributions
defined at high flux densities. Consider a single type of source, and
suppose that for these sources a complete luminosity distribution n(P)
for S > S_0 is available. For a given world model and evolution function
F(P,z), the local luminosity function may be obtained directly:

$$\rho_0(P)dP \quad = \quad n(P)dP \; / \; \int_0^{z(S_0)} F(P,z) \; dV(z)$$

where $z(S_0)$ is the redshift at which a source of luminosity P has an
observed flux density S_0. The count for this type of source is then
computed in the usual manner:

$$N(>S) \quad = \quad \int_0^\infty dP \; \int_0^{z(S)} F.\rho_0.dV$$

A complete luminosity distribution, statistically adequate in size,
therefore makes it possible to proceed directly to a predicted count
for each postulated evolution function F. The requisite input <u>guesses</u>
for our approach are thus (i) a world model and (ii) an evolution
function, while input <u>data</u> supplied are (iii) a luminosity distribution
$n(P,S_0)$ and (iv) an observed source count N(S); the procedure is to
use (i), (ii) and (iii) to calculate the model count directly as above,
and to compare it statistically via a χ^2-test with (iv), the observed
N(S) for this type of object. At the outset we have used simple
analytic forms of evolution functions, whose free parameters are
allowed to vary widely. The evolution models which survive the χ^2-test
may be examined in more detail by comparison with observational data at
lower flux densities.

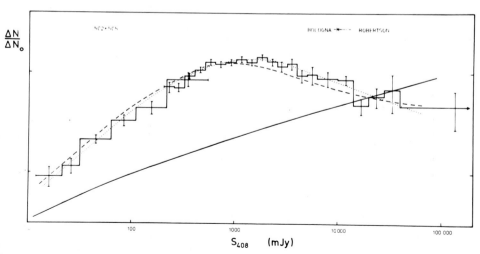

Figure 1.   The 408 MHz source count.   Solid curve: model 1;
dashed: model 2a (optimized);   dotted: model 5 (optimized).

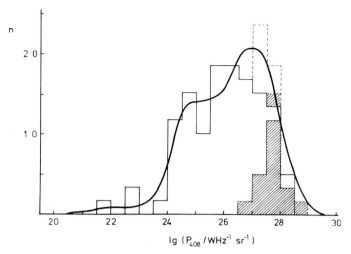

Figure 2.   The 408 MHz luminosity distribution, $S_{408} > 10$ Jy, 5.86 sr.
Hatched area: QSOs; dotted area: no redshift estimates available; solid
curve derived by smoothing the unbinned data.

2. THE 408 MHz SOURCE COUNT

The evolutionary requirements of the 408 MHz count (Fig. 1) are
particularly amenable to exploration by the present method.   Some 90 %
of all sources found at this frequency are of the extended, steep-
spectrum type, so that a valid first approximation is to assume that
the count consists of only these sources, with a spectral index $\alpha = 0.75$
$(S \propto \nu^{-\alpha})$.   Moreover, recent identifications and redshift determinations
yield the well-defined luminosity distribution of Fig. 2, the details
of which will be described elsewhere.

    The guesses of our initial investigation are (i) world models of
Friedmann geometries with $\Omega = 0$ and 1, $H_0 = 50$ km s^{-1} Mpc^{-1}, and
(ii) the following forms for the evolution function F.

1. $F(P,z) = 1$ (non-evolutionary, source-conserving)

2. $F(P,z) = \chi_1(P) + \phi\chi_2(P)$, with $\phi = \exp\{M(1-t/t_0)\}$, and
        $\chi_1 = (P_c/P)^m/\{1+(P_c/P)^m\}$, $\chi_2 = 1/\{1+(P_c/P)^m\}$;
      2a: no redshift cutoff;   2b: $F = 0$ if $z > z_c$.

3.   As for 2b, but with $\phi(P,z) = (1+z)^\beta$ for $z < z_c$.

4. $F(P,z) = \exp\{M(P)(1-t/t_0)\}$, with
        $M(P) = 0$ for $P < P_1$, $M(P) = M_{max}$ for $P > P_2$, and
        $M(P) = M_{max}(\lg P - \lg P_1)/(\lg P_2 - \lg P_1)$ for $P_1 < P < P_2$;

4a: no redshift cutoff;   4b: $F = 0$ if $z > z_c$.

5.   As for 2a, but with $\lg P_c = a \lg z + b$.

The first three are conventional models from the literature. All fail to produce satisfactory models for the 408 MHz source count for all possible variations of the parameters M, m, z_c, β, and P_c. Model 1 (no evolution) produces a count which differs from that observed to an extreme degree of statistical significance, as is evident in Fig. 1. Model 2a has attractively few parameters (M, P_c, m), but after optim-izing these, is rejected at the 99.9% level of significance. The difficulty (Fig. 1) is in obtaining the rapid turnover towards the faint flux densities when enough evolution is supplied to fit the steeply-rising count of bright flux densities; and when this turnover is achieved, falloff is too rapid. Imposing a redshift cutoff (2b) aggravates this behaviour, and model 3 is even worse in this respect because the density enhancement increases rapidly up to the redshift cutoff. The basic problem is that these models produce too many sources at the low-luminosity end of the evolving component. Models 4 and 5 were invented to avoid the difficulty. Model 4 was derived by examining how the evolution function depends on radio luminosity; the value of $<V/V_m>$ as a function of luminosity for sources with $S_{408} > 5$ Jy was used to define M(P), and in this sense there are no free parameters in 4a, and only one (z_c) in 4b. However, we succeeded in finding acceptable models of type 4 only in the $\Omega = 1$ geometry. Model 5 has the transition luminosity between evolving and non-evolving components as a function of redshift, thereby reducing the contribution to the count of weak, high-redshift sources. It produces the best fit to the source count, and bears considerable resemblance to luminosity evolution. The dependence of luminosity function on epoch given by models 4 and 5 is shown in Fig. 3, and the optimum parameters are as follows:

Model 4a: $\lg P_1$ (W Hz^{-1} sr^{-1}) = 26.0, $\lg P_2$ = 27.1, M_{max} = 11.5
      4b: $\lg P_1$ = 25.0, $\lg P_2$ = 27.3, M_{max} = 11.0, z_c = 3.5
      5:    M = 9.8, a = 3.14, b = 26.8, m = 1.06.

These models are very different in character, as exemplified by the redshift distribution which each predicts at low flux densities (Fig. 4); the identification content of such a sample clearly provides a powerful discriminant. For example, the identifications of Richter (1975) for the 5C3 survey suggest to us that $\geq 35\%$ of all sources with $S_{408} > 10$ mJy have redshifts less than 0.6. The proportions of sources with $z < 0.6$ are 26% and 42% for models 4a and 4b, and 28% for model 5, and it is therefore not clear that models 4a and 5 remain tenable. It is clear that identifications and redshifts for even a small sample of sources at this level would provide powerful constraints to the permitted forms of evolution.

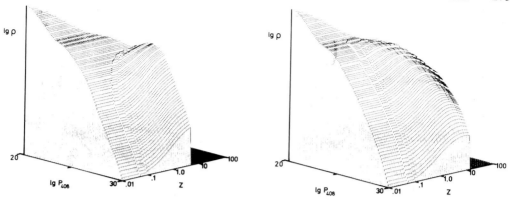

Figure 3.   Dependence of luminosity function on redshift in model 4b (left) and model 5 (right).

3. THE 2700 MHz SOURCE COUNT

In surveys at frequencies above 408 MHz, the proportion of compact objects with 'flat' radio spectra becomes significant and increases as survey frequency increases. Consider for simplicity that <u>two</u> classes of source are represented in the 2700 MHz count: 'flat-spectrum' and 'steep-spectrum' sources. A first attempt to derive evolution models for these two populations might proceed as follows. (1) Assume that the steep-spectrum objects are the same as those in the 408 MHz count. (2) Translate a successful model for the 408 MHz count to 2700 MHz, obtaining the appropriate normalization either from the spectral index distribution at some S_{2700} or from the 'steep-spectrum' component of the 2700 MHz count (Fig. 5 of Wall, this volume). (3) Subtract this from the total 2700 MHz count to obtain a complete count for the flat-spectrum sources. (4) Analyse this with the procedure of section 1 to set limits on the evolution required for these sources.

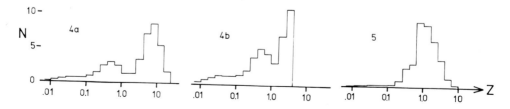

Figure 4.   Predicted distributions of redshifts for sources with $10 < S_{408} < 50$ mJy; models 4a, 4b and 5.

Fig. 5 demonstrates why our first attempt to follow such a procedure
was unsuccessful. The best high-frequency luminosity distribution
available is shown here, divided into flat- and steep-spectrum components. The steep-spectrum component peaks at a lower luminosity than
that predicted by successful evolution models from the 408 MHz count
analysis. This is because the known correlation between radio luminosity
and spectral index for radio galaxies has been neglected (Condon and
Jauncey 1974). The implication is that assumption (1) is wrong – that
the 408 MHz count cannot be simply translated to 2700 MHz to remove the
steep-spectrum component of the count. However, successful evolution
models from the 408 MHz count analysis may be used to calculate a count
for the 2700 MHz steep-spectrum sources, provided that the steep-
spectrum component of the 2700 MHz luminosity distribution is well
defined. The 28 objects of Fig. 5, 12 of which do not have measured
redshifts, are clearly inadequate in this regard. Moreover, the flat-
spectrum component of the luminosity function contains only 20 objects,
and exploration of the evolution function for flat-spectrum sources on
the basis of such a sample is somewhat optimistic.

There is nevertheless some indication that the cosmic history of
the flat-spectrum population differs considerably from that of the
steep-spectrum population. There are differences amongst the source
counts themselves (Figures 4 and 5 of Wall, this volume). These may
be due to differences in luminosity functions rather than evolution
functions, but there is further indication that this is not so.
C. J. Masson has measured magnitudes to $\pm 0^m.3$ for a complete sample of
55 QSOs in unobscured regions of the $\pm 4^\circ$ declination zone of the Parkes
2700 MHz survey. Redshifts for many of these were kindly communicated
to us by R. Lynds and D. Wills, and the $\langle V/V_m \rangle$ for the sample (Masson
and Wall, in preparation) is 0.58 ± 0.03, significantly less than the
values of 0.65 to 0.7 obtained for objects from low-frequency surveys
(Schmidt, this volume). The result suggests that flat-spectrum

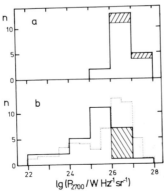

Figure 5. The 2700 MHz luminosity distribution (a) for flat-spectrum
sources, (b) for steep-spectrum sources. Hatched areas: no redshift
estimates available; dotted histogram: prediction from 408 MHz results.

sources show relatively low $<V/V_m>$, and indeed a Spearman rank test on the 55 objects showed a correlation between V/V_m and high-frequency spectral index which is significant at the 95% confidence level. Dividing the sample at $\alpha_{HF} = 0.5$ yields a steep-spectrum subsample of 14 objects with $<V/V_m> = 0.70 \pm 0.05$ whose properties appear to be identical with the (steep-spectrum) 3CR and 4C samples studied by Schmidt and by Lynds and Wills, and a flat-spectrum subsample of 41 objects with $<V/V_m> = 0.54 \pm 0.04$. All 41 of these QSOs have structures dominated by compact components less than 0".1 arc in extent (Bentley et al. 1976). These results do not require a complete lack of evolution for compact QSOs, but they do imply different cosmic histories for QSOs of different radio structures.

4. CONCLUSIONS

Some authors have argued that satisfactory evolution functions are trivial to obtain, firstly because of a wide range of permissible assumptions - geometry, local luminosity function, and of course the form of the evolution itself, and secondly because of highly permissive data - poorly defined source counts, luminosity distributions, background temperature requirements. In the face of modern data this point of view is hard to sustain, and the definition of satisfactory models becomes distinctly non-trivial. Progress is therefore possible on investigating the form of the evolution function for different source populations, and in particular on providing constraints on this evolution. From our examination of the 408 MHz counts we conclude that all conventional models of evolution for steep-spectrum sources are unsatisfactory, and that identifications and redshifts of sources at the flux density level of the 5C surveys provide powerful constraints on models which do result in correct source counts. From our preliminary 2700 MHz results we conclude that existing luminosity distributions permit only the crudest of analyses, but the difference in evolution for compact and and extended QSOs suggested by the $<V/V_m>$ test provides considerable incentive to obtain the necessary observations for this investigation.

Bentley, M., Haves, P., Spencer, R.E. and Stannard, D.: 1976, Monthly Notices Roy. Astron. Soc. 176, 275.
Condon, J.J. and Jauncey, D.L.: 1974, Astron. J. 79, 1220.
Richter, G.A.: 1975, Astron. Nachr. 296, 65.

DISCUSSION

Murdoch: Does the luminosity evolution force a sharp modification to the density law in the vicinity of z=3?

Wall: We have considered basically density evolution; of our successful models, type 1 works (in $q_o = 0.5$ geometry) both with (1a) and without

(1b) redshift cut-offs, while type 2 imposes an effective redshift cut-off by progressively narrowing the range of luminosities allowed to evolve.

Murdoch: Robertson (Sydney) is doing similar work, but instead of parametrized models he uses an array of numbers for $\delta(z)$ which are modified until convergence to the source counts is obtained. The conclusions are similar: $(1+z)^\beta$ must be modified, and in particular it is forced to drop sharply at $z \sim 3$.

Wall: Robertson's approach is one that we are very interested in trying. The difficulty comes in communicating the answers; if you have to invent a parameterized model to describe the resultant array, you might as well have started with this model in the first place. It has always been known that $(1+z)^\beta$ models require a redshift cut-off.

Grueff: You stressed the necessity of optical identifications at the 5C flux density level. Did you compare the predictions of your most successful models with the data already available at ∼1 Jy? The 1 Jy point in the log N - log S is a rather interesting one, since it is there that we observe the largest source excess with respect to Euclidean prediction.

Wall: I confess not; our look at the identification data has been preliminary, and we went for the longest baseline in flux density, namely to what is known of the 5C identifications. Our luminosity distribution is "defined" at 10 Jy; 1 Jy is not very far away considering the breadth of the luminosity function.

E.M. Burbidge: I am still puzzled as to how you estimate redshifts for radiogalaxies which do not have measured redshifts.

Wall: We do what we can from estimated magnitudes and the m-z plot for radio galaxies. But I want to emphasize that extreme assumptions about these objects do not affect the structure of the luminosity distribution markedly, and in particular do not change the conclusions of the analysis.

Roeder: Why did your (1/2) model not work in the empty universe, $q_o = 0$?

Wall: The difficulty is at the faint (5C) end of the 408 MHz count, and is the same as the one which destroys the "conventional" models of the literature. The model in question just fits the counts with $q_o = 0.5$ geometry, because the relatively small volumes, ΔV, corresponding to the Δz's assist in converging the faint source counts gradually; these ΔV's are much <u>larger</u> for $q_o = 0$, and the faint counts take on quite the wrong shape.

Ekers: We know from the bivariate luminosity function analysis presented by Perola that there is a variation of optical luminosity for the lower radio power ranges, so the estimation of distance for these galaxies could be affected. Perhaps you should consider including the bivariate luminosity function in your analysis so that you can correctly include the galaxies with no redshift determination.

Wall: I agree in principle.   However, from several considerations, it is clear that the missing-z objects have high radio powers, off the top of the well-defined part of Perola's bivariate function, and before this refinement becomes necessary, the missing redshifts will probably be available.

THE REDSHIFT-MAGNITUDE RELATION FOR RADIO GALAXIES

Harding E. Smith
Department of Physics
University of California, San Diego

ABSTRACT

   We examine the Hubble diagram for radio galaxies and compare radio galaxies and first-ranked cluster galaxies as cosmological test objects. Radio source identification programs are now producing reliable identifications with galaxies as faint as $V \approx 23$ and spectroscopy of these objects has already resulted in the discovery of galaxies with redshifts as high as 0.75, thus there are great expectations for progress in the near future. As in the past, indeterminate corrections, notably luminosity evolution and a possible correlation between radio power and optical luminosity, preclude the determination of q_0.

1.   RADIO GALAXIES AS STANDARD CANDLES

   Current interest in observational cosmology has centered around the determination of the deceleration parameter q_0 by means of the redshift-magnitude diagram. Most workers have restricted themselves to the use of first-ranked cluster galaxies as test objects since they exhibit uniformity of optical luminosity and are intrinsically bright objects thus detectable at cosmologically interesting distances. Gunn and Oke (1975) and Sandage, Kristian and Westphal (1976a) have both recently analyzed the Hubble diagram for first-ranked cluster galaxies, coming to substantially different conclusions about the interpretation of the diagram in terms of q_0.

   After the identification of Cygnus A as the first extragalactic radio source, it was soon recognized that powerful radio sources are generally associated with optically luminous galaxies (c. f. Bolton 1960), often the first-ranked galaxy in a rich cluster. Sandage (1972b) has subsequently shown that radio galaxies have similar absolute magnitudes, $\langle M_V \rangle = -23.0$ ($H_0 = 50$ km s^{-1} Mpc^{-1}) with a dispersion

D. L. Jauncey (ed.), Radio Astronomy and Cosmology, 279-293. All Rights Reserved.
Copyright © 1977 by the IAU.

about the mean of 0.5 mag. Radio galaxies appear to differ from
first-ranked cluster galaxies in this instance in that the distribution
of absolute magnitudes has a slow tail off to fainter magnitudes rather
than a gaussian about the mean (Sandage 1972b).

Radio galaxies offer several advantages over first-ranked
cluster galaxies for use as standard candles:

i)   The presence of radio emission allows us to identify candidate objects very efficiently. About 3/4 of identified 3C radio
sources are galaxies. By selecting radio sources which show no
optical counterpart to the limit of the Palomar Sky Survey, we have
a high probability of selecting galaxies at high redshift ($z \gtrsim 0.4$), thus
candidates of interest for the Hubble diagram.

ii)   Because one need not identify other galaxies which may
be 0.5-1.0 mag fainter than the candidate to establish the existence of
a cluster, we can identify fainter, thus potentially more distant, radio
galaxies than first-ranked cluster galaxies. Figure 1 shows a direct
red, photograph of 3C 330, a distant radio galaxy ($z = 0.549$) in a rich
cluster of galaxies obtained with the 4-m telescope at Kitt Peak
National Observatory (Spinrad, Liebert, Smith and Hunstead 1976).
Other cluster galaxies are easily detected and the cluster can be
classified Bautz-Morgan II-III. This may be compared with Figure 2
showing 3C 427.1 identified by Smith, Burbidge and Spinrad (1976)
from comparable plate material obtained during the same observing
run.

Our experience indicates that the limiting magnitude for galaxies
in direct photographic studies with fine-grain emulsions is on the
order of $\underline{V} \approx 22$ or under the best possible conditions $\underline{V} \approx 23$. Even to
$V = 23$, the surface density of background galaxies is approximately
1 per square minute of arc, thus for sources with accurate positions
and without large structural peculiarities, unambiguous identifications
may be made to the limits available to the largest optical telescopes.
Kristian has already reviewed the excellent progress in identifying
radio sources with faint optical objects earlier in this symposium.

The Scott effect (bias in the Hubble diagram due to correlation
of brightness of the first-ranked cluster galaxy with cluster richness)
does not operate on the sample of galaxies selected by their radio
properties. There is, however, a corollary effect which may be even
more serious. If optical luminosity is correlated with radio power,
then in any sample limited to a given radio flux density, the more
distant sources will have greater radio and optical brightnesses,

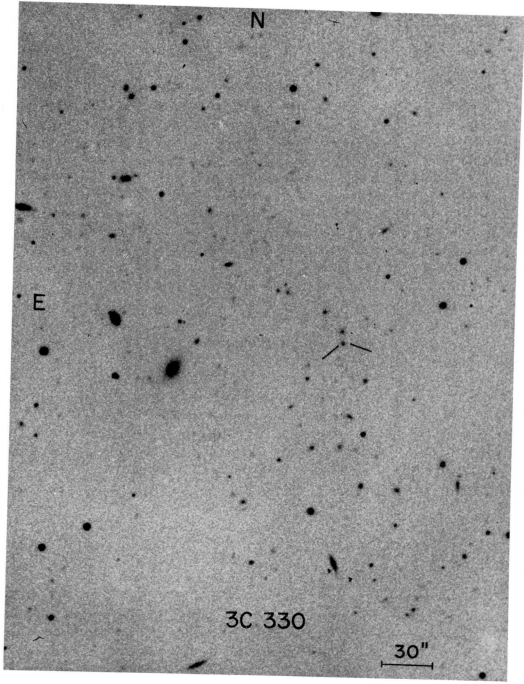

Figure 1.   The radio galaxy 3C 330 (z = 0.549) in a rich cluster of galaxies.

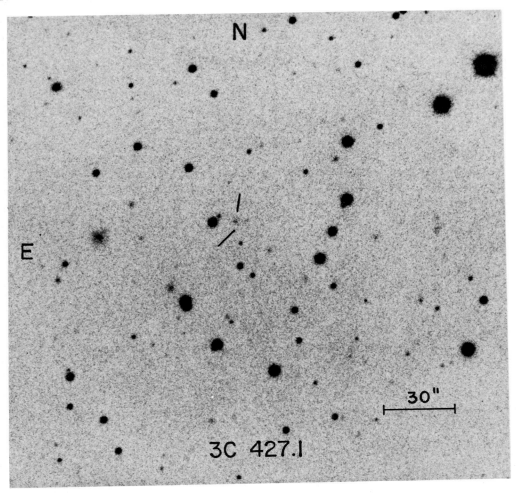

Figure 2.   The galaxy identified with 3C 427.1.

biasing the redshift-magnitude relation toward larger values of q_o.
It is this effect that has caused Gunn and Oke (1975) to question the
use of 3C 295 in their Hubble diagram for first-ranked cluster galaxies
since it was selected by its radio properties rather than the presence
of a cluster.   In a study of the radio/optical bivariate luminosity
function of a complete sample of less powerful Bologna radio sources
($m_{pg} \leq 15.7$, $S_{408} > 0.25$ Jy) Colla et al. (1975) conclude that

$$\langle M \rangle \propto 1.25 \log P_{408}$$

This strong correlation does not, however, continue into the sample
of more powerful 3C radio galaxies.   For the radio galaxies in the

Revised 3C Catalog with $m \lesssim 19$, where the sample has some measure of completeness, the mean absolute magnitude is relatively constant at $M_V = -23$ over more than four decades in radio power at 178 MHz. Nonetheless, until this question is resolved, the "radio Scott effect" poses the most serious challenge to the use of radio galaxies in the Hubble diagram.

2.   SPECTROSCOPY OF FAINT RADIO GALAXIES

There is other justification for obtaining spectroscopic observations of radio galaxies besides simply placing them on the redshift-magnitude plot. Redshifts and spectrophotometric line and continuum measures are necessary to understanding the physics of the non-thermal processes producing the radio emission and the emission-line spectrum. Observations of galaxies at high redshift, where the look-back time is a significant fraction of the "age of the Universe" are important to understanding the evolution of radio sources (and galaxies) and of course it is these that we wish to observe for the Hubble diagram. One must go to a redshift $z \approx 0.5$ before the separation between the $q_o = 1$ and $q_o = 0$ solutions for the redshift-magnitude relation is comparable to the dispersion in absolute magnitude for powerful radio galaxies. Obviously the higher the redshift the better since the separation increases with z. We have estimated that current identification programs can reach down to $\underline{V} \approx 23.0$ under the best possible conditions. For a normal galaxy with $\overline{M_V} = -23$ this corresponds to a redshift $z \approx 0.75$ where the separation between the two solutions is more than 3/4 of a magnitude.

Spectroscopy to such faint optical magnitudes is another matter. With an aperture that subtends only 5 square seconds of arc (appropriate to many modern spectroscopic detector systems), a galaxy at $\underline{V} = 23$ contributes less than 1/10 the signal contributed by the background night sky at even the darkest northern hemisphere sites. Thus high quantum efficiency, sky subtraction detector systems are essential to programs for obtaining spectroscopic information from galaxies at high redshift. Even with such devices integration times may be very long and observations from several nights must be summed to reduce the noise to a level such that spectral features may be detected. Figure 3 shows a section of the average of four nights observations of 3C 123 with the Lick Observatory image-dissector scanner (Spinrad 1975). In some cases obtaining the redshift is easier due to the presence of strong emission features. However the presence of emission lines presents its own set of problems. Apart from the difficulty of obtaining emission free, continuum magnitudes, the presence of strong emission lines is highly correlated with the

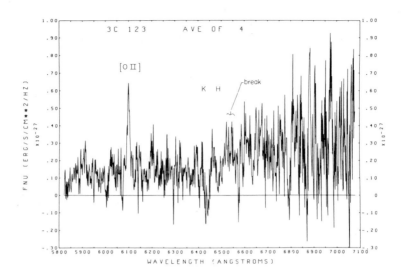

Figure 3.   The spectrum of 3C 123, z = 0.637.

presence of optical non-thermal radiation.   Unless the non-thermal
component can be separated from the stellar galaxy, these systems
must be discarded for use as "standard candles."   An excellent
example is the powerful radio galaxy Cygnus A which was the prototype of the class.   Cygnus A has a very strong emission-line spectrum
and, although it is not classified as an N galaxy, the optical spectrum
is dominated by a non-thermal power law with spectral index α = 1.6
(Osterbrock and Miller 1975).   No trace of absorption features due to
starlight has yet been discovered.   Attempts have been made (Sandage
1973a; Smith, Spinrad and Hunstead 1976) to subtract a non-thermal
power law from the optical spectra of N galaxies in order to isolate
the stellar component.   However it is highly questionable whether
this practice is applicable to the spectra of faint galaxies at high
redshift.   Certainly it must increase the scatter in the Hubble diagram.
More importantly, the subtraction process depends upon reproducing
the observed energy distribution of nearby low-redshift galaxies; if
there has been significant color evolution over look-back times of
several billion years, then the subtraction process will produce
systematic errors in the magnitude of the galaxy as a function of z.
It is alarming to note that even galaxies without strong emission lines
may show optical non-thermal continua.   Ulrich et al. (1975) have
reported non-thermal emission from three galaxies identified with
Bologna radio sources, which do not show line emission.   It is
unlikely that this non-thermal radiation would be recognized in the

noisy spectrophotometric observations of galaxies at high redshift.

Another possible difficulty is van den Bergh's (1975) suggestion that star formation may be triggered by the event that produces the radio emission in radio galaxies. This speculation is based on the relatively blue colors of the stellar population near the absorption band of Centaurus A and by the presence of emission knots which appear to be H II regions in the radio lobes of Cen A (Blanco et al. 1975). The presence of young stars could be very troublesome since at high redshift we are observing the emitted UV where the contribution by young hotter stars would be most significant.

Certainly some high-redshift radio galaxies do show clear evidence for the presence of stars. Both 3C 295 (z = 0.461) and 3C 123 (z = 0.637) show the expected stellar absorption features as well as a single moderate strength emission line identified with [O II] $\lambda 3727$. In fact our experience shows that the faint galaxies identified with radio sources more often show weak emission lines or a faint red stellar continuum without detectable emission lines than a rich Cygnus A type emission-line spectrum. The oft-quoted misconception that "all radio galaxies show strong emission lines" has been caused in large measure by the relative ease of measuring a redshift for an emission-line object and also by the possible ambiguity in identifying a very extended or asymmetric radio source with a faint galaxy showing no spectroscopic peculiarity.

Another difficulty for galaxies without a strong emission-line spectrum is presented by our unfamiliarity with the ultraviolet spectra of galaxies. The classical landmarks Ca II K and H, the 4000 Å discontinuity in the continuum associated with ultraviolet blanketing by metals, the G-band at $\lambda 4300$ begin to be shifted into the region of the spectrum longward of 6900 Å at $z \approx 0.7$. Here the night-sky spectrum is dominated by strong OH emission which varies on very short timescales thus making detection of absorption features very difficult. Morton, Spinrad, Bruzual and Kurucz (1976) have obtained observations of the 2100-3200 Å spectra of α Agl and α C Mi from Copernicus from which they have selected absorption features likely to be strong in the emitted UV spectra of galaxies. A list of absorption/continuum features is given in Table 1 along with the redshift for which that feature is shifted into the atmospheric OH bands.

Despite the difficulties great progress has been made in recent years in obtaining redshifts of radio galaxies at high redshift. Minkowski's (1960) redshift z = 0.461 for 3C 295 stood for nearly

TABLE 1

Galaxy Absorption Features for Redshift Determination

λ_o	Description	z_{6900}
2640	Fe Continuum Discontinuity	1.61
2799	Mg II Resonance Doublet	1.47
3933,68	K, H Ca II	0.75
4000	Continuity Discontinuity	0.725
4303	G-band	0.603
5175	Mg-b	0.333
5895	D Lines Na I	0.170

fifteen years as the largest radio galaxy redshift. Following Spinrad's (1975) result for 3C 123 a number of radio galaxies have now been found with z > 0.461 by the groups at Lick Observatory and at Hale Observatories. Those available as of July 1976 are listed in Table 2.

TABLE 2

New High-Redshift Radio Galaxies

Object	z	V	Remarks
3C 318	0.752[1]	21	N-galaxy, strong non-thermal continuum, weak emission lines
3C 343.1	0.750:[2]	21	1 emission line = [O II] λ3727?
3C 123	0.637[3]	20.5	[O II] λ3727 emission, stellar absorption features
3C 467	0.631[4]	19.5	N-galaxy, strong non-thermal
PKS 0353+027	0.602[4]	18.5	continuum, strong emission lines
PKS 0116+082	0.594[5]	21	Strong emission, cluster
3C 330	0.549[5]	21.5	Strong emission, cluster
3C 19	0.482[6]	20.5	Strong emission, cluster
3C 411	0.467[7]	19.5	N-galaxy, strong non-thermal continuum, strong emission lines.

References to Table 2:

[1]Spinrad and Smith (1976); [2]Spinrad unpublished; emission line confirmed by Sandage, Kristian and Westphal (1976b); [3]Spinrad (1975); [4]Smith, Spinrad and Hunstead (1976); [5]Spinrad, Liebert, Smith and Hunstead (1976); [6]Sandage, Kristian and Westphal (1976b); [7]Spinrad, Smith, Hunstead and Ryle (1975).

3. CONSTRUCTION AND INTERPRETATION OF THE HUBBLE DIAGRAM

Four of the galaxies in Table 2 are N galaxies with strong optical non-thermal radiation. Given the difficulties in interpreting the optical spectrum, we will not consider them further. The remainder are normal radio galaxies and thus are candidates for placement on our Hubble diagram. We will carry 3C 343.1 as tentative (although no more tentative than Minkowski's redshift for 3C 295). Photometry for 3C 343.1, 3C 123, and 3C 19 have been obtained by Sandage, Kristian and Westphal (1976b). For 3C 330, we have taken the spectrophotometric observations of the strong emission-line galaxy associated with the radio source and recent observations of the other bright cluster galaxy by Spinrad. These data have been added to the body of data previously available (largely from Sandage 1972b) to produce the redshift-magnitude diagram for 3CR radio galaxies shown in Figure 4. The following corrections have been applied to the photometric data:

i) The data have been corrected to a standard linear diameter (86 kpc, $q_0 = 1$) according to the method described in Sandage (1972a). This correction depends on q_0, of course, but the difference between $q_0 = 1$ and $q_0 = 0$ is only 0.08 mag (smaller for $q_0 = 0$) at z = 0.5. Sandage's correction is about 1/2 mag larger than that of Gunn and Oke (1975) who choose a smaller standard diameter.

ii) K-corrections have been applied from Oke (1971). His high-quality spectrophotometric observations of 3C 295 have allowed correction to z ≈ 0.72. For redshifts greater than 0.5 this correction is large (> 1.5 mag) and somewhat uncertain, thus photometry at the redshifted emitted V continuum would be preferable, however this requires photometry in the near infrared where detectors are less sensitive and the sky background is brighter.

iii) We have adopted the interstellar extinction correction given in Sandage (1973b), however differing methods of estimating the extinction as a function of galactic latitude (e.g. $E_{(B-V)}$, galaxy

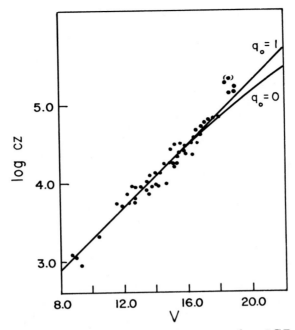

Figure 4.   The Hubble diagram for 3CR radio galaxies.

counts, H I column density) disagree. Heiles (1975) has recently
emphasized this point concluding that extinctions calculated for extragalactic objects must be uncertain by a few tenths of a magnitude.   In
a few cases for galaxies at low galactic latitudes where the extinction
is most patchy and uncertain, other indicators (color, etc.) have been
used to derive the extinction correction. Since we are not dealing
specifically with cluster galaxies, and because we hope to press
beyond the limits where clusters may be effectively classified, we
have not applied corrections for either cluster richness or Bautz-
Morgan type.

It is not the intention of this review to introduce yet another
formal value for q_0 into the literature. We believe that the difficulties
discussed previously, the few galaxies yet available with z > 0.4, and
the unknown correction for luminosity evolution make solution for q_0
premature (let alone the consideration of Λ, the cosmological constant
in Lemaitre models). A few comments may be made, however. The
number of data points in Figure 4 is comparable to the number of
first-ranked cluster galaxies considered by Gunn and Oke (1975) and
Sandage, Kristian and Westphal (1976a). Neither of these studies
have any galaxy with z > 0.4 save 3C 295 (which Gunn and Oke discard).
There is a clear tendency in Figure 4 for the galaxies at high redshift
to lie above the q_0 = 1 line. If none of the selection effects we have

discussed are operating (most of which would make a galaxy appear too bright), this would indicate an <u>evolution-free</u> solution for q_0 in excess of 1.

The correction for galaxy evolution currently presents the greatest uncertainty in the interpretation of the Hubble diagram. Tinsley and her co-workers have been most active in modeling the evolution of elliptical galaxies with a view toward providing corrections for the luminosity and colors of galaxies at previous epochs (c.f. Tinsley 1972; Tinsley and Gunn 1976). The calculated corrections range from a few tenths to over a magnitude at $z = 0.5$, all in the sense that galaxies were brighter in the past. Application of these corrections then revises the value of q_0 downward by as much as 1.5 from that derived without consideration of evolution. The necessity of applying this potentially large and uncertain correction precludes the determination of q_0 at this time. Observational detection of color evolution over the look-back times to distant galaxies would provide a valuable constraint on theories of galaxy evolution. It thus seems that spectrophotometry of galaxies at cosmological distances is necessary first as input for theories of galactic evolution which will then allow proper placement of these objects on the Hubble diagram.

We conclude with the obligatory call for more and better data before substantive conclusions may be drawn from the Hubble diagram. In this case, however, the prospects are excellent for obtaining this data in the near future. The barrier at $z = 0.461$ has been broken and programs of identification and spectroscopy of distant galaxies now going on at Lick and Hale Observatories and elsewhere have likely made this review outdated before it is presented, let alone published in the proceedings of this symposium.

It is a pleasure to acknowledge the contribution of Dr. Hyron Spinrad who initiated and collaborated with me on most of the research discussed here. I would also like to thank Drs. E. M. and G. R. Burbidge for their encouragement and helpful discussions and Drs. A. Sandage and J. Kristian for discussing their data with me before publication.

REFERENCES

Blanco, V. M., Graham, J. A., Lasker, B. M., and Osmer, D. S.: 1975, <u>Ap. J. (Letters)</u> 198, L63.

Bolton, J. G.: 1960, Obs. Owens Valley Radio Obs., No. 5.

Colla, G., Fanti, C., Fanti, R., Gioia, I., Lari, C., Lequeux, J., Lucas, R., and Ulrich, M.-H.: 1975, Astron. & Astrophys. 38, 209.

Gunn, J., and Oke, J. B.: 1975, Ap. J. 195, 255.

Heiles, C.: 1976, Ap. J. 204, 379.

Kristian, J., Sandage, A., and Katem, B.: 1974, Ap. J. 191, 43.

Minkowski, R.: 1960, Ap. J. 132, 908

Morton, D., Spinrad, H., Bruzual, G., and Kurucz, J.: 1976, Ap. J. (in press).

Oke, J. B.: 1971, Ap. J. 170, 193.

Osterbrock, D. E., and Miller, J. S.: 1975, Ap. J. 197, 535.

Sandage, A.: 1972a, Ap. J. 173, 485.

Sandage, A.: 1972b, Ap. J. 178, 25.

Sandage, A.: 1973a, Ap. J. 180, 687.

Sandage, A.: 1973b, Ap. J. 183, 711.

Sandage, A., Kristian, J., and Westphal, J.: 1976a, Ap. J. 205, 688.

Sandage, A., Kristian, J., and Westphal, J.: 1976b, Ap. J. (in press).

Smith, H. E., Burbidge, E. M., and Spinrad, H.: 1976, Ap. J. 210 (in press).

Smith, H. E., Spinrad, H., and Hunstead, R.: 1976, Ap. J. 206, 345.

Smith, H. E., Spinrad, H., and Smith, E. O.: 1976, P.A.S.P. 88 (in press)

Spinrad, H.: 1975, Ap. J. (Letters) 199, L3.

Spinrad, H., Liebert, J., Smith, H. E., and Hunstead, R.: 1976, Ap. J. (Letters) 206, L79.

Spinrad, H., Smith, H. E., Hunstead, R., and Ryle, M.: 1975, Ap. J. 198, 7.

Tinsley, B. M.: 1972, Ap. J. 178, 319.

Tinsley, B. M., and Gunn, J.: 1976, Ap. J. 203, 52.

Ulrich, M.-H., Kinman, T. D., Lynds, C. R., Rieke, G. H., and Ekers, R. D.: 1975, Ap. J. 198, 261.

van den Bergh, S.: 1975, Ann. Rev. Astron. & Astrophys. 13, 217.

DISCUSSION

Jaffe: You can use Dr. Perola's bivariate luminosity function to sort out the radio-optical selection process for radio galaxies even to the extent of constructing a probabilistic H-R diagram of radio/optical color to narrow the dispersion in M of the galaxies observed, whether "radio" galaxies or not.

It may be dangerous to assume a characteristic M of radio galaxies which does not change with z since the radio properties of the galaxies, as well as the optical, evolve with time.

H.E.Smith: Yes, the business of the radio optical luminosity function is very important to the question of using radio galaxies in the Hubble diagram. As I understand it, for powerful radio galaxies there is no correlation between radio and optical luminosity.

We hope to account for optical evolution. It is not clear to me how important the radio evolution will be to $z = 0.5 - 1.0$.

Lewis: You made no reference to the Ostriker & Tremaine correction to be expected from dynamical friction acting on close satellites. Perhaps this might be more severe for a sample of radio galaxies than for the first brightest cluster galaxy, as the radio activity may be a sign of the recent acquisition of a satellite galaxy, with the largest redshift galaxies being liable to have on average accreted more massive satellites.

H.E. Smith: I think that result is speculative and it can influence the diagram either way. Your suggestion seems to be a variation of the colliding galaxies hypothesis for production of radio sources, which has not proven very attractive.

Certainly there are a large number of radio galaxies which are not in clusters of galaxies and for which this effect cannot be important.

Tinsley: Ostriker and Tremaine noted that dynamical evolution (accretion of stars by the galaxy) can be important for any cluster galaxies. Several studies have shown that it will be very difficult to detect dynamical evolution even if it occurs at cosmologically important rates, so it cannot be eliminated as a correction. Moreover, its effect on the Hubble diagram could be of either sign.

It is worth noting that for any evolution (stellar evolution as discussed by Smith, or dynamical evolution), a luminosity change of only 3-5 percent per 10^9 years changes the apparent value of q_o by unity.

H.E. Smith: As I have said, I consider the Ostriker and Tremaine result rather speculative, and of course, unless the theory can make a specific prediction, it cannot be tested.

The corrections are large and the cosmological effects are small. If this work were easy it would have been done a long time ago. Hopefully careful spectrophotometric observations of galaxies at high redshifts will allow us to detect the stellar evolution over times that are a significant fraction of the Hubble time.

Grueff: Do you have any information about the (V-R) color for the new high redshift radio galaxies?

H.E. Smith: Our spectrophotometric observations show weak evidence for bluer colors than might be expected from redshifting a normal galaxy to 0.6 or 0.7. Perhaps Dr. Kristian who has obtained more accurate broad band observations would like to comment.

Schmidt: There is no need for the K-correction to be uncertain. All that is required is spectrophotometry at a given rest wavelength, say at $H\delta$, for each of the galaxies.

H.E. Smith: That's right. Unfortunately good spectrophotometry does not exist for this sample. I expect that to be done in the near future.

Ekers: I wonder if the Cambridge astronomers could provide a 'revised Revised 3C catalogue' - I mean especially the problem of the flux errors. I believe some 20 to 50 objects would be involved.

Shakeshaft: Unfortunately 178 MHz is now a television band - maybe 150 MHz would do.

Longair: It is really rather non-trivial. What we have been doing is using the revised Kellermann, Pauliny-Toth and Williams flux densities which are the best available. We at Cambridge use a complete sample of 166 sources brighter than 10 Jy on this scale. (Appendix I)

Pooley: The radio galaxy 3C123 has the highest known radio luminosity; it is therefore important to be certain that the galaxy is correctly associated with the radio source. The structure of 3C 123 is very asymmetrical and the galaxy does not lie at the radio centroid. When mapped with the 5 km telescope at 15 GHz at a resolution of 0".7 x 1".3, it becomes clear that the geometry of the source is that of a double with a flat spectrum nuclear component coincident with the galaxy, and both compact and diffuse components on either side. The ratio of the flux densities of the two compact components is very large (of the order of 50:1) and the faint features would be very difficult to detect in weaker sources. This emphasises the need for high sensitivity, high resolution radio telescopes to make sure of optical identifications in such cases.

Arp: Have you looked for any optical identifications of the components that are now resolved in 3C123?

Pooley: Yes, and there are none.

THE HUBBLE DIAGRAMS FOR QUASARS

John N. Bahcall, Edwin L. Turner
Institute for Advanced Study, Princeton, New Jersey

We discuss in this talk the optical and radio Hubble-diagrams for the brightest quasars. We shall infer from our results: (1) a strong correlation between redshift and received flux that is consistent with the cosmological interpretation of the emission-line redshifts; and (2) a quantitative upper limit on the permissible amount of pure luminosity evolution. We establish these conclusions twice, by examining the Hubble diagrams both with and without corrections for the selection effects introduced by the flux limits of the quasar catalogs considered.

The sample of 3C R and 4C quasars we use has been described by Schmidt (1976) in these preceedings (see also Schmidt 1968, 1972, 1974, Olsen 1970, Lynds and Wills 1972). We adopt the redshifts, optical and radio fluxes, and radio spectral indices given by Schmidt (1976) as well as the three sets of flux limits he has specified. We note that there are theoretical and observational advantages in defining quasars in terms of their absolute optical luminosity, including observational completeness, independence of observing conditions, and the possibility of isolating one of the most important physical variables (see Bahcall 1971). We eliminate therefore from the sample described by Schmidt those objects that are less than one magnitude brighter than the brightest ellipticals in rich clusters, i.e., we require $\log F_{2500} \geq 22.8$ (F_{2500} in watts per Hz, or $M_B \leq -23.2^m$, $H_o = 55$ km/s Mpc, Sandage(1973)). We are left with a sample of 112 objects. We assume $q_o = 1.0$ for specificity (and because selection effects are less critical for our purposes with $q_o = 1.0$ than, e.g., with $q_o = 0.0$).

The first step in conventional investigations of the Hubble diagram for galaxies is to chose as "standard candles" the brightest galaxies in rich clusters. We make an analogous choice. Following Bahcall and Hills (1973), we use the brightest quasars in equally-populated redshift bins, indicated by horizontal lines in Figure 1. In the quasar case, there are important selection effects that are associated with the flux limits of the radio and optical samples. We shall discuss these selection effects in detail later. A point by point comparison is made by Hills and Bahcall (1973) between the present treatment of the brightest

D. L. Jauncey (ed.), Radio Astronomy and Cosmology, 295-303. All Rights Reserved.

quasars and what is usually done in forming the Hubble diagram of the brightest galaxies in rich clusters.

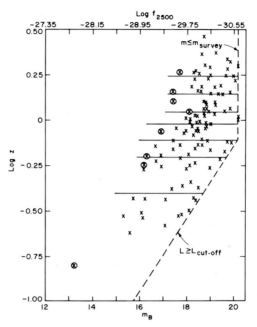

Figure 1.   The redshift-magnitude diagram for the whole sample of 112 quasars

We consider first the optical Hubble diagram constructed from the raw data without any corrections. The results are shown in Figure 2a for the brightest quasars in each of eight redshift bins (the same objects are indicated by circles in Figure 1). The bins were chosen so that they all contain the same number of objects (14) in order to provide a sampling of the luminosity function that, except for flux limits, is unbiased with respect to redshift. There is obviously a strong correlation between redshift and apparent magnitude or flux. The best-fit straight line has the form

$$m_{uncorrected} = (4.43 \pm 0.44) \log Z + 17.3 \qquad , \qquad (1)$$

where we have replaced $\log f_{2500}$ given by Schmidt with an equivalent B magnitude, $m_B = -2.5 \log f_{2500} - 56.375$. The correlation coefficient between $\log Z$ and $\log f_{2500}$ is 0.956, i.e., the probability that this result would have arisen purely by chance is less than one in a thousand. Similar results have been obtained by Bahcall and Hills (1973), Burbidge and O'Dell (1973), Petrossian (1974) and others using inhomogeneous samples. The relation analogous to equation (1) for the brightest radio quasars is $\log f_{500}$ Mhz = -24.9 -(1.1 $\pm$ 0.35) $\log Z$, with a correlation coefficient of -0.73 (less than 10 percent chance of having arisen by accident).

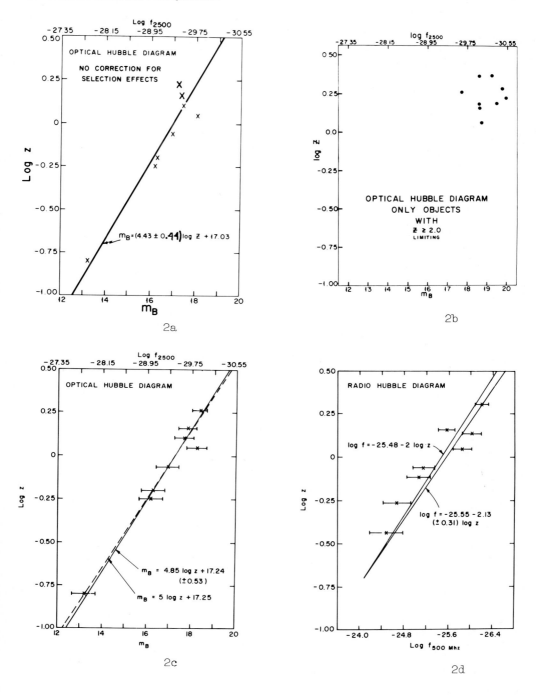

Figure 2.   The Hubble Diagrams for the Brightest Quasars

The strong correlation observed in the raw data between the apparent magnitudes of the brightest objects and their redshifts is consistent with a cosmological origin for the redshifts. No local theory that we are familiar with can explain this correlation. We assume in what follows that the emission-line redshifts of quasars are caused by the expansion of the universe.

We can also calculate an upper limit for optical luminosity-evolution by using the uncorrected luminosity of the brightest object in each of the eight redshift bins. Assuming $L_{brightest}(t) = L_{brightest}(t_o)$ exp $-\beta t$, we find

$$\beta_{upper\ limit} = 1.46 \pm 0.91 \tag{2}$$

where t is the age of the universe in the units of the present age. It should be stressed that the value of β given by equation (2) is an upper limit since we have ignored so far the selection effect in the sample arising from a fixed cut-off in optical flux. This cut-off naturally gives rise to an apparent brightening of objects at small t (large Z) i.e., a positive β. In fact, we shall see below that a value of β equal to zero is consistent with the present data and we shall assume, in what follows, pure density evolution ($\beta \equiv 0$).

Equation (1) implies an important physical result: the optical luminosity of the brightest quasars has not changed appreciably over many generations (possibly hundreds of quasar lifetimes). This suggests that at least the brightest quasars were not very different at early times (t $\simeq$ 0.15) than they are today.

The limits on luminosity evolution for the brightest radio quasars are less informative. One finds, analogous to equation (2), $\beta(radio)_{upper\ limit} = 3.6 \pm 1.4$. The relatively large values for both the variance and the absolute value of β for the radio-brightnesses could be primarily a result of the importance of the radio flux limits in selecting the present sample.

We must now consider what to do about the selection effects introduced by the limiting optical and radio fluxes. The simplest procedure would be generalize a suggestion by Burbidge and O'Dell (1973) and use only objects that are bright enough to be unaffected by either flux limit. They also suggested using the second and third brightest objects in each redshift bin as additional tests. However, Burbidge et al. (1973) ignored all radio selection effects. This certainly cannot be done with the present sample; for 85 out of the 112 quasars, or 75 percent, the maximum distance to which they could be seen and remain in the sample is determined by their radio, not optical, flux. If we require quasars to be intrinsically bright enough to be visible at Z$\geq$2.0 in both radio and optical wavelengths, we find that only 4 objects out of the 3C R sample qualify. There are 9 objects in the 4C Olsen-sample that are bright enough and 7 in the Lynds-Wills 4C sample. With the present approach, we cannot combine objects from the different samples since the optical luminosity function depends upon the radio selection procedure. We also cannot use quasars with $Z > 2.0$ since they would correspond to a greater intrinsic luminosity cut-off and hence require us to eliminate some of the smaller Z objects. We therefore choose the sample with the largest number (9) of intrinsically bright objects, the Olsen 4C sample. The results are shown in Figure 2b. It is obvious that the 9 bright quasars selected in this way are all at such large redshifts that one cannot determine meaningfully a slope in the red-shift-magnitude diagram even if we ignore statistical fluctuations in this small sample. The brightest objects are extremely rare and are naturally found only at the largest redshifts. The total number of objects remaining in the sample of 112 quasars treated as described above is less than were in each of the eight independent redshift bins of Figure 1 and 2a. A formal least squares solution yields $m_b = (0.92 \pm 2.34) \log Z + 18.74$. The bunching effect at Z(limiting) shown in Figure 2b cannot be avoided by using a different value for Z(limiting). For example, if Z(limiting) = 1.0, then 8 out of the allowed 9 objects have $\log Z$ within ± 0.07 of each other. We conclude that further progress requires an explicit calculation of the selection effects so that quasars at smaller redshifts may be used.

Bahcall and Hills (1973) described a method by which a given luminosity function can be used to evaluate both the selection effects associated with the fixed limiting fluxes and the variances expected because of the finite sample size in each bin (14 objects in the present case). The later feature is important because it allows us to make a self-consistent test for goodness-of-fit. In what follows, we use the Bahcall-Hills procedure to construct both the optical and radio Hubble diagrams. The basic assumptions are: (1) the emission-line redshifts are of cosmological origin and (2) the shape of the luminosity function is independent of time (redshift), i.e., there is no luminosity evolution.

The ideas behind the procedure may be most clearly visualized by supposing for a moment that the luminosity function (radio or optical) is a uniform distribution between L(cut-off) and a maximum quasar luminosity, L(MAX). Then for redshifts so small that luminosities less than L(cut-off) could have appeared in the Schmidt-catalog, there is no correction because we are sampling the entire luminosity function equally in each bin. However, for redshifts above some critical value ($Z_c \simeq 0.78$, $\log Z_c = -0.11$ for optical luminosities in the Schmidt sample) only objects with luminosities L(limiting,Z) > L(cut-off) could appear in the quasar catalog. By randomly sampling (14 times in our case) the luminosity function between L(limiting,Z) and L(MAX) we obtain progressively brighter objects as Z increases [L(limiting,Z) approaches L(MAX)]. Also the standard deviation of the brightest object in each bin will decrease as L(limiting,Z) approaches L(MAX). This simple example gives the correct physical picture of the correlations to be calculated; the rest is details. We mention, incidentally, that of the two processes described above, (1) binning into equally populated samples and (2) calculating luminosity corrections, binning is more important and calculating the corrections is more complicated.

We obtain luminosity functions by requiring that (Schmidt 1968) the average value of V/V_{MAX} be equal to one-half when density evolution is included. This does not uniquely or optimally define a luminosity function, but the corrections we calculate are not sensitively affected by small differences in luminosity functions. Some satisfactory evolutionary laws are of the form $\rho(t) = t^{\alpha} \exp \gamma(1-t)$, with $\alpha = 0$, $\gamma = 10^{+4}_{-3}$, $\alpha = 1$, $\gamma = 15$, $\alpha = 2$, $\gamma = 17$, $\alpha = 3$, $\gamma = 20$, and $(1 + Z)^n$ with $n = 5.5^{+2.5}_{-1}$. We prefer evolutionary laws that are expressed in terms of cosmic time since these are simplest to interpret theoretically. Moreover, laws of the form $e^{+\alpha}e^{\gamma t}$ are suggested by quasar theories in which stellar or galactic evolution is important and, for $\alpha > 1$, these laws predict very few quasars with $Z > 2.5$.

We have calculated luminosity functions using several of the derived evolutionaly laws; the results do not depend much on the particular evolutionary law that was chosen. For $q_o = 1.0$ and $\rho(t) = \exp 9.5(1-t)$, we find with

$$\Phi(L)dL = \Phi^*(L/L_*)^{\alpha} \exp{-L/L_*}\, dL/L_* \qquad (3)$$

that for optical wavelengths $\alpha = -1 \pm 1$, $\Phi^* = 0.2 \pm 0.1$ Gpc^{-3}, and $F^*(2500) = (4.5 \pm 0.9) \times 10^{23}$ watts HZ^{-1} ($M^* = -25.3 \pm 0.2$). For the radio luminosity function, $\alpha = -2.4 \pm 0.3$, $\Phi^* = (3 \pm 2) \times 10^{-3} Gpc^{-3}$ and $F^*(2500) = 2.1 \pm 0.8) \times 10^{28}$ watts HZ^{-1}. The errors quoted are equivalent to 1-σ errors. Functions of the general form given by equation (3) have been used by Schechter (1976) and others to describe galaxy luminosity functions (Schechter obtains $\alpha \cong -1.25$ and M^*_B galaxy ≈ -20.85). We used functions of the form given in equation (3) since they provide convenient empirical fits to the average luminosity functions defined by the present sample. The most important feature expressed by equation (3) is that there is a sharp steepening of the slope of the empirical luminosity functions near a characteristic luminosity L_*.

The corrections for the selection effects can be calculated from the empirical luminosity functions if we assume that the N-quasars in each redshift bin are Poisson-distributed over the accessible (at the given Z) part of the luminosity function. We have, up to a common factor,

$$\int_{L_{limiting}(Z)}^{\infty} \Phi(L)dL = N \qquad , \qquad (4a)$$

$$\int_{L_{brightest}(Z)}^{\infty} \Phi(L)dL = \ell n2 \qquad , \qquad (4b)$$

with (in our case) N=14. A slightly more complicated equation determines the variance σ^2 (log L(brightest,Z). For the radio corrections, each of the three samples (3C R, Olsen, and Lynds-Wills) were treated separately in each redshift bin and the properly weighted averages were used in the Hubble diagram (for the optical case, the limits of the three samples are so similar that this averaging was unimportant). We show in Table 1 the numerical corrections and their variances that were calculated for use with the present sample. We also calculated corrections for a specific case treated by Bahcall et al. (N=15, log F_{2500}(minimum) = 22.4) with Monte-Carlo computer calculations on a numerical luminosity function derived by Schmidt (1970) for a sample of optically-selected quasars. Our results are in good agreement with the previously derived corrections, indicating that the computed values of Δm and $\sigma^2(\Delta m)$ are not sensitive to details of the luminosity function. Strictly speaking, we should have used a bi-variate luminosity function (or both optical and radio luminosities, L_O and L_R) to calculate the corrections. However, our approximation is exact if the bi-variate luminosity function can be factored as a product of independent functions of L_O and L_R or (for the optical corrections) as a product of a function of L_O times a function of L_O/L_R. We intend to carry out corrections using bi-variate luminosity functions, but we expect that the improvements introduced by this more complicated procedure will be smaller than the variances $\sigma^2(\log L_B, Z)$.

TABLE I.   CALCULATED CORRECTIONS

Z median (bin)	Δm OPT	$\sigma(\Delta m)$ OPT	$\Delta \log f$ radio	(log f) radio	Brightest in Optical	Brightest in Radio
0.309	0.0	0.51	0.00	0.30	3C 273	3CR 48
0.531	0.0	0.51	0.32	0.25	4C 23.24	3CR 147
0.704	0.0	0.51	0.43	0.21	4C 16.30	3CR 380
0.872	0.05	0.50	0.53	0.19	3CR 454.3	3CR 196
1.031	0.15	0.45	0.62	0.17	3CR 208	3CR 208
1.220	0.25	0.42	0.70	0.17	4C 06.41	3CR 181
1.561	0.43	0.40	0.81	0.15	3CR 298	3CR 298
2.060	0.68	0.34	0.93	0.12	4C 29.1	3CR 9

We note a simple prediction that follows from the joint assumptions of no luminosity evolution and a cut-off (or monotonically-decreasing) luminosity function at large luminosities (cf. equation 3). These ideas imply that the standard deviation of the luminosity of the brightest objects in equally-populated redshift bins should decrease monotonically with increasing redshift if the sample is flux limited (neglecting intrinsic variability). This prediction may be testable since the effect is rather large (see Table I, columns 3 and 5).

Our final results are shown in Figures 3c and 2d. The indicated errors bars are, in all cases, the theoretical standard deviations calculated as described above. Both the optical and the radio Hubble diagrams are good fits (in the χ^2 sense) to the predicted cosmological curve obtained by assuming no luminosity evolution and the calculated (Table 1) corrections and variances. This is satisfying but not surprising since the uncorrected Hubble diagrams (cf. Fig. 2a) already show that the basic assumptions of cosmological redshifts and no

luminosity evolution are good approximations. We note in closing that
the radio and optical Hubble diagrams refer to different objects; in
six of the eight redshift bins the brightest optical quasar is not the
brightest radio quasar.*

REFERENCES

Bahcall, J.N., 1971, Astron. J. 76, 283.
Bahcall, J.N. and Hills, R.E., 1973, Ap. J. 179, 699.
Burbidge, G.R. and O'Dell, S.L., 1973, Ap. J. 183, 759.
Hills, R.E. and Bahcall, J.N., 1973, Annals of the New York Academy of
      Science.
Lynds, R. and Wills, D., 1972, Ap. J. 172, 531.
Olsen, E.T., 1970, A.J. 75, 764.
Petrossian, V., 1974, Ap. J. 188, 443.
Sandage, A., 1973, Ap. J. 183, 731.
Schechter, P., 1976, Ap. J.
Schmidt, M., 1968, Ap. J. 151, 393.
_____   1970, Ap. J. 162, 371.
_____   1972, Ap. J. 176, 273
_____   1974, Ap. J. 193, 505.
_____   1976 (private communication).

*This research was supported in part by National Science Foundation
  Grant No. NSF-GP-40768X.

                              DISCUSSION

Arp: What effect do the violently variable quasars have on your
analysis?

Bahcall: They are very rare and do not influence importantly the statistical averages. The brightest quasars at the epoch studied are shown in
Fig. 1. If quasar variability either in frequency or amplitude were
strongly correlated with redshift, then this should show up in the Hubble
diagram.

Petrosian: The fact that the m-log(z) slope is five after application
of correction is an indication of the correctness of calculation but not
an independent proof of the cosmological hypothesis. Is the difference
between the expected slope of -2 for the $\log(S_{radio})$ - log(z) relation
and the calculated one due to uncertainty in the calculated luminosity
function or some other cause?

Bahcall: The strong correlation observed in the raw data between the
luminosities of the brightest quasars and their redshifts supports the
cosmological hypothesis. The corrected Hubble diagrams are consistent,
to present accuracy, with the joint assumptions of cosmological redshifts
and no appreciable pure luminosity evolution. The difference between the

expected and observed slopes for the radio Hubble diagram is not
significant.

Setti: I am not clear about your limit on the luminosity evolution,
because suppose that the luminosity function steepens with increasing
redshift, then the expected luminosity of the most luminous objects in
the various redshift bins may not change while the average luminosity
per bin would increase with z.

Bahcall: If by "steepen" you mean that the characteristic luminosity L_*
increases with cosmic time, then the expected luminosity of the brightest
objects also increases in the same way; it just scales with $L_*(t)$. This
is the case we have treated in the text. On the other hand, one can
imagine special cases in which the shape of the luminosity function
varies in just such a way that the average luminosity in a bin changes
with time but $L_{Brightest}$ remains constant. These special cases are
probably best studied using a maximum-likelihood estimate that utilizes
all the members of a bin.

Setti: Recently Zamazani and I have derived the shape of the optical
luminosity function for radio selected quasars. This has been done
following a step by step procedure in the redshift-apparent magnitude
diagram of all known radio emitting quasars, taking into account possible
strong selection effects. We find that the shape of the optical luminosity function can be best represented by two power laws with a rather
sharp break at $F_{2500} \approx 23.4$ ($H_o = 100$ km s^{-1} Mpc^{-1}, $q_o = 1$), in very good
agreement with the findings reported by Dr. Bahcall.

McCrea: Does the constant in the m-z relation give any new determination
of the Hubble constant? What is the descriptive meaning of the particular number-evolution that you have used - does it mean that there were
more quasars in the past?

Bahcall: First question: no.
        Second question: yes. It is the same as used by Schmidt and
others.

Jauncey: It will be interesting to see where Malcolm Smith's Tololo
quasars fit on your m-z diagram.

CLUSTERS OF GALAXIES AND RADIO SOURCES

Walter Jaffe
Institute for Advanced Study
Princeton, New Jersey 08540

For some time there have been suggestions that there is a special association of radio galaxies with rich clusters of galaxies, and more recently that the radio galaxies in clusters may show different characteristics from those outside. I will discuss the evidence for three types of such differences, in luminosity function, morphology, and occurance of steep spectrum sources. In each case I will try to connect any difference I find to the cluster environment.

A. THE DATA SET

Much of what I will say is of a very tentative nature because of the small amount of relevant data. Table 1 lists the observations which to date have been most useful for cluster studies. These include two types: 1) Large area surveys, which are the most general way to study any extragalactic radio problem, but in which the percentage of cluster galaxies is small. 2) Cluster specific studies which are much more efficient in sampling cluster properties, but which suffer from various special selection effects, and which usually require another, general, survey to provide a baseline.

Survey	Number Abell Clusters Surveyed	Log Typical Luminosities	Number Detected Cluster Galaxies
General:			
3C	~2000	$39.5 - 42$ erg s^{-1}	10
2d Bologna	~15	$38 - 40.5$	11
Cluster:			
W4	5	$37 - 40$	25
4C/cluster	25	$39 - 41$	17
Owen	~500	$40 - 41$	130

References: 3C; Macdonald et al 1968, Mackay 1969, Elsmore et al 1970: Bologna; Colla et al 1975; W4; Jaffe and Perola 1975: 4C; Riley 1975: Owen 1975

D. L. Jauncey (ed.), Radio Astronomy and Cosmology, 305-316. All Rights Reserved.

B.   THE LUMINOSITY FUNCTION

The first question we ask about radio galaxies in clusters is whether they occur more or less often than those outside. Specifically we ask whether the fraction of galaxies of given optical type and absolute magnitude that are radio sources of given luminosity depends on the cluster environment. As Dr. Perola has reviewed, the answer, at least for elliptical galaxies, seems to be that there is no difference (greater than a factor of two or so) in this luminosity function (LF) for galaxies with $-19 > M_p > -21$ (for $H_o = 100$ kms^{-1} Mpc) and for luminosities of 10^{38} to ^{42} erg s^{-1}.

There is one caveat here and that is the weak end of the LF ($10^{38}-10^{40}$ erg s^{-1}) is determined from a rather small number of clusters, about 5, and could be incorrect if these were unusual in some way. Possible evidence for this comes from Owen's (1975) survey of the total luminosity of a large number of clusters measured with a low resolution telescope. The whole cluster LF he finds, shown in figure 1, shows a well-defined peak at $L \simeq 10^{40.5}$ erg s^{-1} but only accounts for 30% clusters observed. He postulates a second, low luminosity type of cluster, to account for the other 70%.

To explain Owen's result under the hypothesis that the individual galaxy LF is independent of cluster type, we need to assume that the galaxy population of these two types of clusters is drastically different. Since the individual galaxy LF depends very strongly on absolute magnitude, one possibility is that the radio bright clusters contain one superbright cD type galaxy, with $M_p < -21$, which dominates the cluster LF. Figure 1 also shows two whole cluster LF's predicted from the single galaxy LF's for clusters with and without a single superbright galaxy. We see that the inclusion of the one galaxy indeed makes a large difference and that with some mix of the two cluster types we can reproduce the position and amplitude of the peak, but that the faint end

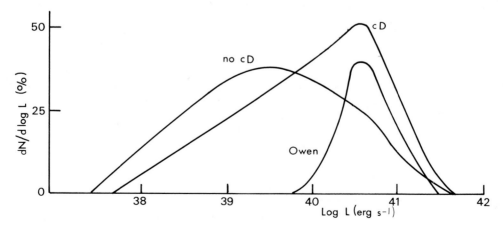

Fig. 1   Whole cluster luminosity function observed by Owen and predicted for clusters with and without a superbright galaxy.

of the LF does not fit that of Owen. It is possible that in this region
there may be systematic errors in Owen's data because the clusters detectable in this range were nearby, and resolved into individual galaxies by his telescope. Solving this problem will require high resolution
measurements of some of these low luminosity clusters; these I expect
will be done soon.

If in fact there is no difference in the cluster/noncluster LF we
are in the unexpected position of having to explain this lack of difference. Either the environments of galaxies have no effect on their
LF's, or the environments around cluster and noncluster radio galaxies
are the same. The second alternative should be taken seriously; present
X-ray measurements do not exclude the possibility that giant ellipticals
outside clusters carry with them a fairly large, say 100 kpc radius,
gas halo of density and temperature similar to that believed to exist
in clusters. The X-ray source around M87 might be evidence for this.

If this is not the case, the lack of change of the LF with density
puts a restraint on any model of radio sources that involve an external
medium. If in a given model the luminosity of a source changes as ρ^α
then the LF, in one of the regions where it follows a power law, should
vary as $\rho^{\alpha(\gamma-1)}$ where γ is the index of the power law. For strong radio
sources $\gamma \approx 2.5$ so we can insist that $|\alpha| \log (\rho_{cl}/\rho_{field}) \lesssim 0.2$, or say
$|\alpha| \lesssim 0.2$ if we allow at least a factor of 10 variation in ρ between
cluster and field.

Whatever the density around isolated galaxies, they do differ from
cluster galaxies in that most of the latter move through the surrounding
gas at speeds of 1000 km s^{-1} or so, and are probably stripped of their
intrinsic interstellar material (Gunn and Gott 1973), while the random
velocity of galaxies in small groups or in the field is about 200 km s^{-1}
(Geller and Peebles 1973). The similarity of the cluster LF to the
field LF seems then to weigh against source models requiring the accretion of interstellar gas by galactic nuclei.

C. SOURCE MORPHOLOGY

Though we find no striking cluster/noncluster differences in terms
of luminosity function, we do better when we look at the morphologies
of individual sources.

The most spectacular sources specifically associated with clusters
are the tailed sources. The two examples in the 3C catalog, 3C129 and
3C82.1B show the defining characteristics of well confined emitting
regions extending to one side only from the parent galaxies. In each
the radio surface brightness peaks a small distance from the galaxy and
decreases irregularly at larger distance. Both also show the double
stranded form, and 3C129 the steepening of spectral index with distance
from the galaxy that are seen in many tails.

We are beginning to get enough examples of tailed sources to discuss their statistical properties. Studies by Owen (1975) and Rudnick
and Owen (1976) of Abell clusters show that the tails make up roughly
10% of cluster radio sources of luminosity $10^{40.5}$ to $10^{41.5}$ erg s^{-1} with
few outside this range. Other surveys (Riley 1975, Colla et al. 1975)
give consistent numbers, with larger uncertainties. The total number
so far detected is about 20. The wide angle tails, intermediate between
normal double and true tails may make up another 10%, (Owen and Rudnick
1976), although it is difficult to distinguish them rigorously from
somewhat asymmetrical doubles. The wide angle tails tend to be more
luminous and associated with more dominant galaxies than the narrow
tails.

Are there tails outside clusters? Maybe. Combining the Bologna
results with a selection from the 3C (3C radio galaxies with z < 0.16
and |b| > 15°, for which cluster identification is fairly complete), I
find one narrow tail source (B1621 + 38) among 22 noncluster sources in
the same luminosity range. The galaxy is however part of the A2197/2199
supercluster, although not in the condensed region of either cluster.
A similar case is 1601 + 17w1 from the W4 survey which is a short tail,
associated with NGC6034, midway between A2147 and A2151 in the Hercules
supercluster. It is not clear whether we should call these cluster
sources or not.

In any case there is a clear cut example of a tailed source in a
sparse group of galaxies, that associated with NGC7385 in the Zwicky
cluster 2247.3 + 1107 (Schilizzi and Ekers 1975) and another in a group
that is possibly part of the A2197/2199 association (B1615 + 35).

There may also be cluster induced distortion of higher luminosity
sources, but the data is as yet too sparse to tell. In Hooley's (1974)
homogeneous subset of the 3C he finds that for monochromatic power
$P_{178} > 10^{25.5}$ W Hz^{-1}, 2 out of 6 well resolved sources in Abell type
clusters are complex, including 1 wide tail, compared to 2/12 of the
sources in poor groups or clusters and 1/4 outside clusters. Hooley
also finds no evidence for a size difference between cluster and non-
cluster doubles but the statistics are equally poor. From this it is
hard to conclude much. Improving on this will require deeper radio and
optical data and a lot of work.

The association of tailed sources with clusters probably arises
from one of the conditions that seems to be necessary to the formation
of tails: that the parent galaxy move through the surrounding medium
at a velocity large compared with the ejection velocity of the radiating
material. Thus cluster galaxies, which move on the average some five
times faster than group or field galaxies, would more likely form tails,
at least if the ejection velocity of the material in these low luminosity sources is around 500 km s^{-1}. The absence of tails among more
luminous sources could arise either because the ejection velocities in
these sources are higher (e.g. the ejection velocity in the very
luminous source Cyg A is estimated at 10,000 km s^{-1} by Hargrave and

Ryle, 1974) or because they are associated with more dominant galaxies whose velocities relative to the surroundings are lower, or both.

The tailed sources also tell us that clusters contain something, most likely a hot gas, which prevents the emitting regions from expanding during periods of 10^8 y or more. All the cluster tails found so far have minimum internal pressure of about 1×10^{-11} dyne cm^{-2}, which agrees quite well with the pressure from a gas of density 3×10^{-27} g cm^{-3} and temperature of a few times 10^7K, such as may be responsible for the X-radiation from clusters (Lea et al. 1973). The one poor cluster tailed source shows a lower minimum pressure, about 10^{-13} dyne cm^{-2} (Schilizzi and Ekers 1975) consistent with the lower maximum temperature that the gas could have and still be bound in a low mass group.

We think then that we can understand the tailed source/cluster connection, although the detailed physics of this formation of the tails is uncertain. The few non-cluster tails raise some special questions. Do those in superclusters mean that there is hot gas at long distances from the cluster centers and that the gas, or the galaxies, have high random velocities? And is the one poor cluster tail just a rare case of a high velocity galaxy in such a group?

D.   SPECTRAL INDICES

The last cluster radio phenomenon I will discuss is that of steep spectrum sources. In Table 2 I list the fraction of 3C radio galaxies with high frequency spectra steeper than 1.0 (H), and low frequency spectra steeper than 1.0 (L) as a function of cluster membership from the sample I mentioned earlier. Here high frequency means from about 1 GHz upward and low frequency from 100 MHz downward.

	Abell Clusters	Poor Groups or Clusters	Outside Clusters
H	2/9	6/22	1/22
L	5/9	6/22	1/22

Four of the "group" sources are counted in both rows because they have steep spectra at all frequencies. In both rows there seems to be a correlation with cluster memberships. The evidence is stronger for the L types and is confirmed by Slingo's (1974) study of fainter and more rigorously selected L sources (with low frequency spectra steeper than 1.2). Here 12 of the 13 identified sources belonged to Abell clusters.

The conventional explanation for steep spectrum sources involves the confinement of relativistic electrons in an emitting region for a period longer than their radiative lifetimes. The L̲ sources, which often have an approximately straight line spectrum down to the lowest frequency observed, about 10 MHz, seem to be cases of a reservoir capable of storing quasi-continuously injected electrons for periods longer than 10^{10} $B^{-3/2}$ y, where B is the source magnetic field in μG. This reservoir need not coincide with the intense "hot spots" seen with high frequency telescopes, but may be a cavity around the intense regions (Scheuer 1974) or may be the entire cluster medium as seems to be the case in the Coma Cluster (Willson, 1970, Jaffe et al. 1976). Some direct observations (Hargrave and Ryle 1974) seem in fact to support the idea that electrons can leak out of the hot spots so quickly that little radiation loss occurs.

The H̲ sources seem to arise from an intermittancy effect, where the source of electrons to an emitting region turns "off" for longer than the radiation lifetimes of the electrons seen at high frequency. This accounts for the features in sources like 3C338 where the source components seem relaxed, without sign of recent motion or injection, and where the source structure in the relaxed regions does not change with frequency (Jaffe and Perola 1974).

The dependence of radiation lifetime on B suggests that the enhanced number of steep spectrum sources in clusters derives from higher magnetic fields in cluster sources, the result of stronger confinement there by the hot gas. This argument is too simple however. In the case of the H̲ sources, for example, it assumes that the old sources do not leak effectively, which may or may not be true. Also this requires that the H̲ sources disappear by some process other than radiation losses. If, to the contrary, radiation losses are responsible for the fading of these sources as well as the spectral changes, shortened lifetimes tend actually to decrease the number found in a given catalog. Without a convincing model of leakage, intermittancy etc. it is difficult to argue in detail along these lines.

A more general, and possibly more important effect is one of luminosity selection. Increasing the magnetic field in remnant emitting regions increases their luminosity and makes them more likely to be included in a catalog. For example, if remnant sources are formed from a cloud of electrons left over from an active source, the luminosity of the remnant will be proportional to $B^{1.7}$ if adiabatic losses are unimportant and to a higher power if they are. As I mentioned in Section B, an increase in source luminosity, due to an external effect, of a factor β increases the LF of such sources by a factor of about $\beta^{1.5}$, so the number of remnant sources found in a given luminosity interval will vary at least as strongly as $B^{2.6}$ due to this effect. This more than compensates the more rapid removal of these sources from view by higher radiation losses.

For the L̲ sources similar luminosity selection applies. Also, as in the H̲ sources, the more rapid aging of the electrons in a higher field would increase the number of L sources observed only if there is a competing, non-radiative electron loss process. Otherwise all halos or cavities would eventually form L type sources, regardless of the value of the magnetic field. If in fact leakage is an important non-radiative loss for the halo sources, then the sheer size of the cluster gas clouds may be instrumental in confining the electrons until they can properly age.

In finishing this section I would like to emphasize that one must be careful in associating a given phenomenon observed in a radio source catalog with a specific physical cause, for example, the occurrence of steep spectrum sources with shorter radiation lifetimes. In addition to the direct physical effect being studied, one must also consider the change in the rate at which sources of a given type enter and leave the luminosity domain being observed.

E.   CONCLUSIONS

There are three conditions affecting radio sources which a priori might account for the differences observed in cluster radio sources: the density of the cluster gas, its pressure, and the fast motion of the cluster galaxies through this gas. The last two seem instrumental in explaining the tail-like morphologies of some cluster sources and the higher occurrence of L and H steep spectrum sources in clusters.

We have on the other hand found no effect directly attributable to a higher density about cluster galaxies. This may be because all bright galaxies are in fact surrounded by a high density halo, or it may be that the effects we looked for, principally changes in the luminosity function, are not sensitive to density differences.

We can conclude by asking what all this does and can tell us about radio sources and clusters that we didn't already know from other lines of study. The most concrete fact is the evidence from tailed sources that there is also hot, dense gas in some smaller groups, and between the clusters in supercluster associations. Also some of the galaxies in these associations seem to show high random velocities.

We can also look for signs of cluster phenomenon in quasars and very distant radio galaxies. Those in catalogs like the 3C are so luminous that comparison with the cluster galaxies we have looked at here on the basis of LF or morphology is impossible. The spectral index measurement can be significant even for very luminous sources. At least one (3C 334, Z = .55) of the 50 or so 3C quasars is an L source, and several of the 4C quasars are. This suggests both that some quasars are surrounded by large hot gas halos and that these quasars have been active long enough, 10^{15} s or so, to build up such a spectrum. I look forward to studying these sources in more detail.

REFERENCES

Colla, G., Fanti, C., Fanti, R., Gioia, I., Lari, C., Lequeux, J.,
    Lucas, R., Ulrich, M.H., 1975, Astron. & Astrophys. 38, 309.
Elsmore, B., Mackay, D.C., 1970, Monthly Notices Roy. Astron. Soc.
    146, 361.
Geller, M.J., Peebles, P.J.E., 1973, Astrophys. J. 184, 329.
Gunn, J.E., Gott, J.R., III, 1972, Astrophys. J. 176, 1.
Hargrave, P.J., Ryle, M., 1974, Monthly Notices Roy. Astron. Soc.
    166, 305.
Hooley, T., 1974, Monthly Notices Roy. Astron. Soc. 166, 259.
Jaffe, W.J., Perola, G.C., 1974, Astron. & Astrophys. 31, 223.
Jaffe, W.J., Perola, G.C., 1975, Astron. & Astrophys. 46, 275.
Jaffe, W.J., Perola, G.C., Valentijn, E.A., 1976, Astron & Astrophys.
    49, 179.
Lea, S.M., Silk, J., Kellogg, E.M., Murray, S., 1973, Astrophys. J.,
    184, L105.
Macdonald, G.H., Kenderdine, S., Neville, A.C., 1968, Monthly Notices
    Roy. Astron. Soc. 138, 259.
Mackay, C.D., 1969, Monthly Notices Roy. Astron. Soc. 145, 31.
Owen, F.N., 1975, Astrophys. J. 195, 593.
Owen, F.N., Rudnick, L., 1976, Astrophys. J. 205, L1.
Riley, J.M., 1975, Monthly Notices Roy. Astron. Soc. 170, 53.
Rudnick, L., Owen, F.N., 1976, Astrophys. J. 203, L107.
Scheuer, P.A.G., 1974, Monthly Notices Roy. Astron. Soc. 166, 513.
Schilizzi, R.T., Ekers, R.D. 1975, Astron. & Astrophys. 40, 221.
Slingo, A., 1974, Monthly Notices Roy. Astron. Soc. 168, 307.
Willson, M.A.G., 1970, Monthly Notices Roy. Astron. Soc. 151, 1.

DISCUSSION

Swarup: Do you have any data on comparison of the angular structure of
the radio sources associated with the cD galaxies located inside and
outside the clusters?

Jaffe: I have no first hand information on this.

Wilson: Vallee, Lari, Parma and I have surveyed Abell clusters A262,
A779 and A1314 at Westerbork at 610 MHz to a flux density level of
~ 8 mJy.   Comparing our results with those of Jaffe and Perola, we find
a radio luminosity function for the E and S_o galaxies in A779 and A1314
that is in agreement with that for field galaxies and galaxies in richer
clusters.   In A262, however, which is a spiral rich cluster, both the
E + S_o and the S + Irr seem over luminous.   The chance probability of
finding as many radio detections as observed is only ~ 1% for both
classes.   Abell 262 has been suggested as a 3U X-ray identification.

Longair: There used to be a strong correlation between finding a head-
tail radio source in a cluster and finding another source in the same
cluster.   What is the story on that now that you have a much larger
sample?

Jaffe: I think it has disappeared.

OBSERVATIONS OF THE MICROWAVE BACKGROUND RADIATION IN THE DIRECTION OF CLUSTERS OF GALAXIES

S. F. Gull

·Northover and I have obtained evidence for the existence of small diminutions in the cosmic microwave background radiation in the directions of several rich clusters of galaxies known to be X-ray sources. We have now accumulated 670 hours of observations on 7 Abell clusters and some control areas of blank sky, using the 25 m telescope at the S.R.C. Chilbolton Observatory. The results are not consistent with the null hypothesis and can best be explained as the Compton scattering of microwave background photons by the hot plasma responsible for the X-radiation.

Cluster	Temperature difference	Cluster	Temperature difference
A 376	− 0.13 mK ± 0.66 mK	A 2218	− 1.94 mK ± 0.54 mK
A 478	+ 0.33 mK ± 0.52 mK	A 2319	− 0.13 mK ± 0.41 mK
A 576	− 0.71 mK ± 0.57 mK	A 2666	− 0.27 mK ± 0.35 mK
Coma	− 1.51 mK ± 0.40 mK	Blank sky	− 0.01 mK ± 0.32 mK

Partridge: For those who like coincidences, how about a comment by Partridge following one by Gull?

During 1975-76, George Lake and I at Haverford have made observations of 8 clusters using a procedure closely resembling that of Gull and Northover. We worked with the 36-foot N.R.A.O. telescope at λ = 9 mm; the beam width was 4', and the separation between the main and reference beams was 19'. The short wavelength was chosen to reduce as far as possible contamination of the signal by weak sources within the clusters. Our preliminary results are as follows:

Cluster	Temperature difference	Cluster	Temperature difference
A 376	+ 0.54 mK ± 0.80 mK	A 1656	
A 401	− 0.39 mK ± 0.61 mK	(Coma)	+ 0.60 mK ± 0.81 mK
A 426		A 2079	− 0.35 mK ± 1.24 mK
(Perseus)	+ 1.91 mK ± 0.82 mK	A 2319	+ 0.27 mK ± 0.77 mK
A 576	− 0.34 mK ± 0.51 mK	A 2666	+ 1.86 mK ± 0.86 mK

We ignore A 426 (Perseus), a known radio source. With this exclusion our data are consistent with the null hypothesis - we have no evidence for the inverse compton cooling of the microwave background. In particular, we do not see the 'dip' in Coma. I should stress, however, that our data analysis is not complete, and that we plan further observations of clusters 1656, 2079 and 2666. In addition, both groups hope to be able to make scans across one or more clusters to attempt to map the radial distribution of hot intergalactic gas.

Bahcall: Do you point your reference beam well outside the limits of the X-ray extent?

Partridge: Yes, even for the cluster of largest angular extent in our group, Coma.

Webster: Are your quoted temperatures antenna temperatures or surface brightness temperatures corrected for sidelobes?

Partridge: They are sky temperatures, properly corrected for side-lobes, antenna inefficiency, etc., and are thus directly comparable to the results of Gull and Northover.

Shaffer: Are they sky temperatures corrected for the atmosphere, especially at 9 mm?

Partridge: They are sky temperatures, corrected for atmospheric extinction.

A METHOD OF DETERMINING Ω FROM A REDSHIFT SURVEY

E.L. Turner and W.L.W. Sargent

The distribution of galaxies in space is approximated by their distribution in a "redshift space" in which their radial coordinate is cz/H_0. Deviations from a smooth and uniform Hubble expansion, due either to perturbations arising from density fluctuations in the distributions of galaxies or to virial motions in bound groups and clusters, cause characteristic distortions in "redshift space". A method of detecting and measuring these distortions (anisotropies) from the relative redshifts and positions on the sky of pairs of galaxies is proposed. An approximate (covariance function) and a more powerful general method of relating these characteristic distortions to their associated density enhancements (and hence Ω) are presented. The limited data presently available are used to illustrate the approximate method, and a very tentative result of $\Omega \approx 0.07$ is obtained. Redshifts accurate to ~ 20 km s^{-1} for a magnitude limited sample of ~ 4000 galaxies (from which a volume limited sample of ~ 1000 galaxies may be extracted) would be required for a strong test of Ω. These requirements suggest, among other possibilities, a 21-cm redshift survey.

Thuan: It would be nice to extend this treatment to a magnitude limited sample rather than a volume limited sample in order not to waste data.

Turner: Extracting a volume limited sample from an observed magnitude limited one requires throwing out $\geqslant 3/4$ of the data. The analysis could be generalised to avoid this but it would require the assumption of a universal luminosity function with a known form. In the ideal case of plentiful, high quality data, the assumptions can be avoided.

COSMOLOGICAL IMPLICATIONS OF THE ULTRAVIOLET SPECTRUM OF GALAXIES

A. D. Code

Based on OAO-2 ultraviolet photometry of bright galaxies in the spectral region from 1550 - 4250 Angstroms, Code and Welch have found significant variations in the ultraviolet flux for early type galaxies with similar energy distributions longward of 4000 Angstroms.

The energy distribution for most giant ellipticals fall below satellite detection threshold shortward of 2400 Angstroms as they should on the basis of models.   However, the energy distribution of the giant elliptical NGC 4486 (M87), which has been measured at 1910 Angstroms is considerably brighter than model predictions.   M87 is the only bright radio galaxy in the sample.   The jet cannot make a substantial contribution to the total integrated flux at 1910 Angstroms and the result suggests that this excess is due to early-type stars.   Sandage (1972, Ap. J. **178**, 25) notes that giant ellipticals that are radio sources are often bluer than those that are radio quiet.   Van den Bergh (1975, Ann. Rev. of Astron. & Astrop. **13**, 217) presents evidence for a burst of star formation accompanying violent events in the nuclei of supergiant elliptical galaxies.

The K-corrections for an energy distribution similar to M87 differ significantly from those for other giant ellipticals and SO galaxies for red shifts in excess of $z = 0.4$.   Finally it is noted that for $z > 0.5$, in the absence of evolution, the late-type spiral galaxies may become photographically brighter than normal giant ellipticals in distant clusters.   The expected increase in the B magnitude relative to the average giant elliptical at $z = 1.0$ would be $-1\overset{m}{.}50$ for M87, $-1\overset{m}{.}75$ for the average Sb galaxy and $-2\overset{m}{.}0$ for the average Sc galaxy.

Tinsley: The evidence for OB stars in M87 is especially interesting since this is one of the relatively few elliptical galaxies in which a supernova has occurred.   Traditionally, the occurrence of supernovae in elliptical galaxies has been taken to mean that very old stars (in particular, white dwarfs) can explode as supernovae, because massive stars are thought to have died out long ago in normal ellipticals. Code's results add further support to the alternative idea that even Type I supernovae (the only type identified in E galaxies) arise from massive stars.

Arp: I am worried about the jet in M87.   At 3700 A the photographic image is dominated by the blue jet. Have you removed specifically the effect of the blue jet in your ultraviolet measures?

Code: No I don't think the jet is an important contribution.   The photometry field of view was 10' arc which includes essentially all the light from the galaxy.   If one extrapolates the contribution of the jet to the total light from the blue to 2000 Angstroms by assuming a flat spectrum the jet contributes only a few per cent.   If the jet were to be the principal source of the ultraviolet excess it would have to have a very steep and uncharacteristic spectrum.

Rowan-Robinson: Can you say anything about these young stars?

Code: The type of stars required to produce the flux shortward of 2000 Angstroms in M87 are O or early B stars; certainly earlier than B5.

Tinsley: Evolution of the stellar populations in galaxies is likely to affect their colors significantly in the redshift range ($0 \leqslant z \leqslant 1$) over which Code has predicted K-corrections and broad-band colors. Calculations of color (e.g. B-V) vs. z are presented here, for a typical elliptical and late-type spiral; further details will be published elsewhere (Ap. J. 1977, in press). In the cases without evolution, the usual effects of K-corrections on colors are reproduced (cf Pence, 1976, Astrophys. J., 203, 39): the steep UV spectra of typical elliptical galaxies causes the colors to increase steeply with z, while for late-type galaxies the effects of redshift are much smaller. Evolutionary models lead to the following predictions:

(i)   the colors of later-type galaxies are not strongly affected by evolution, because a similar distribution of early-type stars always dominates their light at the relevant wavelengths.

(ii) colors of elliptical galaxies at z > 0.5 are significantly affected by the fact that the main-sequence turn-off of a predominantly coeval population is bluer at earlier times. For example, instead of having an observed B-V ~ 2.0 (the approximate value for no evolution) at z = 0.7, a normal elliptical galaxy may have B-V ~ 1.5; exact color-redshift observations would help to evaluate a number of uncertain parameters in the elliptical galaxy models.

VI

MICROWAVE BACKGROUND RADIATION

MICROWAVE BACKGROUND SPECTRUM - SURVEY OF RECENT RESULTS

E. I. Robson and P. E. Clegg
Physics Department, Queen Mary College, London.

Introduction

For those enamoured of the primaeval fireball, the relict
radiation has proved a tantalising mistress. Since the famous
discovery in 1965, by Penzias and Wilson, of an excess antenna
temperature consistent with cosmological expectations, many observers
have succeeded in measuring a flux consistent with a thermodynamic
temperature of ~3K. Until recently, however, no *direct* spectral
measurements had been made at wavelengths shorter than the Planckian
peak corresponding to radiation at this temperature. At such wavelengths atmospheric emission and absorption are overwhelming from even
the highest mountain site and observations must be made from at least
balloon platforms. The pioneering broadband rocket and balloon
measurements covering this wavelength region produced consternation
when excessively high fluxes were reported; successive flights
gradually eliminated the excess, emphasising the practical difficulties
of such observations. A review of this phase of the pursuit is given
by Blair[1]. Nevertheless, it is upon direct submillimetre measurement
of the spectral density I_ν that confidence in the interpretation of the
longer wavelength results must reside. The outcome of the first such
measurement, by Queen Mary College in 1974, seemed completely to
justify such confidence. Unfortunately, the subsequent observations by
Berkeley, *although leading to the same conclusion* about the value of
the thermodynamic temperature, were so discrepant in detail from those
of QMC as once again to raise doubts. We have since been eagerly
awaiting the results of observations by independent groups but these
have been frustrated by instrumental failures. We attempt here to
assess the present situation. We conclude that although present
measurements indicate a flux not inconsistent with a Planckian
spectrum corresponding to a temperature of ~3K, they do not demand such
an interpretation. Moreover, because we may not actually *expect* a
Planckian curve from a fireball, very much more detailed information
is needed to obtain a view of the early thermal history of the universe.

D. L. Jauncey (ed.), Radio Astronomy and Cosmology, 319-325. All Rights Reserved.
Copyright © 1977 by the IAU.

Review of Observations

A total of 16 direct monochromatic (radiometer) measurements has been made[2-16] from the ground, mountain peaks and an aircraft platform. (A recent, quasi-monochromatic measurement at ~1.2mm[17] sets an upper limit and is discussed below). The weighted mean of these results, if interpreted in terms of a Planck spectrum, gives a thermodynamic temperature of 2.73 ± 0.08K. The value of χ^2 for these is 9, compared with an expected value of 15 ± 5. Although it is well within the 95% confidence level, the low value may reflect, as has been remarked by Peebles[18], a justifiable caution on the part of observers.

Three groups, Cornell, MIT and Los Alamos, have flown rockets or balloons to make broadband measurements of the background flux; the detailed history of these is discussed by Blair[1]. In spite of the very high submillimetre fluxes initially reported, apparently supported by observation of submillimetre line emission[19] which has certainly not been substantiated[20],[21],[22] the latest broadband measurements are mutually consistent and consistent with a temperature of 2.7K. However, as Blair points out, the Cornell and Los Alamos results are equally consistent with a temperature of zero and only MIT provides positive evidence of a 2.8 ± 0.2K equivalent flux.

Finally, a series of temperatures and upper limits have been deduced from observation of interstellar molecular absorption lines (see review by Thaddeus[23]), the most precise of which gives a temperature of 2.83 ± 0.15 K at 2.62mm. Such observations are, of course, measures of excitation and only by interpreting this excitation as being directly due to background continuum radiation does one arrive at a temperature. Although arguments against other excitation mechanisms are persuasive, the submillimetre upper limits do not impose very severe restraints upon the background. However it should be stressed that the early interstellar molecular line observations, giving upper limits to a cosmic background flux at wavelengths of 1.32mm, 0.56mm and 0.36mm were historically most important. During the same period, broadband rocket and balloon experiments were reporting extremely high fluxes in just this wavelength region. For the molecular line measurements and the broadband results to be at all consistent, one possibility was for the "excess" flux to be extremely local in origin; another featured the presence of a spectral line, leading to the 11.7 cm^{-1} emission feature saga mentioned above. The 1.32mm upper limit has now been replaced by the assignment of a definite temperature of 2.9 $^{+0.4}_{-0.5}$ K [29].

The shorter wavelength monochromatic, and the latest broadband and interstellar excitation measurements are summarised in figure 1, which emphasises the need for the subsequent short-wavelength spectral measurements.

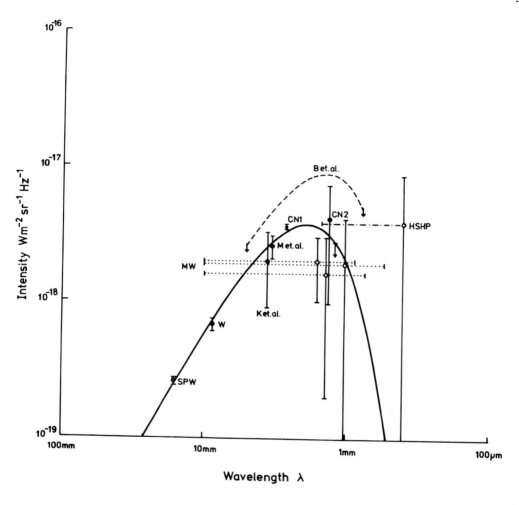

Figure 1 Short-wavelength monochromatic, broadband and interstellar
excitation observations of the background flux. References
as follows: SPW (8); W (11); K et al. (13); M et al. (15);
MW (23); B et al. (25); HSHP (26); CN1 (27); CN2 (28).

Balloon-borne Spectrometric Observations

In spite of having been made over two years ago, the balloon-borne
spectrophotometric observations of QMC[30] and Berkeley[31] remain,
unfortunately, the only such measurements. A recent re-flight by
Berkeley had instrumental difficulties and added little new data. The
long-awaited announcement of results by an independent group has not
appeared, at least one experiment having suffered mechanical failure

under the severe conditions experienced in the stratosphere.

At first sight, the temperatures deduced by QMC and Berkeley, 2.7 $^{+0.2}_{-0.3}$ K and 2.99 $^{+0.07}_{-0.14}$ respectively, are in remarkable and encouraging agreement. Detailed comparison of the observed spectra (cf. discussion in Rowan-Robinson[32]) quickly cools enthusiasm. In spite of similarities of instrumentation and flight conditions, the observed spectra differ considerably in the contribution of the residual atmosphere, this appearing to be much more significant in Berkeley's result. While QMC believed that they could distinguish directly the turnover of a 2.7 K Planck spectrum at ~1mm wavelength, Berkeley obtained their temperature by fitting an emission model to their spectrum. Their model atmosphere is isothermal, has an exponential pressure profile and constant mixing ratio and uses tabulated line parameters, some of which (Oxygen lines:- Gebbie, private communication) are somewhat uncertain. As free parameters, they used the column densities of oxygen, ozone and water vapour and the temperature of a Planck spectrum.

Assessment of Present Position

All the present direct observations of the background flux are summarised in figure 2, where the logarithm of antenna temperature, T_A, is plotted against the logarithm of the wavelength λ. Antenna temperature, defined by

$$I_\nu \equiv \frac{2\nu^2}{c^2} kT_A \ ,$$

where I_ν is the specific intensity at frequency ν, has been chosen as a direct measure of I_ν, avoiding the prejudice of deducing a thermodynamic temperature. Filled circles are results of references 2-16. The bounded region above 2.5mm represents the 2σ limits of the Berkeley group *when they fit a Planck spectrum*; the error bars on the open circles (QMC) are 1σ. The solid line corresponds to a Planck spectrum of 2.82 ± 0.06 K, the weighted mean of all results shown. (χ^2 = 12.6, compared with an expected 16 ± 6). You will see that the curve does not fit entirely within either the QMC 1σ or Berkeley 2σ limits; *neither does a 2.99 K* curve, corresponding to Berkeley's best estimate of the temperature. In fact, although one can claim that present results indicate a dramatic drop of antenna temperature at short wavelengths, they cannot be said to be strong evidence for a *Planck* spectrum. Also, the upper limit of 2.7 K at 1.2mm imposed by the recent Italian mountain based observation[17] is rather embarrassingly low. Although this measurement is extremely difficult and should therefore be treated with some caution, it well indicates the uncertainty in this spectral region.

Finally, the dotted line shows a fairly extreme Compton-distorted[33] spectrum (adapted from Chan and Jones[34]). It is clear that the precision of present measurements is completely inadequate to determine such distortions with any confidence; we believe that this

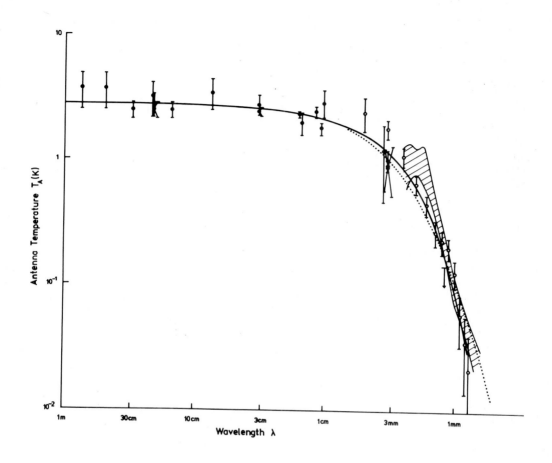

Figure 2   Monochromatic and spectral measurements of the background
antenna temperature. References: Filled circles (2-16);
open circles (QMC, 30); hatched region, Berkeley 2σ limits
(31); upper limit (17); Solid curve, 2.82 K Planck
spectrum; dotted curve, Compton-distorted spectrum (see text).

will remain true for observations at balloon altitudes.

Conclusions

The considerable effort devoted to observing the microwave to
submillimetre background flux has not, in my opinion, unambiguously
demonstrated the Planckian nature of its spectrum. Indeed, because of
the discrepancy between the two submillimetre spectral observations,
even the point of turnover of the spectrum is open to question,
although this matter should be settled by future balloon observations.

Even further are we from investigating thermal events in the early universe by measuring departures from a Planck curve. We are convinced that only by making observations from space platforms, clear of the contamination by atmospheric radiation, may we hope to obtain this information.

REFERENCES

Blair, A.G. in "IAU Symposium No. 63", ed. M.S. Longair, D. Reidel, 1974, p. 143.

Howell, T.F. and Shakeshaft, J.R., Nature 216, 753 (1967)

Penzias, A.A. and Wilson, R.W., Ap.J. 72, 315 (1967)

Howell, T.F. and Shakeshaft, J.R., Nature 210, 1318 (1966)

Pelyushenko, S.A. and Stankevich, K.S., Soviet Astron. 13, 223 (1969)

Penzias, A.A. and Wilson, R.W., Ap.J. 142, 419 (1965)

Roll, P.G. and Wilkinson, D.T., Phys.Rev.Lett. 16, 405 (1966)

Stokes, R.A., Partridge, R.B. and Wilkinson, D.T., Phys.Rev.Lett. 19, 1199 (1967)

Welch, W.J., Keachie, S., Thornton, D.D. and Wrixon, G., Phys.Rev.Lett. 18, 1068 (1967)

Ewing, M.S., Burke, B.F. and Staelin, D.H., Phys.Rev.Lett. 19, 1251 (1967)

Wilkinson, D.T., Phys.Rev.Lett. 19, 1195 (1967)

Puzanov, V.I., Salomonovich, A.E. and Stankevich, K.S., Soviet Astron. 11, 905 (1968)

Kislyakov, A.G., Chernyshev, V.I., Lebskii, Yu.V., Mal'tsev, V.A. and Serov, N.V., Soviet Astron. 15, 29 (1971)

Boynton, P.E., Stokes, R.A. and Wilkinson, D.T., Phys.Rev.Lett. 21, 462 (1968)

Millea, M.F., McColl, M., Pedersen, R.J. and Vernon, F.L., Phys.Rev.Lett. 26, 919 (1971)

Boynton, P.E. and Stokes, R.A., Nature 247, 528 (1974)

Dall'Oglio, G., Fonti, S., Melchiorri, B., Melchiorri, F., Natale, V., Lombardi, P., Trivero, P. and Sivertsen, S., Phys.Rev.D. 13, 1187 (1976)

Peebles, P.J.E., "Physical Cosmology", Princeton, 1971

Beery, J.G., Martin, T.Z., Nolt, I.G. and Wood, C.W., Nature 230, 36 (1971)

Nolt, I.G., Radostitz, J.V. and Donelly, R.J., Nature 236, 444 (1972)

Beckman, J.E., Ade, P.A.R., Huizinga, J.S., Robson, E.I., Vickers, D.G. and Harries, J.E., Nature 237, 154 (1972)

Mather, J.C., Werner, M.W. and Richards, P.L., Ap.J. 170, L59 (1971)

Thaddeus, P., Ann.Rev.Astron.Astrophys. 10, 305 (1972)

Muehlner, D.J. and Weiss, R., Phys.Rev.Lett. 30, 757 (1973)

Blair, A.G., Beery, J.G., Edeskuty, F., Hiebert, R.D., Shipley, J.P. and Williamson, K.D., Phys.Rev.Lett. 27, 1154 (1971)

Houck, J.R., Soifer, B.T. and Harwit, M., Ap.J. 178, L29 (1972)

Bortolot, V.J., Clauser, J.F. and Thaddeus, P., Phys.Rev.Lett. 22, 307 (1969)

Heygi, D.J., Traub, W.A. and Carleton, N.P., Phys.Rev.Lett. 28, 1541 (1972)

Heygi, D.J., Traub, W.A. and Carleton, N.P. Ap. J. 190, 543 (1974)

Robson, E.I., Vickers, D.G., Huizinga, J.S., Beckman, J.E. and
    Clegg, P.E., Nature 251, 591 (1974)
Woody, D.P., Mather, J.C., Nishioka, N.S. and Richards, P.L.
    Phys. Rev. Lett. 34 1036 (1975)
In "Far Infrared Astronomy", ed. M. Rowan-Robinson, Pergamon, 1976
    p.143 ff.
Zel'dovich, Ya.B. and Sunyaev, R.A., Ap. and Space Sci. 4 301 (1969)
Chan, K.L. and Jones, B.J.T., Ap. J. 198, 245 (1975)

DISCUSSION

Partridge: Would you comment further on the apparent differences between your results and those of the Berkeley group? In particular, could you say why your atmospheric spectrum differs from theirs?

Robson: The crucial points are that in the QMC spectra the atmosphere does not dominate the cosmic background spectrum, whereas in the Berkeley results the atmosphere appears as a continuum spectrum which must first be subtracted before the cosmic background spectrum can be measured.

COSMIC BACKGROUND EXPERIMENT FROM SPACE

S. Gulkis

The United States space agency, NASA, has recently appointed a team of six scientists to study high sensitivity measurements of the diffuse infrared and microwave background radiation from a dedicated satellite in space. The study will last for one year and produce a preliminary plan for a spacecraft experiment. Presently, three groups of instruments are being considered which will measure the diffuse infrared and micro-wave background over the broad range from 5 microns to 2 cm wavelengths. The three groups of instruments consist of a cryogenic far infrared spectrometer which covers the frequency range from 3 to 30^{-1}, to measure the spectrum of the 3 K background; a group of differential radiometers using microwave and infrared techniques to cover the frequency range 0.5 to 20 cm^{-1} (15 to 600 GHz), to measure anisotropy of the 3 K background; and finally a large beam absolute radiometer which covers the range from 100 to 2000 cm^{-1}, to measure zodiacal dust emission, galactic dust, and extragalactic diffuse light. The spatial resolution of the anisotropy measurements will be $\sim 10^{\circ}$ and be able to measure temperature differences of approximately 0.0001 K on that scale. The spectrum experiment is planned to be able to detect a spectral intensity which departs from a best-fitting blackbody by one part in 10^{4} with a resolution of 1 cm^{-1}.

SMALL SCALE FLUCTUATIONS IN THE MICROWAVE BACKGROUND RADIATION ASSOCIATED WITH THE FORMATION OF GALAXIES

R.A. Sunyaev
Space Research Institute, USSR Academy of Sciences,
Moscow, USSR.

According to current ideas, massive extragalactic systems such as galaxies and clusters of galaxies formed as a result of the growth of small fluctuations in density and velocity which were present in the early stages of expansion of the Universe under the influence of gravitational instability. According to the hot model of the Universe at the epoch corresponding to a redshift $z \approx 1500$, recombination of primaeval hydrogen took place and as a result the optical depth of the Universe to Thomson scattering decreased abruptly from about 1000 to 1 - the Universe became transparent. Therefore the observed angular distribution of the microwave background radiation (MWBR) contains information about inhomogeneities in its spatial distribution at a redshift $z \sim 1000$. Silk (1968) was the first to note that this "photograph" of the Universe at the epoch of recombination must be enscribed with fluctuations associated with perturbations in the space density and velocity of motion of matter which will later lead to the formation of galaxies and clusters of galaxies.

Detailed investigations of this question show that the real picture is much more complicated and, in particular, that the fact that recombination is not instantaneous (Zeldovich et al 1968, Peebles 1968) changes significantly the result; the recombination of the primaeval plasma takes place over a period of time and so a real "photograph" is not obtained. Furthermore, the possibility of secondary non-equilibrium heating of the matter in the Universe at redshifts $z \ll 100$ leading to re-ionisation of the matter could lead to a large optical depth to Thomson scattering and to damping of primaeval temperature fluctuations. Nevertheless, observations of small-scale fluctuations in the MWBR seem to be one of the few methods of determining the amplitude and nature of perturbations of velocity and density at the epoch of recombination. Knowledge of the amplitude of the fluctuations and the laws according to which they evolve with time enable the epoch of formation of galaxies and clusters to be determined - and this is one of the most important questions of contemporary cosmology.

D. L. Jauncey (ed.), Radio Astronomy and Cosmology, 327-334. *All Rights Reserved.*
Copyright © 1977 by the IAU.

1. PRIMAEVAL FLUCTUATIONS IN THE MICROWAVE BACKGROUND RADIATION

In the following discussion we will introduce the mass $M = \frac{4\pi}{3} \rho \lambda^3 = \frac{4\pi}{3} \Omega \rho_{cr} \lambda^3 z^3$ corresponding to the mass of a fluctuation of scale λ ; the angular diameter of this fluctuation Θ for $z \gg 1$ is

$$\Theta = \Omega z \frac{4\lambda H_o}{c} \approx 6' \Omega^{2/3} h^{7/4} \left(M / 10^{15} M_o \right)^{1/3}$$

Here and below $\Omega = \rho / \rho_{cr} = 8\pi G \rho / 3 H_o^2$; $h = H_o / (50 \text{ km s}^{-1} M_{pc}^{-1})$ where H_o is the present value of the Hubble constant.

1.1 Adiabatic perturbations

Before the epoch of recombination these fluctuations were stationarysound waves in which the fluctuation in density is uniquely related to the perturbation in the temperature of the background radiation $\Delta T / T = \frac{1}{3} \Delta \rho / \rho$. If recombination were instantaneous, we would observe such fluctuations of brightness temperature $\Delta T / T(\Theta)$. The fact that recombination is not instantaneous leads to an abrupt decrease in ΔT on small scales as was shown by Sunyaev and Zeldovich. The point is that during recombination small scale perturbations have optical depth less than 1 to Thomson scattering at a much earlier stage than the Universe as a whole because $\lambda \ll ct$. As a result the fluctuations in the MWBR are much reduced. By the epoch of recombination this effect gives rise to the damping of adiabatic temperature fluctuations on all scales $M < 10^{15} M_\odot$. The other reason why relatively small temperature fluctuations are expected on small scales is damping up to and during the epoch of recombination of density and velocity perturbations with masses $10^{12} - 10^{14} M_\odot$ because of the effects of radiative thermal conductivity and viscosity (Silk 1968, Chibisov 1972, Weinberg 1975).

The principal effect which leads to temperature fluctuations on scales $10^{12-15} M_\odot$ is scattering of photons by moving electrons (Sunyaev and Zeldovich 1970). According to the equation of continuity $\partial(\Delta\rho/\rho)/\partial t = - \text{div } \underline{u}$ perturbations of density are uniquely related to velocity perturbation which in this case are a function of spatial coordinates. As a result of Doppler scatterings,

$$\frac{\Delta T}{T} = \int \frac{u(z)}{c} \cos\Theta \, e^{-\tau(z)} d\tau(z)$$

the factor $\cos\Theta$ taking account of the fact that only the radial component of velocity gives contributions to the fluctuations. Detailed computations have been made by Sunyaev and Zeldovich (1970) and by Peebles and Yu (1970). The results of the calculations of Doroshkevich et al are as follows. The spectrum of initial fluctuations is taken to be of power-law form

$$\overline{\left(\frac{\Delta\rho}{\rho} \right)_k^2} \propto k^n$$

Damping and phasing of the perturbations leads to spectrum at the moment of recombination of the form

$$\overline{\left(\frac{\Delta\rho}{\rho}\right)^2_k} = A\,(kR_c)^n\,e^{-kR_c}\,\frac{\sin^2 kR_J}{k^2 R_J^2}\,R_c^3$$

where R_J is the Jeans' wavelength and $R_c = \frac{6\times10^{24}}{\Omega h^2(1+z)}\,\left(1 + (40\,\Omega h^2)^{-3/4}\right)^{1/2}\,cm$ is the dissipation length at the epoch of recombination. Combining this result with observation provides a method for estimating the epoch of formation of clusters of galaxies. Estimates of the magnitude of the expected temperature fluctuations in the MWBR suggest that $\Delta T/T \approx 10^{-4}$ on angular scales $\approx 1 - 10$ arc min. If the spectrum of fluctuations is not of power law form but is rather narrow, the predicted amplitudes of the fluctuations for the same $\Delta\rho/\rho$ are less than 10^{-4}.

1.2 Entropy fluctuations

In the case of entropy fluctuations the adiabatic relation $\Delta T/T = \frac{1}{3}\Delta\rho/\rho$ does not apply and temperature fluctuations arise because of Doppler scatterings by inhomogeneities in the velocity distribution (Sunyaev and Zeldovich 1970). Velocity perturbations generate density inhomogeneities after perturbations on a given scale have become transparent on a time scale of the order of the hydrodynamical timescale. The temperature fluctuations $\Delta T/T$ should be of the same order of magnitude as in the case of adiabatic fluctuations if the amplitude and initial spectra of the perturbations are the same. They may be much greater on small scales because there is no damping of density or velocity perturbations before or during recombination.

1.3 Whirl perturbations

In this model the turbulent "whirl" velocities are much greater than the velocities predicted according to the theory of adiabatic and entropy fluctuations. Again the most important effect in producing temperature fluctuations is Doppler scattering. The computations of Chibisov and Ozernoi (1969) and Anile et al (1976) show that the fluctuations expected according to this theory exceed by a considerable factor the existing experimental limits of Conklin and Bracewell (1967), Parijskij (1973), Carpenter et al (1973) and Stankevich (1974). Agreement with the observations can, however, be obtained if it is supposed that secondary heating of the Universe took place which leads to damping of the fluctuations. However, Anile et al (1976) note that the large velocities of matter during the period of secondary heating also lead to significant fluctuations.

2. OTHER SOURCES OF FLUCTUATIONS

2.1 Radio sources

Longair and Sunyaev (1969) noted that radio sources radiating at short wavelengths (1 - 10 cm) lead to fluctuations in the MWBR. In addition there should also be present a large number of compact radio sources associated with the nuclei of galaxies and quasars. A simple extrapolation of the Cambridge counts of radio sources at 408 MHz to

a wavelength of 4 cm with a radio source spectrum $F_\nu \propto \nu^{-0.75}$, ie $\alpha = 0.75$
shows that the radio sources contributing to the source counts guarantee
fluctuations at a level $\approx$ 1 to 3 x 10^{-5} on angular scales 1 arc min to
1 degree. The most important contribution to the fluctuations is due
to sources which have surface density roughly one per beam-width of
the radio telescope.

2.2 Clusters of galaxies

Sunyaev and Zeldovich (1970, 1972) showed that in the direction
of rich clusters of galaxies containing hot intergalactic gas, there
should be a decrease in the brightness temperature of the MWBR

$$\frac{\Delta T}{T} = - 4 \frac{kT_e}{m_e c^2} \sigma_T N_e R$$

Such an effect was discovered in the Coma cluster by Parijskij (1973)
and recently confirmed by Gull and Northover (1976); $\Delta T/T$ is roughly
3 x 10^{-4}. If it is supposed that the effect does not depend on redshift
and that the number of rich clusters (with richness classes greater or
equal to that of Coma) along the line of sight to a redshift $z \approx 4 - 5$
is roughly 1, it is clear that clusters of galaxies can give a significant contribution to fluctuations of the MWBR on angular scales of the
order of a few min arc.

2.3 Perturbations originating during secondary heating of the inter-
galactic gas

The formation of observed objects in the Universe is apparently
accompanied by the ionisation and reheating of the intergalactic gas.
Inhomogeneities in the heating must lead to fluctuations in the MWBR
for the same reasons described in 2.2. As is well known, under the
influence of gravitational instability, velocities corresponding to
slowly growing perturbations of density also increase with time. Therefore scattering by moving ionised matter must also lead to temperature
fluctuations of the MWBR.

3. REHEATING OF THE INTERGALACTIC GAS

The optical depth of the Universe to Thomson scattering is equal to

$$\tau_T = \int_0^{z_{max}} \sigma_T N_e(z) c \, dt = \sigma_T \Omega N_{cr} c H_0^{-1} \int_0^{z_{max}} \frac{(1+z)}{(1+\Omega z)^{1/2}} x(z) \, dz$$

where $x(z) = N_e/(N_e + N_H)$ is the degree of ionisation of hydrogen. If for
small $z < z_{max}$ the hydrogen is completely ionised and at $z > z_{max}$ it is
neutral, $\tau_T = 0.025 \, \Omega^{1/2} h \, z_{max}^{3/2}$. In order to damp out fluctuations effectively, it is necessary that $\tau_T \sim 5$, corresponding to weakening of
the fluctuations by a factor of 150. Evidently, effective damping of
temperature fluctuations can only take place if reheating of the inter-
galactic gas occurs sufficiently early at $z > 40 \, \Omega^{-1/3} \left(\frac{\tau}{5}\right)^{2/3} h^{-2/3}$
At redshifts $z > 10$, the most important cooling process for heated
intergalactic gas is Compton scattering of photons by hot electrons,
leading to an increase in the mean square frequency of photons and cooling of the electrons.

3.1 Shock waves

One possible method of secondary heating is dissipation of turbulent motions. Subsonic turbulent velocities decrease during the expansion as $(1 + z)$. Turbulence which was subsonic before recombination becomes supersonic after recombination because of the abrupt drop in the velocity of sound i.e. shock waves are formed immediately after recombination. The time scale for Compton cooling of the plasma at a redshift $z \sim z_{rec} = 1000$ is much shorter than the cosmological time-scale.

$$\frac{t_c}{t_{cosm}} = \frac{3}{4} \frac{m_e c}{\sigma_T \varepsilon_r} H(z) \approx 20 \frac{\Omega^{1/2}}{z^{2.5}}$$

Consequently, hot ionised gas is only present in the shock front and behind it the gas cools rapidly and recombines. It is difficult to conceive that in this situation it is possible to obtain $\tau_T > 1$. Besides, the gas in the shock front has a large velocity and temperature and this must lead to strong fluctuations of the intensity of the MWBR. We also note that the rapid cooling must lead to the early appearance of dense objects of large mass. Apparently, early heating by shock waves does not seem to be an effective mechanism for damping primordial fluctuations.

3.2 Photoionisation

The other, more probable, possibility is that the secondary heating is due to the presence of strong sources of ulta-violet radiation.
The optical depth of the Universe for ionising photons is

$$\tau_{ph} = \sigma_o' N_H c t_{cosm} \approx 10^5 \Omega^{1/2} h \, z^{3/2} \left[1 - x(z)\right]$$

and is very high. Here and below $\sigma_{ph} = \sigma_o' \left(\frac{\nu_o}{\nu}\right)^3 = 6.3 \times 10^{-18} \left(\nu_o / \nu\right)^3 \, cm^2; \, h\nu_o = 13.6 eV$ is the ionisation potential of hydrogen. Therefore we can investigate the local ionisation balance

$$4\pi N_H \int_{\nu_c}^{\infty} \frac{J_\nu}{h\nu} \sigma_{ph}^r(\nu) \, d\nu = (\alpha_t - \alpha_1) N_e^2 \qquad (1)$$

where $(\alpha_t - \alpha_1) = 2.5 \times 10^{-13} T_4^{-1/2} \, cm^3 s^{-1}$ is the recombination coefficient to all levels except the first and $T_4 = T/10^4 \, K$.

The emissivitiy of sources per unit volume may be determined as follows

$$j_\nu = \frac{1}{4\pi} N(z) L_\nu \quad erg \, cm^{-3} s^{-1} Hz^{-1} sr^{-1}$$

where $N(z)$ is the mean space density of sources; $L_\nu = L_o \left(\frac{\nu}{\nu_o}\right)^{-\alpha}$ is the mean spectral luminosity close to the Lyman limit. The total luminosity of the source is $L = \int L_\nu \, d\nu$. Absorption of photons takes place to a distance $D(\nu) \approx (\sigma_{ph} N_H)^{-1}$ from the object, corresponding to an optical depth $\tau_{ph} \approx 1$. The intensity of radiation at any point closer than $D(\nu)$ is

$$J_\nu = j_\nu D(\nu) = \frac{1}{4\pi} N(z) L_\nu \frac{1}{\sigma_{ph} N_H}$$

The temperature of a hydrogen plasma under the combined effects of photoionisation and cooling due to recombination and the excitation of lines by electron collisions cannot greatly exceed 10^4 K. At large redshifts Compton cooling of the plasma plays an important role but at redshifts $z < 30$ and for a power-law spectrum of ultraviolet radiation,

the temperature of the electrons does not fall significantly below 10^4 K.
Substituting for J_ν into (1), we find

$$\int_{\nu_c}^{\infty} \frac{N(z) L_\nu \, d\nu}{h\nu} \approx (\alpha_t - \alpha_1) N_e^2(z)$$

Complete ionisation of the gas takes place when the number of ionising
photons being created per unit volume per unit time exceeds the number
of recombinations in the same volume. Taking $N(z) = 0.03 (1+z)^3$ Mpc^{-3},
a density of the order of that of galaxies such as our own, and
$L_\nu \approx 3 \times 10^{29} erg \, s^{-1} Hz^{-1}$ $(L \approx 10^{45} erg \, s^{-1})$, we see that such sources could lead to
ionisations of the intergalactic gas at any redshift

$$z \leqslant 27 \, \Omega^{-2/3} h^{-4/3}$$

In this case the optical depth of the gas to Thomson scattering must be
less than $\tau \leqslant 4 \Omega^{-1/2} h^{-1/2}$

   We note that the total number of quasars which were active at any
time in the Universe is of the order of the total number of galaxies
such as our own (Lynden-Bell 1969, Komberg and Sunyaev 1971). If the
models of Doroshkevich et al (1971) are correct, in which it is proposed
thatthe cosmological evolution of quasars and radio galaxies are more or
less simultaneous, both in their birth-rate an subsequent decline in
activity, then quasars can be responsible for the secondary ionisation
of the gas. This was noted by Arons and Wingert (1971). Another natural
sources of ionising photons are young galaxies because, just as in the
case of quasars, they are anomalously bright at ultraviolet wavelengths
(Weymann 1966). Therefore the secondary ionisation of intergalactic gas
can lead to effective washing-out of fluctuations in the MWBR if
observed galaxies and quasars formed at $z \approx 20$ to 30 and went through
a bright phase at that time.

   It should be noted that the early formation of galaxies contradicts
the adiabatic model of primaeval fluctuations in which first clusters
of galaxies form at a redshift of 4 - 10 and then later fragment into
separate galaxies (Sunyaev and Zeldovich 1973, Doroshkevich et al, 1974).
On the other hand in models of entropy fluctuations (Rees and Gott, 1976)
and of whirl perturbations (Ozernoi and Chernin, 1968), early formation
of galaxies is completely natural.

   It should be noted that at that period the velocities of matter on
scales $10^{13} - {}^{16}M_\odot$ were very large which must inevitably give rise to
fluctuations in the MWBR because of Doppler scatterings. However, at
that time the optical depth to scattering for a single object in this
mass range was many times $(10^2 - 3)$ smaller than the optical depth of
the Universe. Naturally, for a completely random distribution of
velocities, the effect corresponds to

$$\frac{\Delta T}{T} \approx \frac{1}{\sqrt{n}} \sigma_T N_e \lambda \, \frac{v}{c} \langle \cos^2\theta \rangle^{1/2}$$

and the region within which such an effect is expected corresponds to
$c\tau \approx 1$.

Literature

Anile A.M., Danese L., De Zotti G. and Motta S., 1976. Astrophys. J, 205, L59.

Arons J. and Wingert D., 1972. Astrophys. J., 177, 1.

Carpenter R.L., Gulkis S. and Sato H., 1973. Astrophys. J., 182, L61.

Chibisov G.V., 1972. Astron. Zh., 49, 286.

Chibisov G.V. and Ozernoi L.M., 1969. Astrophys. Letters, 3, 189.

Conklin E.K. and Bracewell R.N., 1967. Phys. Rev. Letters, 18, 614.

Doroshkevich A.G., Longair M.S. and Zeldovich Ya.B., 1970. Mon. Not. R. astr. Soc., 147, 139.

Gull S.F. and Northover K.J.E., 1976. Nature (in press).

Komberg B.V. and Sunyaev R.A., 1971. Astron. Zh., 48, 235.

Longair M.S. and Sunyaev R.A., 1969. Nature, 223, 719.

Parijskij Yu.N., 1972. Astron. Zh., 49, 1322.

Parijskij Yu.N., 1973. Astron. Zh., 50, 453.

Peebles P.J.E., 1968. Astrophys. J., 162, 815.

Peebles P.J.E. and Yu, I.T., 1970. Astrophys. J., 162, 815.

Silk J., 1968. Astrophys. J., 151, 459.

Stankevich K.S., 1974. Astron. Zh., 51, 216.

Sunyaev R.A. and Zeldovich Ya.B., 1970. Astrophys. and Space Science, 7, 20.

Sunyaev R.A. and Zeldovich Ya.B., 1972. Comments Astrophys. and Sp. Phys.,4, 173.

Weinberg S., 1972. "Gravitation and Cosmology", Wiley, New York.

Zeldovich Ya.B., Kurt V.G. and Sunyaev R.A., 1968. ZhETP., 55, 278.

DISCUSSION

Bonometto: I would like to comment on the possibility that isothermal fluctuations may cause fluctuations of BB radiation over scales $\sim 10^{12} M_\odot$ or so.

We must remember that the surface of last scattering for BB photons is at $z \sim 1000$, while any significant dragging effect between matter and radiation on those scales has stopped at z ~1300 to 1200. During the Δz interval between 1200 and 1000, radiation density fluctuations will be completely damped because of the residual ionisation. All that is, however, strongly dependent on the rate of residual ionisation in the late recombination period. Previous statements implied no extra energy input. If such inputs took place, fluctuations would be even more strongly damped.

Longair: I have concentrated mainly upon scales M $\sim 10^{15} M_\odot$ where fluctuations $\Delta T/T \sim 10^{-4}$ are expected, taking into account the effects you mention.

Lynden-Bell: Is there any evidence for reheating?

Longair: Sunyaev has adopted the conventional view that there must be some intergalactic gas because the process of galaxy formation cannot

have been 100 per cent efficient in mopping up intergalactic gas and
there is no evidence for depression of the continuum in the spectra of
quasars beyond redshift Ly-α. The cooling rate of the intergalactic
gas at redshifts z $\sim$ 30 - 100 is less than cosmological timescales and
therefore ionisation and consequent reheating seem inevitable.

FLUCTUATIONS IN THE MICROWAVE BACKGROUND CAUSED BY ANISOTROPY OF THE UNIVERSE AND GRAVITATIONAL WAVES

I.D. Novikov
Space Research Institute,
Academy of Sciences of the USSR, Moscow.

The problem described in the title has already been theoretically analysed several times (see for example Dautcourt, 1969 and Novikov, 1974). Recently, however some important new aspects of the problem have been discovered; they are discussed briefly in this report which is based upon the calculations of Doroshkevich, Lukash, Novikov and Polnarev. First consider the influence of primordial gravitational waves on the microwave background. It is natural to assume that in the Universe, in addition to acoustic (or adiabatic) density perturbations which result in galaxy formation and corresponding metric perturbations, there also exist metric perturbations in the form of gravitational waves with wavelengths of the same order of magnitude as the acoustic perturbations. The amplitude of such gravitational waves could in principle be quite arbitrary. Their amplitude can be estimated by comparing the theory of such waves in the expanding Universe with the observed fluctuations in the microwave background which are now available or will be in future.

Gravitional radiation – as well as any non-stationary metric perturbation – effect the microwave background radiation (we will refer to it as the relict radiation) and results in the observed temperature T of the relict radiation being different in different directions.

The amplitude h of the gravitational waves decreases in the course of the cosmological expansion. For this reason the major contribution to the fluctuations ΔT are made by h values at the earliest observable epoch, i.e. at the epoch of recombination. The calculations of Doroshkevich, Novikov and Polnarev which I report here include two important factors that were not taken into account accurately enough or in a consistent manner in the previous investigations. First, allowance is made for the fact that the recombination of the primordial plasma is not instantaneous, but rather it gradually becomes transparent over a time interval of the order of $0.1\ t_{rec}$ where t_{rec} is the moment at which the recombination begins. This circumstance is important for the wavelengths noticeably smaller than the horizon at the moment t_{rec}. The depth of the plasma layer that contributes to the observed microwave

D. L. Jauncey (ed.), Radio Astronomy and Cosmology, 335-339. All Rights Reserved.
Copyright © 1977 by the IAU.

background extends over a number of gravitational wavelengths, i.e.,
we see simultaneously a number of shells contributing with opposite
signs to the distortion of the relict radiation. This effect strongly
blurs fluctuations of the background.

Second, when calculating the expected mean square amplitude of
ΔT and the correlation function, one should take account properly of
the polar diagram of the radio antenna. Below results are given for a
number of different polar diagrams.

The fluctuations in the electromagnetic relict radiation due to
the primordial gravitational waves are quite different from perturbations of the acoustic (or adiabatic) type discussed in the paper by
Sunyaev (this volume). The most important features that enable us to
distinguish between them are as follows.

The fluctuations $\frac{\Delta T}{T}$ caused by adiabatic density perturbations are
of the same order of magnitude as $\delta\varepsilon/\varepsilon$ (provided the wavelength of the
perturbations is much less than the scale of the horizon, ct). The
metric perturbations in this case are given by $\delta g = \left(\frac{\delta\varepsilon}{\varepsilon}\right)\left(\frac{\lambda}{ct}\right)^2$. The
perturbations $\frac{\Delta T}{T}$ caused by δg can be neglected. For gravitational
waves the quantity $\frac{\Delta T}{T}$ is of the order of δg, while $\delta\varepsilon$ vanishes
identically in this case. Furthermore, when gravitational waves
propagate through matter, the particles move perpendicular to the
direction of wave propagation, while in acoustic waves the particles
move parallel to the direction of propagation.

Finally, the velocity of gravitational waves is the velocity of
light, which is not the case for acoustic waves. These are the main
differences between the process discussed here and that considered in
previous work.

The calculations include the solution of the kinetic equation for
photons in a Friedmann metric with perturbations in the form of
gravitational waves; both factors mentioned above were taken into
account.

The results are as follows. For the amplitude of fluctuations we
have

$$\overline{\left(\frac{\Delta T}{T}\right)^2} = \frac{\pi}{4} \int_0^1 d\mu \int_0^\infty \mathcal{D}_K^2(\mu) K^2 e^{-\alpha^2 K^2 r^2 (1-\mu^2)} dK \qquad (1)$$

were $K = \frac{2\pi a}{\lambda_{GW}}$, α - is the scale factor of the Universe, μ is the cosine
of the angle of observation. Other quantities are defined below. To
represent the observational data, observers use the function

$$F(\Theta) \;=\; \overline{\left(\frac{T(\mu_1) - T(\mu_2)}{T}\right)^2} \tag{2}$$

where the angle between the directions μ_1 and μ_2 is equal to θ. r is the angular diameter effective distance of the epoch for recombination. $F(\theta)$ we have

$$F(\Theta) \;=\; \overline{\left(\frac{\Delta T}{T}\right)^2} \, (1 - f(\Theta)) \tag{3}$$

$$f(\Theta) = \frac{\pi}{4\,\overline{(\Delta T/T)^2}} \int_0^1 d\mu \int_0^\infty K^2 \mathcal{D}_K^2(\mu)\, J_0\left(\Theta K r (1-\mu^2)^{1/2}\right) e^{-\alpha^2 K^2 r^2 (1-\mu^2)}\, dK \tag{4}$$

$$\mathcal{D}_K^2(\mu) \;=\; \frac{1}{4}(1-\mu^2)^2 \left[|A_+|^2 + |A_-|^2\right] h_K \tag{5}$$

$$A_\pm \;=\; e^{-\frac{\Delta^2 K^2 (1\pm\mu)^2}{2}} \int_{\eta_{rec}}^\infty e^{iK\eta(1\pm\mu)} \frac{3 \pm 3iK\eta - K^2\eta^2}{\eta^4}\, d\eta \tag{6}$$

J_0 is a Bessel function. It is assumed here that the spectrum of gravitational waves takes the form

$$h_K = h_0 K^n$$

for all relevant wavelengths, α is beam width of the radio antenna, $d\eta = \frac{dt}{a}$, and Δ is the duration of the process of recombinations in η-time. The results for different beam widths are shown in Fig. 1.

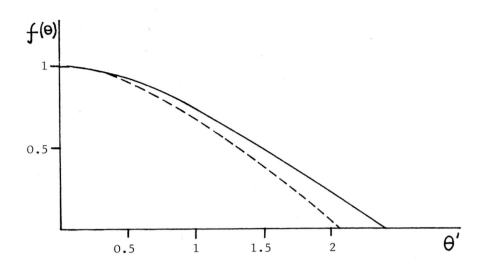

Fig. 1.   The function $f(\theta)$ for $n = 0$ and for $\alpha = 1'$ (solid line) and
for $\alpha = 2'$ (dotted line).

These formula should be used in analysing the implications of
future observations.

Comparison with the observational data now available enables us to
establish an upper limit for the energy density of long gravitational
waves.   This method is most sensitive for gravitational waves with
$\lambda_{GW} \stackrel{\sim}{\sim} ct_{rec}$.   The fluctuations $\frac{\Delta T}{T}$ due to these waves have scale $\stackrel{\sim}{\sim} 0.03$
radian.   If, according to modern observations, we take $\frac{\Delta T}{T} < 10^{-4}$, then
$\varepsilon_{GW}/\varepsilon_{\gamma} < 10^{-8}$ for those gravitational waves which have $\lambda_{GW} = 5.10^{26}$ cm
today where ε_{γ} is the energy density of relict radiation.   The
fluctuations $\frac{\Delta T}{T}$ due to long gravitational waves with $\lambda_{GW} = ct_{today} =$
10^{28} cm, are quadrupole.   For these fluctuations the observations give
$\frac{\Delta T}{T} < 3.10^{-4}$ and so $\left[\dfrac{\varepsilon_{GW}(\lambda = 10^{28} \text{ cm})}{\varepsilon_{\gamma}} \right] < 10^{-3}$.

——I wish to emphasize that having constructed the spectral functions
$\left(\frac{\Delta T}{T} \right)^2_{K}$ we can in principle make a clear distinction between   fluctuations
due to gravitational waves and fluctuations of the acoustic (or adiabatic)
type which result in galaxy formation.   This possibility is closely
related to the facts mentioned above that (i) the velocity of propagation of gravitational waves is the velocity of light, and that (ii)
the gravitational waves are transverse.   As a result, if the spectrum

of δg for adiabatic perturbations and gravitational waves are identical
the spectral function $\left(\frac{\Delta T}{T}\right)_K^2$ at short wavelengths drops exponentially
(due to blurring) in the case of adiabatic perturbations and only as
a power law in the case of gravitational waves (the spectral index α_1
changes from α_1 to (α_1-1). This fact makes it possible to distinguish
between these types of fluctuation.

I turn now to the anisotropy in the relict radiation T caused by
anisotropy of the Universe as a whole. This problem had been
extensively discussed by Novikov (1974), by Doroshkevich, Lukash and
Novikov (1974) and by Zeldovich and Novikov (1975). I shall not
reproduce the formulae and their derivations which are given in these
works. I only recall that in the case of $\rho < \rho_{crit}$ there should be a
spot on the sky in which $\Delta T/T$ greatly exceeds its value over the rest
of the sky. The angular scale of the spot θ is of order of ρ/ρ_{crit}.

If one considers, for example, a perturbation that is a superposition of two homogeneous anisotropic models, say, of Bianchi type V, and
each model differs from an isotropic Friedmannian model only slightly
today, then one arrives at an inhomogeneous model with an inhomogeneity
scale of the same order of magnitude as the curvature radius. The
observational appearance of such a structure will be the superposition
of two spots on the sky. This example demonstrates that we cannot say
that anisotropic models of Bianchi type V is the limit when wavelength
of perturbations tends to infinity. The analysis of perturbations with
the wavelengths greater than the curvature radius in a curved Universe
is a non-trivial problem that requires special consideration.

REFERENCES

Dautcourt, G., 1969. Mon. Not. R. astr. Soc., 144, 255.
Doroshkevich, A.G., Lukash, V.N. and Novikov, I.D., 1974. Astr. Zh.,
    51, 940.
Novikov, I.D., 1974. "Confrontation of Cosmological Theories with
    Observational Data", ed. Longair, M., 273, Reidel, Dordrecht.
Zeldovich, Ya.B. and Novikov, I.D., 1975. "Structure and Evolution of
    the Universe", Nauka, Moscow.

VII

MORE INTERPRETATION OF COSMOLOGICAL INFORMATION ON RADIO SOURCES

THE PHYSICS OF RADIO SOURCES AND COSMOLOGY

P.A.G. Scheuer
Mullard Radio Astronomy Observatory, Cavendish Laboratory,
Cambridge, U.K.

There are two important questions in which the physics of radio
sources impinges upon cosmology. The first is whether the large
apparent expansion velocities of certain compact sources can be
explained satisfactorily within the hypothesis that their red-shifts
are due to the Hubble expansion. The second is the whole broad
question of the evolution of the radio source population with epoch.
I do not have a new and convincing answer to the first, and the second
is pretty nebulous, since we do not even understand radio sources at
the present epoch very well. So I shall not present a general survey:
instead, I shall use my allotted time to discuss a smaller question to
which one can now give a fairly definite answer.

Malcolm Longair's first notable contribution to science was to
point out that the radio source counts require not only that the source
population should evolve, but that powerful sources should evolve much
faster than weak sources. Ever since then he has been trying to
define more quantitatively how one must fill up the P - z plane, and
indeed much of this symposium has been devoted to that and closely
related problems. The theoretician in each of us cannot help also
wondering why. There are plenty of explanations for the greater
abundance of radio sources in the past; all sorts of exciting things
could have happened when the world was young and galaxies first shone
forth out of the primaeval turbulence. There are fewer explanations
of the fact that weaker radio sources weren't nearly as overabundant
(relative to the present epoch) as powerful ones. However, there is
one natural and elegant explanation, which depends on the idea that old,
weak, diffuse sources are extinguished because of inverse Compton
losses on the microwave background. For example, in the halo of M87
one would expect inverse Compton losses to be important over time scales
10^8 years, and indeed this and some other weak sources (e.g. 3C 465,
3C 129) have very steep radio spectra in their outermost regions.
Since the energy density of the microwave background rises as $(1+z)^4$,
weak sources would have been snuffed out much younger at large redshifts; hence the density of weak sources does not increase with z as
fast as that of strong sources. This idea was worked out rather

D. L. Jauncey (ed.), Radio Astronomy and Cosmology, 343-352. All Rights Reserved.
Copyright © 1977 by the IAU.

thoroughly by Rees and Setti (1968), and has been mentioned again in more recent papers (e.g. van der Laan and Perola 1969, Christiansen 1969, Wardle and Miley 1974).   Rees and Setti took expanding spheres as models of radio sources, and computed how the luminosity function should evolve with eposh.   We now have a far more profound and sophisticated ignorance of the physics of radio sources than we had in 1968, but we cannot yet make a decisive improvement on Rees and Setti's work by using better models.   We also have a wealth of observations of the structures of radio sources, and can even discern some correlations between morphology and radio power P; but to check Rees and Setti's theory directly we chiefly need to know how long the fast electrons interact with the microwave background, and we do not really know the ages of radio sources to within a factor 10.   Nevertheless, I think the new observational information can be used to perform a test which may be good enough to exclude the theory.

I shall make no assumption about the age of any source.   I merely note that inverse Compton losses must be unimportant so long as synchrotron losses are greater, i.e. so long as the magnetic energy density is greater than two thirds of the radiation density.   For the magnetic energy density I adopt the equipartition value, worked out separately for each component of a source. . I then take a complete sample of radio sources (it is the subset with $S_{178} > 20$ Jy of the complete sample of 166 sources described by Longair, see final day's discussion in this volume); all the sources in this sample are either very compact or have been mapped with reasonably good angular resolution (most of them at 2" arc resolution).   That complete sample provides a luminosity distribution (admittedly a rough one!) for $z = 0$, which is shown in the top histogram of Figure 1.   (The most powerful sources in the sample are in fact at appreciable red-shifts, but, as we shall see, their distribution is not important for the present exercise since nearly all their flux density is in components of small angular size.) For each component of each source I then find the equipartition magnetic energy density, and hence the red-shirt beyond which inverse Compton loss would exceed synchrotron loss.   I then make the most optimistic assumption I can about the importance of inverse Compton losses: as soon as inverse Compton loss exceeds synchrotron loss, I extinguish that component utterly.   Thus I can find out how a set of sources with the same intrinsic properties would look at various red-shifts z. Some results are shown in the lower histograms of Figure 1.   Some sources have vanished altogether, others have only lost some of their components.   Diffuse components vanish first because they have lower equipartition magnetic fields.   The resulting distributions are shown by the full lines.   Some of the sources still lack red-shift measure- ments; if these are arbitrarily given $z = 0.3$ and included, the histogram rises to the upper dashed line.   But what I have done so far is not quite fair; some of the sources, on losing one or more components, would have dropped below the flux limit $S_{178} = 20$ Jy of the sample.   If we discard a source completely as soon as it falls below that limit (which I believe to be the correct procedure) we are left

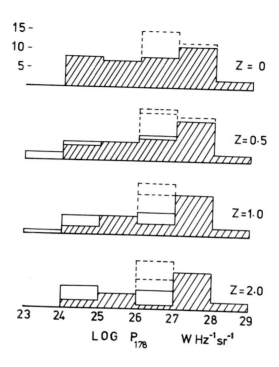

Fig. 1   Luminosity distributions at various red-shifts, derived
from that at z = 0 assuming the greatest possible effects of
intense Compton losses.

with the shaded histogram (or the lower dashed line, if we include
sources with unknown distances as if they had z = 0.3).   Now I divide
each histogram by the z = 0 histogram to get the ratio by which weak
sources are underevolved, obtaining Figure 2.   Figure 2 again shows,
qualitatively, the behaviour that we want.   Now we come to the crucial
comparison.   Is the maximum possible effect of inverse Compton losses,
shown in Figure 2, enough to account for the requirements of the source
counts and red-shift distributions?   The most recent estimates of the
evolution of the radio luminosity function that I know are those that
Wall described at this symposium, and two of his models are sketched in
Figure 2.   It is clear that inverse Compton scattering on the microwave background is quantitatively inadequate, by several orders of
magnitude, to suppress the density evolution of weak sources to the
required extent.

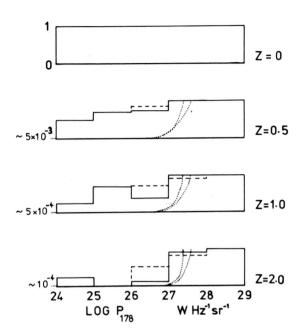

Fig. 2   The density of sources, as a fraction of what it would
have been if the source density had evolved in the same way for
all radio powers P.   The histogram is derived from Fig. 1, i.e.
assuming that variation with P is due to inverse Compton losses
only.   The dotted lines indicate the variation with P required
according to Wall, Pearson and Longair;   these curves fall to
the small fractions written by the left sides of the histograms.

How reliable is this result?

(i)   I have rather arbitrarily taken equipartition magnetic fields.
If the magnetic energy were systematically much less than the fast
particle energy, inverse Compton losses would exceed synchrotron
losses (though not necessarily other e.g. adiabatic losses) at smaller
z.   If $B = \alpha\, B_{equipartition}$, we ought to use the histogram for z* at
red-shift z, where $(1+z^*) = \alpha^{-\frac{1}{2}}(1+z)$.   Thus we can see that the sources
have to be grossly out of equipartition to make the "inverse Compton"
explanation consistent with Wall's estimates of the evolution of the
luminosity function.   But we must bear in mind that we can't prove
that the sources are anywhere near equipartition.

(ii)   Some of the maps do not have enough resolution to show all of the structure.   But any finer structure would represent components with larger $B_{equipartition}$, and so strengthen my conclusion.

(iii)   Figure 2 does not compare theory directly with observation, but with models fitted to the observations.   The weakness of evolution for weak sources is determined essentially by the convergence of the source counts at low flux densities;   my impression is that this feature of the models cannot be shifted very much without contradicting the observations, but obviously it is a question that needs to be looked into.

(iv)   The calculations presented here are preliminary;   the statistics could be improved by using a rather larger sample.

Let me summarise the argument.   Detailed mapping has shown that even fairly weak sources often have a large part of their flux density in small components.   Unless these components are very far indeed from equipartition, their magnetic fields are such that synchrotron losses exceed inverse Compton losses even at red-shifts of 1 or 2.   When one looks at the argument in this way, I think it becomes clear that the conclusion is not likely to be changed by minor tinkering with the radio source parameters.

I conclude, rather reluctantly, that we have to look for something intrinsic to the sources to account for the weaker evolution of weak sources.

I am indebted to Ann Simon for permission to use her computations of magnetic field strengths in radio sources, without which this talk would certainly not have been ready in time for the symposium, and to Drs. Wall, Pearson and Longair for permission to use their estimates of the evolution of the radio luminosity function before publication.

REFERENCES

Christiansen, W., 1969. Mon. Not. R. astr. Soc., 145, 327
Rees, M.J. & Setti, G., 1968.   Nature, 219, 127.
van der Laan, H. & Perola, G.C., 1969.   Astr. Astrophys., 3, 468.
Wardle, J.F.C. & Miley, G.K., 1974. Astr. Astrophys., 30, 305.

DISCUSSION

Conway: Is it possible to turn your argument right round and argue that since the Inverse Compton effect has not completely wiped out every radio source, B must be within some factor of the equipartition value. If so would that set an upper or lower limit to $B/B_{equipartition}$?

Scheuer: No, I can't turn the argument around, because I am using an inequality.   Something other than synchrotron losses could (and in most cases probably does) extinguish sources before inverse Compton losses affect them.

Trimble: How far out of equipartition do the sources have to be to make inverse Compton losses important enough to account for the required evolution?

Scheuer: Looking at Figure 2, one sees that inverse Compton losses at $z = 2$ are still insufficient to fit the luminosity function for $z = 0.5$. Thus one has to decrease the field by a factor exceeding $(2/0.5)^2 = 16$ below equipartition before inverse Compton losses become sufficiently effective. With a more detailed analysis I think one could sharpen that statement quite significantly.

Rowan-Robinson: Some time ago (MN, 150, 389, 1970) I tried to fit models of this type to the radio source-counts, with density evolution combined with this inverse Compton snuffing-out to give a luminosity-dependent cut off. I would agree that equipartition doesn't work, but a model in which the electron energy is the same in all sources, with the magnetic field varying, can fit everything. You have to be careful not to exceed the integrated X-ray background with the inverse Compton radiation.

Scheuer: Thank you for reminding us of this work. In the calculations described here, my purpose was to use observations rather than models; this has only become practicable fairly recently, thanks to the availability of more detailed radio maps and more measured redshifts.

van der Laan: It seems to me that spectral index distributions demonstrate that the sources are snuffed out, not by radiative losses, but by processes that uniformly suppress the emission throughout the radio window. The fact that for much deeper samples $g(\alpha)$ does not show any flux dependence, reinforces this conclusion. (See P. Katgert, this volume).

Scheuer: I have used synchrotron losses only as a lower bound to other losses, a lower bound that is easy to compare with inverse Compton losses. Personally, I should agree with you that sources probably fade chiefly because of processes, such as adiabatic loss and field line reconnection, which do not change the radio spectrum (cf. Jenkins, C.J. & Scheuer, P.A.G., 1975, M.N., 174, 327), though I doubt whether this is generally accepted. However, there are no firm estimates of the rates of such processes in radio sources, so I can't use them here.

Schmidt: Optically selected quasars show a cosmological evolution similar to that of strong radio galaxies, suggesting a common explanation. Since inverse Compton losses are unlikely to affect the very compact optical quasars, it is perhaps agreeable to see them ruled out, too, as a cause of radio galaxy evolution.

Okoye: Is it not true to say that not knowing the actual values of the magnetic field strength in the source components, it is not then reasonable to assume that inverse Compton losses are less than intrinsic synchrotron losses (or if inverse Compton losses are greater, then some

disappear) while at the same time assuming that the source magnetic
field strength is the equipartition value.

Scheuer: Yes, I agree. I am sorry if I did not make that sufficiently
clear. You can keep inverse Compton losses as the cause of differential
evolution between strong and weak sources, if you assign magnetic fields
far below equipartition values to the compact parts of weak radio
sources. That does not appeal to me, because the more compact components
seem to have quite normal radio spectra, but you are quite right to point
out that my calculations do not rule out the possibility.

THE ORIGIN OF THE EXTRAGALACTIC RADIO BACKGROUND AT LOW FREQUENCIES

Ann J.B. Simon

At low frequencies the isotropic extragalactic component of the
radio background peaks at about 3 MHz and then decreases. Independent
evidence that the component from outside the Galaxy peaks at about this
frequency is provided by A.H. Bridle (1968, Nature, 219, 1136) in his
interpretation of the lack of an absorption feature in the direction of
the Magellanic Clouds in observations by G. Reber (1968, J. Franklin
Inst., 285, 1) at 2.1 MHz. The brightness temperature at 178 MHz is
23 ± 7 K for $\alpha = 0.80$ and can largely be attributed to the sum of
contributions from extragalactic radio sources. I shall outline an
attempt to explain the turnover at 2 MHz in terms of absorption in the
individual sources which make up the background.

A complete sample of radio sources (See Appendix) is taken and a
model built for each source. The only absorption effect found to be
significant was synchrotron self-absorption. The frequency at which
the luminosity of a source is a maximum (the synchrotron self-absorption
frequency) is found to be directly related to the source luminosity,
ranging from 1 MHz for low luminosity sources to over 100 MHz for high
luminosity sources. The integrated background radiation was computed
for $\Omega = 1$ for an evolving Universe of the type described by Wall (this
volume) and a non-evolving Universe for comparison. Both spectra
peaked at about 1 - 2 MHz and obeyed the power law $B(\nu) \propto \nu^{-0.8}$ at 178 MHz.
The evolving model predicted $T_b = 15$ K at 178 MHz. Calculations of the
background radiation from sources in different luminosity ranges showed
that for a non-evolving universe the peaks of each contribution ranged
from 1 MHz ($P_{408} \leqslant 10^{24}$ W $Hz^{-1}sr^{-1}$) to 15 MHz ($P_{408} \geqslant 10^{28}$ W $Hz^{-1}sr^{-1}$).
The background for an evolving universe is therefore bound to turn over
somewhere in this frequency range, whatever model of evolution is used.
Even for the evolving Universe, the non-evolving low brightness sources
contribute a very large fraction of the background at all the frequencies considered.

Investigation of the spatial origin of the background radiation
showed that at least half of the background at all frequencies in the
range 0.5 - 178 MHz originates at redshifts less than 1.0, and that the
contributions become progressively smaller for regions of higher

redshift. This minimises the effects of possible absorption by either intergalactic gas or normal galaxies. It can therefore be concluded that the superposition of individual sources leads to a turnover in the radio background at about 2 MHz due to synchrotron self-absorption in the sources. A large fraction of the background comes from low-luminosity sources. The contribution to the background from sources at redshifts greater than about 2.0 is very small.

Feldman: What percentage of the extragalactic background at about 1 MHz is due to clusters of galaxies?

Simon: I don't think this can be meaningfully evaluated since the number of sources associated with Abell clusters in the sample I have used is very small.

Jaffe: I wouldn't take the absence of thermal absorption in the direction of the Magellanic Clouds as typical of the Galaxy as a whole since the gas in the Galaxy is very patchy, and the area of the Magellanic Clouds is known to be an area of low optical absorption, indicating a low gas content in that direction.

Simon: I would agree that it cannot be taken as a hard fact but it does suggest that the extragalactic component has decreased by 2.1 MHz. It is impossible to make any more definite deduction from the data available.

Conway: I believe your graphs refer to a universe with $\Omega = 1$. Have you evaluated the case with low Ω? I would expect that the chief difference would be to make the most important z-range further away.

Simon: No evolving model is yet available for low Ω. If the computation is done for a non-evolving $\Omega = 0$ model, the contribution from the range $0.0 \leqslant z \leqslant 1.0$ still dominates.

Gulkis: Can you please explain how the R.A.E. satellite data are separated into a galactic and radio source component, and in particular how you know that the radio source spectrum turns over abruptly in the 1 - 10 MHz spectral region?

Simon: The original paper (Clark, Brown and Alexander, 1970, Nature, **228**, 847) explains how the spectrum was broken down into two components by the best fit method, which showed that the extragalactic component arriving here has a sharp turnover at about 3 MHz. This could be due either to absorption of a power law by thermal electrons equivalent to a free-free optical depth of at least 3 at 1 MHz, or to a turnover in the true extragalactic spectrum with smaller optical depth inside the Galaxy. In view of the probable patchiness of the electrons inside the Galaxy, I think the latter interpretation is more likely.

THE DISTRIBUTION OF INTRINSIC SEPARATIONS OF 3CR DOUBLE
RADIO SOURCES FOR RADIO GALAXIES

K.C. Jacobs

For the 3CR radio galaxies with measured redshifts the observed angular separation between the components of each radio source may be converted into an apparent separation, d (kpc) [$H_o = 50$ km s^{-1}Mpc^{-1}]. The usual histogram of number of cases with d in logarithmic (decade) bins is sharply peaked at several hundred kpc. However, this "log-binned-log-histogram" is misleading, since it is a severely distorted representation of the quantity of physical interest, namely the probability density function, f(d). We invert the histogram to obtain the observed f(d), which decreases monotonically with increasing d. Since uniform-spherical angular projection effects transform the intrinsic probability density, F(d), into the apparent f(d) via

$$f(d) = d \int_d^\infty \frac{F(D) \, dD}{D\sqrt{D^2 - d^2}}$$

we may deduce F(D). The distribution of intrinsic separations, D, is essentially a decaying exponential function with an e-folding parameter, $D_o \approx 300$ kpc.

Three interpretations of the resulting F(D) are given:

(i) If the double components are separating with uniform speed, v_o, then the source lifetimes must be roughly exponentially distributed with the characteristic lifetime, $t_o \approx D_o/v_o$.

(ii) If the components are decelerated as they separate then the lifetimes are distributed even more steeply than exponential,

(iii) Finally, and most interestingly, if the lifetimes of all sources are approximately equal (and our observations have sampled them fairly uniformly), then F(D) implies that the components of the radio doubles are accelerating apart! Should this interpretation be the correct one, then the only extant theory of double radio sources which can survive is some version of the so-called "beam models".

TWO TESTS OF THE EXPANSION OF THE UNIVERSE

W.H. McCrea

Discussion at the symposium indicates that the following tests may be on the verge of feasibility:

Peculiar motion   In a region of the Universe at distance giving redshift z, let $\bar{v}$ be the mean speed of peculiar motion of gravitationally indepen- dent systems (field galaxies, clusters). Then expansion requires

$$\bar{v} \propto 1 + z$$

(at any rate, roughly).   A speaker quoted 200 km/s as a possible estimate of v in our cosmic vicinity;   this would imply $\bar{v} \approx$ 400 km/s at
z = 1 and so on.   The test is to look for any effect that depends upon
peculiar motion and to see how it varies with z, or any parameter depending on z.

<u>Hubble motion</u>   For a region of the Universe at redshift z, let θ be
the observed angular size of objects of some standard class, and let ϕ
be the observed angular distance to some object of some standard class
that is gravitationally independent of the first object.   Then, for
some reasonable definition of the means < >

$$\frac{<\phi>}{<\theta>} \propto R(t) \propto \frac{1}{(1+z)}$$

The test is to get statistics of θ, ϕ, z.   By using 3 parameters in
this case we get, in principle, a test of the expansion in any region
that is model-independent.

    Conversely, the relation might be used as a test to see whether
objects of certain classes are at the same distance - a test that
would be independent of redshift.

Wittels: For the Hubble-velocity test of R(t) dependence on z, does
one need to pair sources at the same z.

McCrea: Yes, but with sufficient ingenuity I think you could invent a
statistical way of doing it without that restriction.

YOUNG GALAXIES, QUASARS AND THE COSMOLOGICAL EVOLUTION OF EXTRAGALACTIC
RADIO SOURCES

M.S. Longair
Mullard Radio Astronomy Observatory, Cavendish Laboratory,
Cambridge, England.

R.A. Sunyaev
Space Research Institute, USSR Academy of Sciences,
Moscow, USSR.

The V/V_{max} test for quasars and the counts of radio sources show
that the most powerful extragalactic radio sources exhibit strong
evolutionary changes with cosmological epoch (see M. Schmidt and J.V.
Wall et al., this volume). It should be emphasised that these are very
large changes indeed. Schmidt, for example, has shown that for the
world model with $\Omega = 0$, evolution functions of the form $F(t) \propto \exp(-10t/t_o)$
can account for the observations, t_o being the present epoch and t
cosmic time. Since the quasar population from which this law is derived
extends at least to $Z = 2.5$, corresponding to $t = t_o/3.5$ if $\Omega = 0$, the
comoving space density of quasars at $Z = 2.5$ must have been about 1300
times greater than it is at the present epoch. It is not known whether
or not this law continues to hold at larger redshifts but even if it
does, the increase in comoving space density from $Z = 2.5$ to infinity
is only a further factor of 17. Therefore the bulk of the evolution
occurs within the range of observationally accessible redshifts. Notice
that a characteristic time-scale for the decay of the quasar population
of $\approx t_o/10 \approx 10^9$ years comes out of this analysis.

These evolutionary changes are empirical results derived from
observation but there has been little interpretation of their cosmological significance. This is because neither the theory of the
origin and evolution of galaxies nor that of radio sources is in a
state where it can account for the properties at a single epoch, let
alone how they might change with time. Nonetheless, speculation about
the signficance of these results may be useful if it suggests potentially
fruitful lines of observational and theoretical investigations.

1. THE NATURE OF THE COSMOLOGICAL EVOLUTION OF RADIO SOURCES AND QUASARS

The radio structures and properties of extragalactic radio sources
bear little relation to the properties of the parent galaxy or quasar.
It is therefore important to investigate whether the evolutionary changes
should be associated only with the radio properties or whether they must

D. L. Jauncey (ed.), Radio Astronomy and Cosmology, 353-360. All Rights Reserved.
Copyright © 1977 by the IAU.

be associated with the parent object which is believed to be ultimately responsible for the source of energy. Rees and Setti (1968) for example showed that it is possible to account for part of the inferred cosmological evolution by embedding a given type of double radio source in an intergalactic medium with increases in density with increasing redshift. Some aspects of this model are discussed by Scheuer (this volume). Two pieces of evidence suggests that the evolution is more likely to be associated with the parent object. First, there appears to be little change in the physical properties of the extended radio structures of quasars with redshift (J.M. Riley et al., this volume). Second, the same cosmological evolution laws are found for quasars which are powerful radio sources and for radio-quiet quasars. These statements are based upon rather small statistical samples but they both suggest that the evolutionary changes are associated with the parent objects and therefore that understanding the evolutionary problem must be related to the origin and evolution of galaxies and quasars.

2. OTHER EVIDENCE ON THE ORIGIN AND EVOLUTION OF GALAXIES

2.1. The formation of the heavy elements in the Galaxy

According to the well-known argument of Schmidt (1963), the uniformity of the heavy element abundances in red dwarf stars which have ages as old as the Galaxy suggests that most of the heavy elements were formed in a burst of star formation when the Galaxy was about one tenth of its present age. Models to account for these observations have been made by a number of authors and characteristically it is found that such "young galaxies" are expected to be 10-100 times more luminous than our own Galaxy for periods $\gtrsim 10^9$ years.

2.2. The Epoch of Galaxy Formation

None of the arguments concerning the epoch of galaxy formation are unambiguous. Perhaps the strongest comes from comparison of the sizes and densities of clusters of galaxies with their mean separation and the mean intergalactic density. The average density of matter in the Universe was equal to the mean density of clusters at a redshift $Z \sim 10$-30, implying that the epoch at which they became distinct physical systems must have occurred at a somewhat later epoch, probably $Z \sim 5$-10. The latter redshifts correspond to timescales $\gtrsim 10^9$ years, similar to the timescales found from the evolutionary models and the argument of 2.1.

Whether or not the epoch corresponding to $Z \gtrsim$ of 5-10 also corresponds to the epoch of galaxy formation in general, depends upon the model of galaxy formation. In the model of primaeval adiabatic fluctuations, galaxies form by fragmentation of the clouds associated with clusters and hence the epoch of galaxy formation corresponds to $Z \gtrsim 5$-10. In the models of isothermal and whirl perturbations, however, galaxies form much earlier at redshifts Z $\gtrsim$ 30 and clusters of galaxies form later.

We do not enter into all possible scenarios for the origin and evolution of galaxies but consider only the case of the adiabatic model in which galaxies and clusters form at redshifts Z $\approx$ 5-10. The reasons for this choice are: (i) the inferred evolution of radio galaxies and quasars requires strong evolution at small redshifts and this is expected to occur if the redshift of formation of galaxies is small; (ii) at such small redshifts, young objects soon after the epoch of formation may be directly observable and therefore make the hypotheses susceptible to observational verification; (iii) adopting this model, we do not violate constraints set by the observed lack of fluctuations in the microwave background radiation and we alleviate the problems of ionising and reheating the intergalactic gas.

3. YOUNG OBJECTS

We adopt a scenario in which galaxies form at redshifts Z $\sim$ 5-10 by collapse on a hydrodynamical time-scale which is expected to be $\sim$ 10^8 years for a galaxy such as our own. In the first few hydrodynamical timescales, the first generations of stars form, the most massive ones evolve rapidly, explode and return material enriched in heavy elements to the interstellar gas. This general picture agrees with the requirement that the bulk of the heavy elements be formed in the first 10^9 years. The galaxy is therefore expected to be much brighter at these epochs than it is now, (i) because the luminosity of a galaxy is proportional to its rate of heavy element formation and this rate must have been at least 10 times the present rate and (ii) because during the first generations of star formation, there are many more hot young massive blue stars in the luminosity function in comparison with the function at the present day. Various estimates suggest that young galaxies should be $\sim$ 10-100 times brighter than our own Galaxy.

It is interesting to speculate further what the observable properties of young galaxies might be.

3.1. Optical Wavelengths

The optical luminosity of the young galaxy is of the same order as that of quasars. Probably most of the luminosity originates within a sphere of diameter $\sim$ 10 kpc. Therefore adopting a Hubble constant of 50 km s^{-1} Mpc^{-1}, the observed angular size of the galaxy will be less than 1" arc at all redshifts Z > 1 and hence it is indistinguishable from a star (cf Weymann 1967). The optical spectrum is likely to be dominated by the integrated light of young blue massive stars. If the birth rate function follows the Salpeter law, the superposition of their spectra will not result in a Planckian distribution but a spectrum more akin to a power-law extending into the ultraviolet region of the spectrum because of the presence of very hot stars. Whether or

not this is the observed spectrum depends upon the amount of dust and
interstellar gas present which will absorb strongly the ultra-violet
photons. Because of the high inferred rate of star formation, there
must also be many giant HII regions present emitting strong recombination and forbidden line radiation.

As noted by Field in 1964, such a model of a young galaxy resembles
in many ways the properties of quasars. Let us look at the statistics
of quasars at large redshifts in comparison with the numbers of galaxies.
At the present day roughly 1 in 100 quasars is a powerful radio source
and the space density of radio quiet quasars is about 10^{-5} that of
galaxies such as our own. If we adopt the exponential evolution law for
the quasar population, at large redshifts, quasars of all types were
more common by a factor of 2×10^4 and hence, within the uncertainties
in these numbers, the space density of quasars was of the same order
as young galaxies (Lynden-Bell 1969, 1971; Komberg and Sunyaev 1971).
It is also clear that this population of quasar-like young galaxies must
decay rapidly with time, a time-scale of 10^9 years appearing naturally
in this scheme, which might account for the results of the V/V_{max} test
and the counts of radio sources.

In this type of model, there is no danger of exceeding the upper
limits to the extragalactic optical background as may be seen from the
following simple order of magnitude estimate. When primordial hydrogen
is transmuted into heavy elements, a binding energy of $\sim 0.007 \ m_p c^2$
per nucleon is released which is eventually converted into light.
Therefore the ambient energy density in starlight U_{rad} due to the
formation of an abundance of x% of heavy elements is

$$U_{rad} = \frac{0.007 \ x \ \Omega_{gal} \rho_{crit} c^2}{100(1+Z)} \quad eV \ cm^{-3}$$

where ρ_{crit} is the critical cosmological density, $3H_0^2/8\pi G \approx 5 \times 10^{-30}$
g cm^{-3}, and Ω_g is the density parameter corresponding to the mass in
galaxies; the factor (1+Z) takes account of the effects of redshift
upon the energy of photons. Taking, as an example, x = 5%, Ω = 0.03,
Z = 5, we find $U_{rad} = 5 \times 10^{-3}$ eV cm^{-3}, a figure of the same order of
magnitude as the upper limit to the energy density of the optical
background of 3×10^{-3} eV cm^{-3}.

Although the above models resemble quasars in many ways, they are
quite distinct in three respects:

(i)   they have angular sizes $\sim 0\overset{''}{.}1-1''$ arc;

(ii)   they should not exhibit strong optical variability because
the integrated light is the superposition of the light of many stars.
Although supernovae might make a contribution to the total light,
individual supernovae are incapable of changing the luminosity of an
object having $M_V = -26$ by a factor of 2.

(iii) they should not exhibit optical polarisation because the light is the superposition of the emission of many stars.

These properties are different from those normally attributed to quasars but one wonders whether a significant number of quasars already catalogued might not turn out to be this second type of quasar. We note that the "typical" properties of quasar are based upon intensive studies of relatively small numbers of the brightest objects and that some quasars are much more stable in optical output. It would seem to be an observational programme of great interest to discover if any of the catalogued quasars turn out to be "quasars of the second type". A possible programme might consist of selecting those quasars which are optically stable, measuring their optical polarisations and then studying their angular sizes by, say, speckle interferometry.

3.2. Radio Wavelengths

According to this model supernova activity must have been much more intense during the first 10^9 years of a galaxy's life than it is at the present day. The question of how much brighter the galaxy would have been as a radio source depends upon many intangibles, in particular the question of how long the particles are confined within the young galaxy. To produce the observed abundance of heavy elements, we estimate that $\sim 10^9$–10^{10} supernovae must have exploded within the first 10^8–10^9 years, corresponding to supernova rates ~ 30 to 3000 times greater than that of our own Galaxy at the present time. Therefore, we might expect the radio luminosity of a young galaxy to be ~ 1000 times stronger as a radio emitter than our Galaxy, putting it in the class of weak radio galaxies. Radio sources of these luminosities can now be observed at cosmological distances in the deep surveys made by Earth-rotation synthesis radio telescopes. The radio structures are expected to be similar to those of normal galaxies, rather than double sources, and their angular sizes $\sim 1"$ arc. Thus an investigation of the quasars found in deep radio surveys may provide candidates for young galaxies.

3.3. X-ray wavelengths

Because there are many O and B stars in the first generations of stars, there are likely to be many binary X-ray sources consisting of O and B stars and a compact companion. These are expected to be intense X-ray emitters like those observed in our Galaxy. If a Salpeter mass function is adopted, there should be $\sim 10^3$–10^4 times more binary sources in a young galaxy than in our own Galaxy. Therefore, the X-ray luminosity of a young galaxy should be $\sim 10^{42}$–10^{43} erg s^{-1}.

Because there are more supernovae, there will also be more supernovae like the Crab Nebula which is an intense X-ray emitter at soft and hard X-ray energies ($L_x \sim 10^{38}$ erg s^{-1}). If there are 10^9–10^{10} supernovae in 10^8–10^9 years, there will be $\sim 10^3$–10^5 supernovae with ages less than 10^3 years at any time. If a significant fraction of these are Crab-like, the luminosity of the young galaxy will be Lx $\sim$ 10^{41}-10^{43} erg s^{-1}, even at hard X-ray energies.

Both of these types of X-ray source may be observable at cosmological distances with the next generation of X-ray telescopes.

3.4. Summary

Naturally, these are crude estimates but they make the important point that if the epoch of galaxy formation occurred at redshifts Z $\lesssim$ 5-10, young galaxies are potentially observable in many wavebands. Indeed, some young galaxies may already be present in catalogues of quasars and radio sources. Although the programmes to identify those objects are time consuming and arduous, the importance of the results would seem to justify this effort.

4. THE COSMOLOGICAL EVOLUTION OF QUASARS AND RADIO SOURCES

Do the above considerations have anything to do with the problem of the cosmological evolution of quasars and extragalactic radio sources? We may answer with a guarded "yes" in the sense that we can make a case for enhanced astrophysical activity of all types on a time-scale $\sim$ 10^9 years which was most intense when galaxies first formed and which has decayed since that epoch. We have argued that we do <u>not</u> have to account for the appearance of radio sources because it is the parent object which evolves cosmologically. What we have <u>not</u> explained is why real quasars, rather than quasars of the second type, exhibit cosmological evolution. Quasars and radio galaxies require the formation of a compact nucleus and we have given no discussion of how this might come about. A deeper understanding of the nature of galactic nuclei is required.

We give only one example of how real quasars and radio galaxies might fit into the overall picture. We adopt a model in which the nucleus consists of a massive black hole and the source of energy is accretion of interstellar matter. Many authors have discussed how the liberation of energy in an accretion disc can lead to optical phenomena which can account for the observed properties of quasars. During the early stages of evolution, the time-scale for mass loss from stars must have been short to account for the thorough mixing and build-up of the heavy elements. At the present epoch, however, the time-scale for mass exchange is $\sim$ 10^{10} years. If the luminosity of the quasar depends upon the available ambient interstellar gas density, as it does in accretion models, quasars should be much more active during the first 10^9 years. Speculation along the lines of accretion models of this type is encouraged by the observation of small scale archetypes of quasar-like phenomena in our own Galaxy. The X-ray binary system Sco X-1 is believed to be a typical accretion binary system. In addition it is known to "flicker" at optical wavelengths, possesses a highly variable compact radio component coincident with the binary system and has a double radio source on either side of the nucleus with angular separation ≈ 3' arc. In other words, this accretion system possesses all the characteristics of the most powerful quasars but on a very small Galactic scale. It may be that further studies along these lines will provide further insight into the nature of quasars, double radio sources and their cosmological evolution.

REFERENCES

Field, G.B., 1964. Astrophys. J., 140, 1434.
Komberg, B.V. and Sunyaev, R.A., 1971. Astron. Zh., 48, 235.
Lynden-Bell, D., 1969. Nature, 223, 690.
Lynden-Bell, D., 1971. Mon. Not. R. astr. Soc., 155, 119.
Rees, M.J. and Setti, G., 1968. Nature, 219, 127.
Schmidt, M., 1963. Astrophys. J., 137, 758.
Weymann, R., 1967. Unpublished preprint.

DISCUSSION

Tinsley: I would like to draw attention to a recent letter and paper in the Astrophysical Journal by D.L. Meier of the University of Texas at Austin. Meier has constructed synthetic spectra for young galaxies in the early stage of evolution discussed by Longair, with predictions of the colors, angular sizes, surface brightnesses, etc. that one would observe, as a function of the redshift of galaxy formation. Meier's calculations are based on Larson's dynamical models for galaxy formation, in which the metals do indeed form early, presumably with the high supernova rate mentioned by Longair and Sunyaev. In his first paper (Astrophys. J. Letters, 203, L103, 1976), Meier notes that primeval galaxies could be masquerading as red quasars. Of course, as Longair remarks, the general properties of primeval galaxies predicted by Meier do not depend strongly on the specific galaxy models used.

Longair: I would emphasise Dr. Tinsley's last remark. The hyperluminous nature of young galaxies at optical and radio wavelengths and their mimicing quasars would seem to be a very general property of young systems if the epoch of formation is z ~ 5 to 10 and does not depend very strongly on the assumed model.

Madore: Considering the large supernova rate necessary in your model is this not contradictory to your non-variability criterion to be used in selecting quasars to be studied? More specifically, if the number of supernovae is proportional to the number of early-type stars, won't the supernova contribution to the light (and variability) always be significant?

Longair: The integrated light due to supernovae may be significant but it is not possible to produce several magnitudes variability due to a single supernova explosion - it is just not bright enough in comparison

with the total luminosity of the quasar, say $m_v \sim -26$. Supernova
activity may produce statistical fluctuations of, say a few hundreds or
tenths of a magnitude but not the very large fluctuations.

Kafka: I wonder whether your 100 supernovae per year wouldn't dominate
the whole picture. Especially if you think of the nucleus being embedded
in gas and dust it will be difficult to predict the observable features.
I think you might as well expect to observe what we call quasars now.

Longair: I have covered the first part of this question in my answer to
Dr. Madore. There are very many possibilities when we know so little
about young galaxies. I would prefer to isolate those aspects which
would help us to distinguish genuine young galaxies from quasars where,
for the various reasons expressed in our paper, we believe that young
ordinary galaxies should be quite distinct from quasars.

Schmidt: Labeyrie has recently attempted to measure the angular diameter
of 3C 293 with the 200-inch telescope. Although no definitive results
are known yet, it is most likely that he did not resolve the object.
Hence the diameter is probably less than 0.02 seconds of arc, i.e. less
than 100 parsecs.

Longair: This is exactly the type of work which we would like to see
extended to faint quasars to find out if our type of quasar is lurking
in the present catalogues.

Oort: The subject of the symposium is inextricably tied up with the
problem of the formation of galaxies in an expanding universe, and this,
in turn, cannot be separated from that of the evolution of the unevenness
in the distribution of galaxies and the birth of galaxy clusters.
Important progress has been made in this domain by the Soviet astro-
physicists, and recently by Gott and by Silk. I wish there would have
been time to discuss these investigations properly in the context of the
present symposium. A thing that is equally fundamental for the discussion of evolution effects is insight into the process of the formation
and maintenance of radio components of galaxies and quasars. Though we
know nothing for certain as yet, a discussion of the work on this subject
would have been useful.

Longair: I agree completely with Professor Oort's sentiments.

POLARIZATION MEASUREMENTS OF DISTANT SOURCES

R.G. Conway
University of Manchester, Jodrell Bank

Polarization measurements, in addition to their interest for deciphering the internal physics of radio sources, are relevant to cosmology because estimates of n_e, the thermal electron density, may be obtained from measurements of the Faraday effect. We have:-

Position Angle $\quad \chi = \chi_o + R.\lambda^2$

Rotation Measure $\quad R = K.\int n_e B_{\parallel} d\ell$

$\qquad\qquad\qquad\quad K = 0.8 \times 10^6 \text{ rad.m}^{-2}/(\text{cm}^{-3}\text{G.pc})$

The foreground plasma, along the line of sight to the observer, can be investigated using R (see the following contribution by P.P. Kronberg). The degree of polarization, m, is not affected by foreground rotation, but is decreased if there is a contribution to $\int n_e B_{\parallel} d\ell$ arising within the source. The rotation from the back will be rotated relative to that from the front at long wavelengths but not at short wavelengths ("front-back depolarization"). If the Faraday dispersion Δ is the rotation measure of the equivalent uniform slab, then the polarization falls to one half at

$$\lambda_{\frac{1}{2}} = (1.90/\Delta)^{\frac{1}{2}}\text{m}$$

Hence one may estimate n_e within the source from measurements of $\lambda_{\frac{1}{2}}$, provided i) that side-side variations of R can be neglected ii) that the source is transparent.

Conway et al. (1974) estimated n_e in quasars this way, using only those sources which on spectral evidence seemed to contain no opaque components. Assumption i) regarding side-side depolarization rested on rather slender evidence (see below).

The values of n_e ranged from 3.10^{-5} to 3.10^{-2}cm^{-3}. (Note that because of an error the published values were too low by x10). The values show considerable scatter, but do increase on average with z, as

$$n_e = 10^{-4} \times (1+z)^{3.4 \pm 3.7}$$

D. L. Jauncey (ed.), Radio Astronomy and Cosmology, 361-366. All Rights Reserved.
Copyright © 1977 by the IAU.

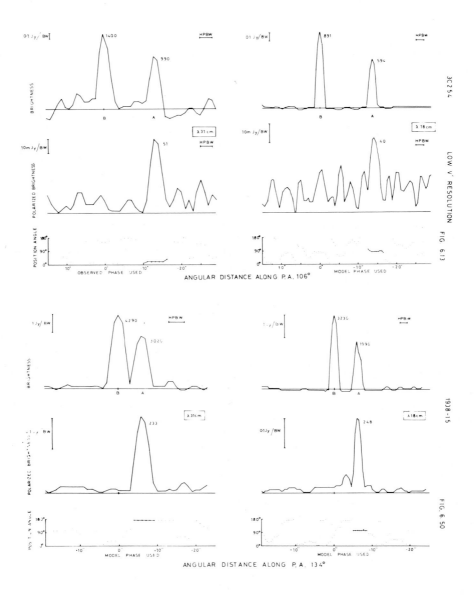

Figs 1 and 2   Polarization measurements at λ18 and 31 cm made with
a 23 km interferometer by R.J. Davis. Upper 3C254, lower PKS1928-15.
For each source the top strip distribution gives I(θ), the middle
strip gives mI(θ) and the lowest strip gives χ(θ).

Clearly the data agree with a wide range of cosmologies. One might expect $n_e \propto (1+z)^3$ since the ambient density presumably has this form. The mean $n_e/(1+z)^3$ is $1.4 \times 10^{-4} \mathrm{cm}^{-3}$ compared with a closure density of 3×10^{-6} cm^{-3}, an apparent enhancement over the ambient of at least a factor of 50. Since ram-pressure models, e.g. De Young and Axford (1967) no longer seem satisfactory, we do not appear to have a means of predicting this enhancement factor. However, if there were a satisfactory theory of radio sources which predicted the enhancement, we could explore the ambient density of the Universe as a function of z by means of these measurements.

More evidence is now available on the distribution of polarization, which bears on the assumption that "side-side depolarization" is negligible. Hogbom and Carlssen (1974) have published maps of polarization at 21 cm obtained with the WSRT in Holland. They comment that regions of low T_b are more polarized than those of high T_b: bridges may be ~20% but hot spots are usually <10%. This is clearly relevant for source physics, but we should remember that the sources studied have large angular size, low luminosity and low redshift. At high redshift, surface brightness decreases as $(1+z)^{-4}$ and high-z sources tend to show only hot spots. Hence Hogbom and Carlssen's remark about low T_b regions may not be important for distant sources.

Hogbom and Carlssen selected extended sources. R.J. Davis at Jodrell Bank (unpublished measurements) selected a sample of 20 sources with very compact components. The observations with a 23 km interferometer at λ18 and 31 cm give strip-distributions along the major axis (Figs 1 and 2).

Although the intensity ratio of the components is near unity, the polarizations are often very dissimilar, especially at λ31 cm. Effectively all the polarized flux comes from one component. When this is the case the integrated polarization follows that of the polarized component, and "beating", or the side-side depolarization of Component A by component B does not occur. There seems no rule as to which component will be the more polarized: it does not depend on intensity, spectral index, nor on proximity to the optical object. We are investigating this effect further in larger samples complete to defined selection limits. We suspect that the effect may not be universal but may relate to this particular sample, possibly because the components are so compact.

Conway, Burn and Vallee (in press) mapped 72 sources from the 4C catalogue using the WSRT at 6 cm and 21 cm (Fig. 3). Out of the 72 sources, 45 are double with $\theta \geq 12"$, and 4 are triple with a central component. Taylor and Conway (unpublished) have measured the same set of sources at 18 cm and 31 cm, giving polarizations at four wavelengths of each component of the 45 double sources. The optical fields have been studied by Padrielli at Bologna (Padrielli and Conway, in press), who finds 19% to coincide with a blue object or quasar, 24% with a Galaxy, while 57% are unidentified. The majority are near or beyond

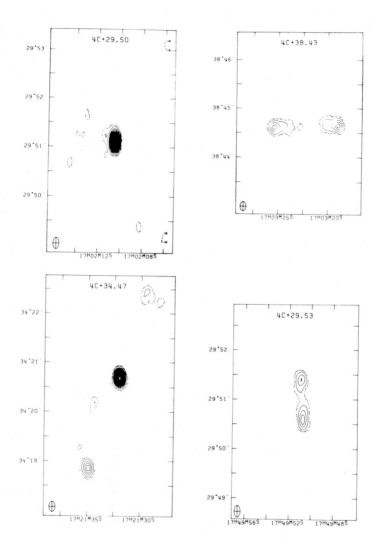

Fig. 3   Maps of 4 sources from the sample studied by Conway, Burn and Vallee. The beam of the WSRT at λ6 cm, 7 arcsec in RA, is shown in the lower left corner of each map.

the P.S.S. plate limit, and we suggest $z \sim 0.25$ for G, $z \sim 0.5$ for U and $z \sim 1.0$ for B objects. For calculating n_e, only very approximate z-values are required, as the dependence on both z and q_0 of the derived n_e is very low. For fixed $\lambda_{\frac{1}{2}}$, S α and θ, the derived values of n_e do not vary by more than a factor 1.9 within the ranges $0.1 < z < 2.0$ and $\theta < q_0 < 1$. Though the n_e values obtained for radio sources will not help to decide q_0, they will, when z is measured, be useful for studying $n_e(z)$. The n_e values of these sources, or rather for the components, range from 10^{-4} to 3×10^{-2} cm^{-3}, very similar to those of the known quasars studied previously.

Even with 4 wavelengths of measurement, the solution for rotation measure may be ambiguous (Haves, 1975). We believe that for 33 sources in the 4C sample unambiguous values for RM can be obtained from the integrated polarizations, giving the intrinsic position angle and hence the orientation of the B-vector on the sky. For 16 of these sources, unambiguous solutions can be obtained for both components. The orientation of the B vectors in the components is usually but not always close to the major axis of the source. If θ_B is the angle projected on to the sky between these two directions, the most probable value of θ_B is $15^\circ \pm 10^\circ$. Approximately one third of the sources have θ_B greater than 45°, i.e. The B vector lies more nearly across the major axis than along it. These statistics are very similar to the results collected by Haves and Conway (1975).

In all cases (even where RM is ambiguous) it is clear that the rotation measures of the two components of a double source are always closely equal. The median value of $RM_A - RM_B$ is 8 rad.m^{-2}. We may draw two conclusions:

i) the differential Faraday rotation between two components produces negligible "beating" even when both are polarized. Hence the assumption by Conway et al. (1974) that "side-side depolarization" can be neglected is supported by the subsequent data.

ii) if a double source were embedded in a cocoon of plasma, one would expect the more distant component to show an extra contribution in rotation measure. If one assumes a separation of 100 kpc, and a field of 10^{-7} Gauss, the value of 8 rad.m^{-2} suggests that the density of such a cocoon is not more than $10^{-3}(1+z)^2$ electrons.cm^{-3}.

References

Conway R.G., Haves P., Kronberg P.P., Stannard D., Vallee J.P. and Wardle J.F.C. (1974). M.N.R.A.S. 168, 137

De Young D.S. and Axford W.I. (1967). Nature 216, 129

Haves P. (1975). M.N.R.A.S. 173, 553

Haves P. and Conway R.G. (1975). M.N.R.A.S. 173, 53P

Hogbom J.A. and Carlssen I. (1974). Astron. and Astrophys. 34, 341

DISCUSSION

Ryle: I think it may be dangerous to deduce too much information from
the integrated polarization from components of double sources.   The
polarization structure of Cygnus A observed with 2" arc resolution shows
a highly complex situation in the tail, with 8 or so components each
having strong percentage polarization and considerable complexity in
the heads.   Similar complexity is evident in many of the 48 3C sources
observed by Pooley and Henbest.   At larger red-shifts, where the components cannot be so well resolved, the integrated polarization may
have a magnitude and position angle which varies in a way not attributable in a simple way to Faraday rotation.

Conway: I entirely agree about Cygnus A, which I am sure is very complex.   I would differ in regarding it as typical: Cygnus A is better
described as a rogue elephant among sources.   It has always been a
puzzle to explain why integrated polarizations behave with the regular-
ity that they do - $m(\lambda)$ always decreases, position angle varies as λ^2
and so on - when one expects a random drunkard's walk behaviour from
the addition of many domains.   However, the regularities are there,
and need explanation. My working hypothesis is that in each component
there is one dominant domain.   My present results suggest that the
dominant domains in double sources are connected, and have the same
position angle.   Clearly, as you say, we need the maximum resolution
at as many wavelengths as possible.   A.J. Kerr, a student at Jodrell
Bank, is at present studying 96 3CR sources at 31 cm at resolutions
going down to 1 arcsec.   This is a pilot project for polarization work
with Jodrell Bank's new Multi-Element Radio-linked system at present
being put together.

ROTATION MEASURES AND COSMOLOGY

Philipp P. Kronberg
Max-Planck-Institut für Radioastronomie, Bonn, F.R.G. and
Scarborough College, University of Toronto, Canada

1. FARADAY ROTATION IN AN INTERGALACTIC MEDIUM

1.1 General remarks

The very low upper limits on distributed intergalactic (i.g.) HI (Gunn and Peterson 1965, Wampler 1967), and H_2 (Field et al. 1966) have made it clear for some time now that if a smooth distributed i.g. gas exists in significant amounts ($\Omega_{ig} \gtrsim 10^{-3}$), it must be ionized. X-Ray emission from rich clusters of galaxies such as Coma, indeed show that an intracluster gas can be seen (cf. Field 1974). Also persuasive in this connection are the head-tail radio galaxies in clusters, whose prototype is NGC 1265 (Miley et al. 1972, Jaffe and Perola 1973).

The use of Faraday rotation data to test for i.g. gas has the advantage that it is insensitive to the temperature of a non-relativistic ionized gas (unlike the X-ray emission), and the disadvantage that we are measuring only the product $nB_{\parallel}$ since a magnetic field is also required. The rotation measure due to an intergalactic magneto-ionic medium (RM_{ig}) can be written in its most general form as follows:

$$RM_{ig} = C \int_{0}^{z_s} \frac{n(z)\ B_{\parallel}\ (z)}{(1+z)^2}\ dr(z) \qquad (1.1)$$

where z_s is the redshift of the source, $n(z)$ and $B_{\parallel}(z)$ are the average values of electron density and line-of-sight component of magnetic field respectively, and C is a constant. If the local values of n and $\underset{\sim}{B}$ increase as

$$n(z) = n_0 (1+z)^3 \qquad (1.2)$$

$$|\underset{\sim}{B}(z)| = |\underset{\sim}{B}_0| (1+z)^2 , \qquad (1.3)$$

where n_0 and B_0 are the present-epoch values (B is assumed to decrease adiabatically as the universe expands), then the sharp increase of $nB_{\parallel}$

D. L. Jauncey (ed.), Radio Astronomy and Cosmology, 367-377. All Rights Reserved.
Copyright © 1977 by the IAU.

overcomes the watering down effect of the Doppler shift $(\alpha(1+z)^{-2}$ —
see eq. 1.1). Under these conditions our sensitivity to an intergalactic magneto-ionic medium improves with increasing redshift. It will be
noted that our simple assumptions in 1.2 and 1.3 ignore accretion to
galaxies or protogalaxies over the redshift range investigated — i.e.
no local evolution up to the maximum look-back time. One might alternatively assume that the density of intergalactic gas clouds is determined by local gravity rather than the cosmological scale factor (cf.
Rees and Reinhardt 1972), in which case the average values of n and $\underset{\sim}{B}$
are independent of z (again ignoring evolution).

Two basic kinds of intergalactic magnetic field may be envisaged
— a uniform, aligned primordial field, and a turbulent, random i.g.
field which changes on the scale of i.g. gas clouds or smaller. Since
the effect of each case on the observed rotation measures is different,
we shall discuss them separately below, and briefly review previous
attempts using RM data to detect an intergalactic RM.

1.2 Effect of an aligned, primordial magnetic field associated with an early phase of the universe

The possibility of such a field has been discussed by Woltjer
(1965) and Zel'dovich (1965). Given the presence of an intergalactic
plasma we would see a prevailing component of rotation measure (RM_p)
whose intensity would vary with direction and redshift in a Friedmann
universe with zero cosmological constant as follows

$$RM_p(\theta,z) = \frac{e^3\,B_{p_0}\,n_0\,\cos\theta}{H_0 \cdot 2m_e^2\,c^3\,q_0^2}\left\{(1+2q_0z)^{3/2} + (6q_0 - 3)(1 - 2q_0z)^{1/2} +(2 - 6q_0)\right\}$$

$$(1.4)$$

(Brecher and Blumenthal 1970), where n_0, B_{p_0} are the local values of
electron density and prevailing field, θ the angle between the line of
sight and the primordial field direction, m_e, e the electron mass and
charge and c the velocity of light. Plots of RM_p (Figure 1) show that
the prevailing component of RM increases with redshift, given the
assumptions of (1.2) and (1.3).

Sofue et al. (1968) and Kawabata et al. (1969) have interpreted
earlier RM data as supporting the existence of a large-scale magnetic
field, whereas Reinhardt (1972), Vallée (1975), and I (1976, unpublished) using successively larger RM data sets have concluded that
there is no positive evidence for such a field. Any large scale
asymmetry in the distribution of rotation measures on the sky due to a
primordial field has to be disentangled from similar variations on a
large angular scale due to Faraday rotation in the Galaxy. This separation is difficult to make. Ideally we could test for B_p by comparing
the redshift dependence of the mean rotation measure for separate
directions in the sky. Unfortunately, a three-dimensional subdivision
of the limited RM sample (e.g. l, b, z) for highly redshifted QSO's

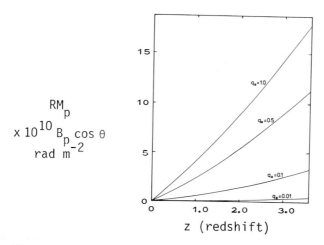

$$RM_p$$
$$\times 10^{10} \, B_{p} \cos \theta$$
$$rad \ m^{-2}$$

z (redshift)

Figure 1. The variation of intergalactic rotation measure with redshift in the presence of a uniform i.g. gas and a primordial magnetic field aligned over $z \gtrsim 3.5$ in a Friedmann universe.

drives us into the unhappy realm of small-number statistics. We shall need to have a larger and more homogeneous sample of RM data in order to conduct a sensitive test for any primordial magnetic field which is uniform over scales of up to $z \sim 3$.

1.3 Faraday rotation in a random, intergalactic magneto-ionic medium

In this case a polarized radio wave performs a random walk in Faraday depth as it traverses independent cells of $n_i \, B_{\parallel i} \, l_i$ (cf. Rees and Reinhardt 1972, Nelson 1973). This will cause a redshift dependence of the scatter, or variance of the distribution of rotation measures of extragalactic radio sources. Using a sample of sources in which ~ 12 had $z > 1$, Mitton and Reinhardt (1972) found a systematic decrease with redshift in the RM scatter: Nelson (1973), whose data included 8 sources having $z > 1.1$, found the variance ($V(z)$) of RM to increase with redshift, and Vallée (1975) using the RM list of 251 sources of Vallée and Kronberg (1975) (of which 21 have $z > 1$) found no significant redshift-dependent effect. A more recent analysis by myself and M. Simard-Normandin (1976) will be described in §2. Before doing this, I shall briefly describe two simple intergalactic medium models in a Friedmann universe with zero pressure and zero cosmological constant ($\Lambda = 0$).

In the first case (model 1) we assume that the average values of $n(z)$ and $B_\parallel (z)$ scale according to (1.2) and (1.3) as before, and that the cell size, $l(z)$, also follows the cosmological scale factor $R(t)$, and that the number of cells to a distant QSO is large. Kronberg, Simard-Normandin and Reinhardt (1976) have shown that the variance $V(q_0, z)$ of RM can be expressed as

$$V(q_0, z) = 9.4 \times 10^{11} \, l_0 \, \eta^2 \, q_0^2 \, H_0^3 \, |B_{||_0}|^2$$

$$\left[\int_0^{z_s} (1+z)^{3/2} (1+2q_0 z)^{-1/4} \, dz \right]^2 \, \text{rad}^2 \, \text{m}^{-4} \, , \qquad (1.5)$$

where z_s is the source redshift, η the fraction of matter in the form of intergalactic ionized gas (assumed here to be 100% hydrogen and independent of epoch), H_0 (km s^{-1} Mpc^{-1}) the Hubble constant, and l_0 (Mpc), $|B_{||_0}|$ (gauss) the average zero-epoch values of cell size and magnetic field. The cells are assumed in this case to be contiguous, that is l_0 represents both the average size and separation of the clouds.

In the second model, (2), the values of n $B_{||}$ and r the cell radius are assumed independent of cosmological epoch; only the average separation scales as $R(t)$. The variance in this case is given by

$$V(q_0, z) = 9.4 \times 10^{11} \, l_0 \, |B_{||_0}|^2 \, f_1^{2/3} \, \eta^2 \, q_0^2 \, H_0^3$$

$$\left[\int_0^{z_s} (1+z)^{-2} (1+2q_0 z)^{-1/4} \, dz \right]^2 \, \text{rad}^2 \, \text{m}^{-4} \qquad (1.6)$$

(Kronberg et al. 1976). For this model, the Faraday rotating cells must occupy only a fraction, f_1, of i.g. space, so that they do not overlap up to the maximum redshift, i.e. $f_1 < (1+z_s)^{-3}$.

Figure 2 shows plots of $V(z)$ for different q_0 values. In model 1 (equation 1.5) the variance (solid lines) increases rapidly with redshift, and hence also the prospects for detecting an intergalactic medium. A plot of the solution of $V(z)$ for model 2 for $q_0 = 0.5$ on the other hand shows that if this more aptly represents the true intergalactic medium the variance is relatively insensitive to redshift above $z \simeq 1$. It is also important to note that, for a given q_0, the variance increases as l_0 in both models, and is a sensitive function of H_0, η and $|B_{||}|$.

2. NEW EVIDENCE ON THE ORIGIN OF ROTATION MEASURE

2.1 The relative galactic and extragalactic contributions

It has been well established that the interstellar medium imposes a systematic variation of RM in (l,b) coordinates (Gardner and Davies 1966, Gardner, Morris and Whiteoak 1969, Mitton 1972, Vallée and Kronberg (1975). The extragalactic contribution has been commonly thought

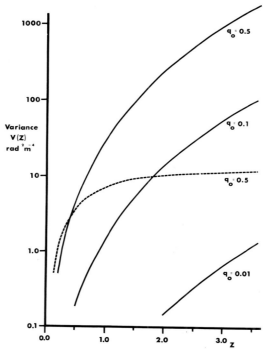

Figure 2. The calculated variation of V(z) for models 1 (solid lines) and 2 (dashed line) over the redshift range $0 < z < 3.6$. The following values were assumed: $B_0 = 1.8 \times 10^{-8}$ gauss, $\eta = 1$, $H_0 = 75$ km s^{-1} Mpc^{-1}, $l_0 = 1$ Mpc, and $f = 1/64$ for model 2. Model 1 is shown for q_0 (= $\Omega/2$) values of 0.5, 0.1 and 0.01.

to be small; $\lesssim 10$ rad m^{-2}. From more recently published data I have assembled a list of 450 rotation measures, which is substantially larger than any sample previously available. I shall briefly summarize the results of an examination of the new data, which shed some new light on the nature of the extragalactic contribution to RM (Kronberg and Simard-Normandin 1976).

2.2 A revised interpretation of the RM data

When we compare the RM distributions of the 450 sample at high and low ($|b| \gtrsim 30°$) galactic latitudes (Fig. 3), we see that the fraction of very large RM's (> 200 rad m^{-2}) is to first order independent of galactic latitude (compare Figs. 3(a) and (b)), whereas the very small RM's (< 25 rad m^{-2}) occur almost exclusively at the high latitudes. Comparison of the distributions in Fig. 3(a) and (b) (see especially the insets) shows that the effect of the galaxy is to "smear out" the sharp peak near $|RM| = 0$, but only to ~ 120 rad m^{-2} at most. It is very likely that the galactic RM commonly exceeds this limit at

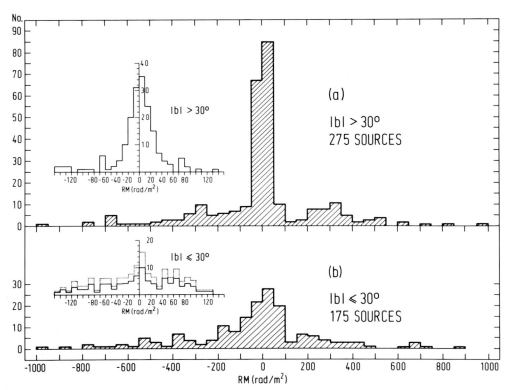

Figure 3. Histograms showing the distribution of the rotation measures for (a) $|b| > 30^\circ$ and (b) $|b| < 30^\circ$. The insets show the central peaks with intervals of 10 rad m^{-2}. The dotted profile in (b) shows the same distribution scaled by 275/175 for comparison with that in (a). (Reproduced from Kronberg, P.P. and Simard-Normandin, M., Nature 263, 653, 1976.)

very low latitudes ($|b| \gtrsim 5^\circ$), however our sample contains too few sources at $|b| < 5^\circ$ to reliably test the galactic contribution in this region. Figure 3 is instructive, and leads us to some new and interesting conclusions. The first is that at $|b| \gtrsim 5^\circ$, the very large RM's ($\gtrsim 200$ rad m^{-2}) must be produced outside of the galaxy, since the fraction of sources with $|RM| > 200$ is independent of path length through the galactic disc. It follows that the galactic contribution to RM is usually less than ~ 120 radians in the range $5^\circ < |b| < 30^\circ$, and very much smaller at the higher latitudes. The distribution for $|b| > 30^\circ$ (Fig. 3(a)) therefore closely approximates the distribution of extragalactic RM's. This brings me to the third conclusion, namely that there appear to be two quite distinct RM populations; those with very low extragalactic RM which I shall call class I which form the narrow peak in Fig. 3(a), and a class II with RM $\gtrsim 120$ rad m^{-2} which is largely extragalactic in origin.

The separation into class I and class II RM's is both good and bad
for the purpose of analysing their origin. For the galaxy it is good,
in that by simply omitting all RM's > 150 rad m^{-2} the galactic contribution can be seen more clearly than hitherto. For our present purpose,
the existence of class II (high extragalactic RM) tends to raise the
level at which we can expect to see an intergalactic contribution.

Kronberg and Simard-Normandin (1976) have investigated the
variance of RM with redshift for 108 of the 450 which are QSO's of
known redshift up to z $\simeq$ 2.5. The variances (after removing the
galactic contribution) are 3 - 6 x 10^4 rad^2 m^{-4}, and we detected no
convincing redshift dependence of V(z). We repeated the investigation
for a control sample of 90 radio galaxies over their (much smaller)
range of z. The variances again showed no clear z-dependent variation,
furthermore collectively they are indistinguishable from those of the
more distant QSO's. Although the samples are different, this result
suggests that the large RM's are not a sensitive function of intergalactic path length and must be generated around or in the sources
themselves. This result is not entirely surprising (it was also surmised by Reinhardt (1972)), in that the very rapid depolarization rates
in some sources suggest a *priori* an associated large RM at the source.
However an attempt to correlate depolarization rate and RM shows a
poor correlation. If we could succeed in isolating class I sources by
some independent measured parameter (e.g. surface brightness or source
size), it is clear that this class of source would make a much more
sensitive probe of intergalactic Faraday rotation than the present
total sample of QSO's.

2.3 Application of the RM data to i.g. medium models

When we compare the formal variances of various sub-groups of our
sample of 108 QSO RM's we can place an approximate upper limit of
1 x 10^4 rad^2 m^{-4} on any systematic increase of V(z) over the range
0 < z < 2.5. This, for q_0 = 0.5 and z = 2.5, gives an upper limit for
model 1 of $\eta n_0 |B_{||_0}| 1\frac{1}{2} \lesssim$ 2.4 x 10^{-13} gauss cm^{-3} Mpc$^{\frac{1}{2}}$ which is only a
factor of 4 above the value of 6.3 x 10^{-14} gauss cm^{-3} Mpc$^{\frac{1}{2}}$ which
corresponds to the curve for q_0 = 0.5 (model 1) in Figure 2. The
present data can rule out a model 1 - type i.g.m. in which q_0 = 1,
l_0 = 10 Mpc and $|B_{||_0}|$ = 5 x 10^{-8} gauss, and η = 0.8. On the other
hand if model 2 is more nearly correct, the RM data are much too insensitive (by 2 or more orders of magnitude in the variance) to detect
an i.g. medium.

Ignoring the formal variances discussed above we can also test
for the presence of a sub-group of high-z QSO's with very small RM's
— by analogy with Class I in Figure 3(a). Figure 4 shows the
distributions of residual RM (RRM) for the QSO's after subtracting the
galactic contribution, and that there may be such a sub-group visible
at high z. The numbers in this sample are too small to give reliable
results, but if more data were to reveal the presence of the class I
component (Fig. 3a) over 0 < z < 2.5 our sensitivity to i.g. rotation

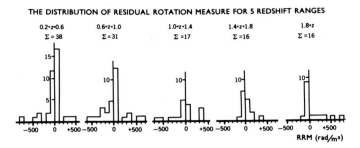

THE DISTRIBUTION OF RESIDUAL ROTATION MEASURE FOR 5 REDSHIFT RANGES

Figure 4. Histograms showing the distribution of the
residual rotation measures (RRM) for 5 equal redshift
intervals beginning at z = 0.2. (Kronberg, P.P. and
Simard-Normandin, M., Nature 263, 653, 1976.)

measure would improve by at least an order of magnitude and provide a
much more stringent upper limit on the quantity $(q_0 \eta |B_{||0}| \, 1_0^{\frac{1}{2}} \, H_0^{\frac{3}{2}})$.

Finally I should remark that in our model 1 we did not assume
clumping of the i.g. gas. If the i.g. gas clumps conserving mass and
magnetic flux, the variance increases as f_2^{-2} where f_2 is the volume
filling factor. In other words, one can make plausible versions of
models 1 and 2 in which the same mass of i.g. HII clouds can produce
a larger Faraday rotation.

3.   CONCLUDING REMARKS

We have shown that most of the source-to-source scatter in the RM
of extragalactic sources is produced at or in the sources themselves,
and that the contribution by the galaxy or intergalactic Faraday
rotating clouds is usually small by comparison. This makes it difficult
to look for variations of V(z) unless class I type can be independently
isolated, in which case they would constitute a sensitive probe for an
intergalactic medium. RM data on QSO's may in the near future provide
an important test for intergalactic matter and on the deceleration
parameter.

I wish to thank the Alexander von Humboldt-Foundation (Federal
Republic of Germany) and the National Research Council of Canada for
support of this work.

REFERENCES

Brecher, K.I. and Blumenthal, G.R.   1970, Astrophys. Letters 6, 169
Field, G.B.   1974, IAU Symp. No. 63 "Intergalactic Gas", p. 13
Field, G.B., Solomon, P.M. and Wampler, E.J.   1966, Astrophys. J.
        145, 351
Gardner, F.F. and Davies, R.D.   1966, Australian J. Phys. 19, 129
Gardner, F.F., Morris, D. and Whiteoak, J.B.   1969, Australian J. Phys.
        22, 813
Gunn, J.E. and Peterson, B.A.   1965, Astrophys. J. 142, 1633
Jaffe, W.J. and Perola, G.C.   1973, Astron. Astrophys. 26, 423
Kawabata, K., Fujimoto, M., Sofue, Y. and Fukui, M.   1969, Publ.
        Astron. Soc. Japan 21, 293
Kronberg, P.P. and Simard-Normandin, M.   1976, Nature 263, 653
Kronberg, P.P., Simard-Normandin, M. and Reinhardt, M.   in preparation
Miley, G.K., Perola, G.C., van der Kruit, P.C. and van der Laan, H.
        1972, Nature 237, 269
Mitton, S.   1972, Monthly Notices Roy. Astron. Soc. 155, 373
Mitton, S. and Reinhardt, M.   1972, Astron. Astrophys. 20, 377
Nelson, A.H.   1973, Publ. Astron. Soc. Japan 25, 489
Rees, M.J. and Reinhardt, M.   1972, Astron. Astrophys. 19, 189
Reinhardt, M.   1972, Astron. Astrophys. 19, 104
Sofue, Y., Fujimoto, M. and Kawabata, K.   1968, Publ. Astron. Soc.
        Japan 20, 388
Vallée, J.P.   1975, Nature 254, 23
Vallée, J.P. and Kronberg, P.P.   1975, Astron. Astrophys. 43, 233
Wampler, E.J.   1967, Astrophys. J. 147, 1
Woltjer, L.   1965, "The Structure and Evolution of Galaxies" (Inter-
        science, New York)
Zel'dovich, Ya.B.   1965, Soviet Phys. JETP 21, 656

DISCUSSION

Conway: The z-dependent variance of rotation measure in your "narrow peak" is equivalent to $[3 \text{ rad m}^{-2}]^2$. In other words, sources at z = 2.5 suffer rotation measure equal to one tenth of that imposed by the Milky Way. This seems to argue rather strongly against absorption in quasars being due to intervening galaxies, unless these are unusual, or else exactly face-on.

Goldstein: How many frequencies go into a typical Faraday rotation measurement?

Kronberg: The average is about 5.5 different frequencies, but it is never less than 4.

THE POLARIZATION OF 3C123 AND 3C427.1

R.G. Strom

P.P. Kromberg and I have made high resolution studies of the luminous radio galaxies 3C123 and 3C427.1 with the NRAO interferometer at 8085 MHz and the Westerbork telescope at 4995 MHz. Both objects, which appear to be very distant (for 3C123, z=0.637), have an integrated degree of polarization which drops very rapidly with increasing wavelength. The 8085 MHz map of 3C123 shows a weak radio bridge extending from the more compact component to the edge of the optical galaxy. At both frequencies we find no emission from the galaxy itself stronger than 0.1 Jy, which combined with the 15 GHz flux density of 0.1 Jy reported by G. Pooley suggests $\alpha \geqslant 0$, much flatter than any other emission within the source.

At 4995 MHz, our good sensitivity to extended emission reveals structure well south of, and connected with, the high brightness component. This gives an overall size of nearly 1' arc, or a linear extent of 320 kpc ($90=0.1$, $H_o=75$ kms^{-1} Mpc^{-1}). Comparing the polarization structure in the two maps when convolved to the same resolution, we find the projected magnetic field is directed along the axis of the extended component. Changes in rotation measure exceeding 400 rad m^{-2} along the component can be explained by the projection of a unidirectional magnetic field running from one end of the component to the other. In 3C427.1 we also observe rotation measure changes exceeding 200 rad m^{-2}. Radio emission from the optical galaxy associated with 3C427.1 does not exceed 1% of that from the outer components.

McEllin: C.J. Jenkins and I have been working on the classification of a complete sample of sources in terms of the proportion of flux density coming from 'hot spots'. We find a strong tendency for high luminosity sources to have very prominent hot spots and weak or non-existent bridges, and, as Strom suggests, this is also the case for the unidentified sources. The population of unidentified objects is in fact very similar to the population of identified sources with powers at 178 MHz greater than 5 x 10^{26} W Hz^{-1}sr^{-1}, and quite different from the population of objects with powers less than this. We conclude that most of the unidentified objects must be at redshifts larger than about 0.5.

INTERGALACTIC FARADAY ROTATION

Yoshiaki Sofue

Statistical analyses are made of Faraday rotation measures RM and redshifts z of extragalactic radio sources. Correction for local contribution from the galactic disk has been made by using the polarization data of pulsars. Absolute magnitude of RM of extragalactic sources increases with redshift. A correlation coefficient r between the corrected RM and z·cos θ is calculated for various assumed directions (l_o, b_o) of a uniform intergalactic magnetic field, where θ denotes an angle between the source direction and the field direction. The correlation coefficient attains its maximum of r=0.65 for 30 sources with $z > 1.0$, when the field is assumed to run toward $(l_o, b_o) \cong (64°, 12°)$. Whereas the maximum correlation coefficient for sources with $z < 1.0$ is only 0.12. These facts indicate the existence of Faraday effect due to a large scale and ordered magnetic field, uniform at least up to $z \cong 2$. The uniform field is of strength of $\sim 10^{-9}$ gauss, if the intergalactic electron density is $10^{-5} cm^{-3}$.

If we take into account the effect of the space curvature and its evolution in a Friedmann universe, we may be able to get a better fit to the observed RM than by using the straight line fitting. In fact, we have discussed magnetic field configurations which may occur in a Friedmann universe when there is no electric current, and have further derived theoretical expressions for the Faraday rotation measure of distant objects in the Friedmann universes (open, flat and closed), assuming that the field lines are geodesics. These expressions are then used to determine the deceleration parameter q_o by use of the χ^2 - fitting method for the observed RM of distant QSOs $(z > 1.0)$. The best χ^2 fit was obtained for $q_o \gtrsim 1$, suggesting a closed universe, although the conclusion may be rather tentative because the amount of data seems not yet sufficient.

[References: SOFUE, Y., FUJIMOTO, M. and KAWABATA, K. 1976, submitted to Publ. Astron. Soc. Japan; and GAFFET, B. and SOFUE, Y. 1976, submitted to Publ. Astron. Soc. Japan]

COSMOLOGICAL INFORMATION FROM NEW TYPES OF RADIO OBSERVATIONS

William C. Saslaw
University of Virginia, and National Radio Astronomy Observatory,* Charlottesville, and Institute of Astronomy, Cambridge, England.

To have squeezed the Universe into a ball
To roll it toward some overwhelming question.

T. S. Eliot

     So far, we have mainly discussed what might be called classical methods for investigating cosmological models using radio observations. These mostly involve the number counts as a function of measured flux, angular diameters, and spatial distribution of radio sources at cosmological distances, as well as the measurements of the microwave background. In this review, I have been asked to give a brief description of some new, or non-classical, measurements that might be made. It seems that such measurements fall into one of three categories: Observations which are being made now and which may bear fruit in the next several years, observations which may be possible in the intermediate future several decades from now, and observations which may be possible in the far future - if ever. Naturally I'll try to concentrate on the first two categories since they will be of more interest to most living astronomers during their working lifetimes. Also, since time is short and it is very difficult to predict the future, I'm afraid that this review will be somewhat impressionistic rather than exhaustive. I hope people will describe additional methods during the discussion.

I.   OBSERVATIONS POSSIBLE NOW

     There are a number of new types of observations which have been made recently and which will be extended in the next several years to give cosmological information. One of the most rapidly developing methods is the radio analogue of the optical magnitude-redshift-diagram.

* The National Radio Astronomy Observatory is operated by Associated Universities, Inc. Under contract with the National Science Foundation.

D. L. Jauncey (ed.), Radio Astronomy and Cosmology, 379-387. All Rights Reserved.
Copyright © 1977 by the IAU.

This is being pursued by Fisher and Tully at NRAO and Marseille. The
basic idea is to measure the 21 cm. redshifts of a large number of
spiral galaxies – which typically requires only several minutes per
redshift with a large low noise telescope – and then to find an indicator of each source's intrinsic 21 cm. radio magnitude. Measuring the
apparent 21 cm. magnitude would then, in principle, enable one to plot
the standard Hubble diagram and determine H_O and possibly q_O (without
the important evolutionary radio corrections, of course). Alternatively,
if the 21 cm. measurement provides an indicator of a sources intrinsic
optical luminosity, its apparent optical magnitude could be used.

The chief problem with this method therefore is finding a reliable
indicator of the source's intrinsic luminosity. There are several
possibilities. It may be that for galaxies in a narrow morphological
class, the intrinsic 21 cm. luminosity can be shown to depend mainly on
mass. To do this one would examine galaxies whose distance is known
independently (e.g. members of the Virgo Cluster) and whose mass can be
found from the width of the 21 cm. profile. Since these results would
also depend on the size and orientation of the galaxy, it is necessary
to measure a large statistically homogeneous sample. An exploratory
start in this direction has been made by Fisher and Tully (1975) using
241 dwarf galaxies catalogued by the David Dunlap Observatory. They
have found some weak evidence for a slightly simpler relation: a dependence of the hydrogen M_H/L ratio (which does not depend on distance) on
the absolute photographic magnitude of the irregular dwarf DDO galaxies.
Here the hydrogen mass is determined directly from the 21 cm. flux density corrected for beam dilution, and L is the intrinsic photographic
luminosity. And they have also found stronger evidence for a relation
between a measure of the total mass of a galaxy and its absolute photographic magnitude. The total mass measure of the galaxy is taken to be
proportional to (distance).(Holmberg angular diameter).(full line width
at half intensity)^2.(cosec^2 inclination). Thus if these two relations
were to hold, it would be possible to measure the distance independently
of the redshift and have another determination of H_O. Another possibility is that the shape of the 21 cm. line profile is itself related to
the intrinsic luminosity of the galaxy for a given morphological type.
This whole area is just beginning to be understood as large statistical
samples accumulate and the absolute calibrations of the telescope are
determined. Moreover, very little has been done in deriving detailed
theoretical models of evolving 21 cm. line profiles to be expected from
different types of galaxies. In future such studies will prove very
useful.

Another contribution which radio observations make to cosmology is
through measurements of the total mass density, ρ_O, of the Universe.
For a Friedman universe dominated by a non-relativistic fluid, the ratio
of the actual density to the critical density, ρ_c, needed to close the
universe is just $2q_O$, when the cosmological constant is zero. If the
energy density is dominated by a relativistic fluid, then $\rho_O/\rho_c = q_O$.
There have been many attempts to determine the amount of intergalactic
neutral hydrogen in ρ_O. Most recently Lang (1976) has searched for

neutral hydrogen in the directions of four radio galaxies and found upper limits $\rho_H < \rho_c$ in all cases. Relativistic ionized hydrogen in clusters has also been searched for by its effect on the microwave background. Here the idea is to detect any diminution of the background radiation at microwave frequencies in clusters caused by its inverse Compton scattering to higher frequencies by local relativistic electrons. Both Partridge and Gull (described at this meeting) have looked at 9 mm. and 3 cm. respectively in several rich clusters. So far there is no strong indication of this effect at a sensitivity of about 10^{-3} $^{\circ}$K. Although the interpretation of these results is model dependent, it suggests that the mass of ionized gas is considerably less than the "virial mass" and thus much less than ρ_c .

The mass density of matter in galaxies may also be found by measuring redshifts for a large complete sample of galaxies. This is an extension of the old virial theorem argument for groups to the statistics of velocity distributions in the sample. The galaxies in this sample need not be divided into groups. These statistical virial measures are based on the departure of the velocity distribution from a smooth Hubble flow (Peebles, 1976; Fall, 1976; Sargent & Turner, submitted to Ap. J.). Other things being equal, larger correlated departures from a smooth Hubble flow at present indicate a greater density of clumped matter. Since the dispersion around the Hubble flow may be quite small, it is necessary to have accurate redshifts of the galaxies, and this can be done best by measuring the 21 cm. emission line. Thuan and Knapp are doing this for a complete sample of about 1100 galaxies in the Zwicky catalogue with $M_{pg} \geq 14$, $|b| \geq 40^{\circ}$ and $\delta > 0^{\circ}$. Some 90 per cent of these galaxies are spirals and their HI radial velocities can be measured on the NRAO 300-foot to an accuracy of about 15 km s^{-1} in about five minutes apiece.

While this will provide a major important estimate of the mass density, this technique also has its difficulties. One is that it does not include any matter such as hot gas, radiation, or neutrinos which is distributed uniformly throughout the Universe. Another is that it assumes a cold initial Hubble flow with no random or correlated velocities. Although the process of galaxy formation remains a mystery, it is quite likely that the initial perturbations from which they arose also included longer wavelength perturbations which started them off with a clustered (i.e. non-random) distribution. If they had a positive initial correlation, this technique would lead to an overestimate of ρ_0/ρ_c , while if they were anti-correlated initially this would underestimate ρ_0/ρ_c .

Next we turn from observations of parameters of Friedman models to other questions of cosmology. One of the most basic is whether the fundamental constants of laboratory physics have changed during the evolution of the Universe. The best known of these is Dirac's speculation that the constant of gravitation may decrease as the Universe ages so as to preserve the equality between the dimensionless ratio of electric and gravitational forces and the dimensionless age of the universe.

However the a priori question of whether these quantities change is a
broader one, to be settled observationally unless one can show that
such constancy is required for our conceptual structure of physics to
be internally consistent.

Recently a radio observation has put strong constraints on temporal
changes of atomic and nuclear physical constants (Wolfe, et al. 1976).
The quasar AO 0235+164 has both hydrogen hyperfine absorption lines and
MgII fine structure lines at the same redshift; z = 0.5239 $\pm$ .0001. By
comparing the terms responsible for the hyperfine splitting of the hydrogen ground state with the fine structure splitting of the MgII doublet,
one finds that the product $\alpha^2 g_p m/M$ has changed by less than two parts in
10^{14} per year over the roughly 10^{10} years which have elapsed since the
radiation was emitted. Here α is the fine-structure constant, g_p is the
nuclear g factor for the proton, and m/M is the electron/proton mass
ratio. Presumably this limit also applies approximately to each of the
factors of the product unless there is some fortuitous cancellation.
Discovery of more examples at higher redshifts would improve this limit
and also check the expectation that the effect is isotropic.

Another constraint on changes of a fundamental constant may be
provided by the microwave background. Noerdlinger (1973) has pointed
out that if Planck's constant is taken to be potentially variable, the
black-body spectrum of the radiation requires the fractional change of
hc with redshift z to be less than three parts in 10^4 since decoupling.
This is derived by noting that the temperature of the background derived
from the Rayleigh-Jeans part of the spectrum, which does not depend on
h, is the same (to within about 30 per cent) as the temperature derived
from the turnover of the spectrum, which does depend on h. The main
question here, however, is whether one would really expect h to change
in principle, since it is a dimensional quantity and its value is
really just a convention. Much as the speed of light is a conventional
conversion factor between length and time, so h would be a conversion
factor between, say, energy and frequency.

Measurements at millimeter wavelengths of the polarization of the
microwave background could provide especially interesting information
about any early anisotropy of the universe (Rees, 1968). Anisotropic
radiation becomes linearly polarized when it is scattered by free
electrons, so if the universe expanded anisotropically, meaning that
the radiation had a different temperature along different axes of
expansion, part of the microwave background would be polarized. For a
given amount of anisotropy, the fractional polarization depends mainly
on the ionization history of the gas during and after decoupling. Under
the most favourable conditions, if the present density of intergalactic
gas is greater than 10^{-5} cm^{-3} and it was reionized at z$\gtrsim$7, the fractional
polarization can be several times the fractional temperature anisotropy.
Inhomogeneities in the matter distribution will also produce some polarization, especially if they have large proper motion relative to an
observer and therefore scatter anisotropically. At present these
effects have not been detected at the ∿1 per cent level, but more sensitive experiments should be possible within the next few years.

II. OBSERVATIONS WHICH MAY BE POSSIBLE IN THE INTERMEDIATE FUTURE.

If we look beyond the near future, the only safe prediction, of course, is that any predictions we make will be covered by surprises. Nevertheless, people have made several suggestions for new observations.

There is an interesting area which may show the effects of the microwave background on radio galaxies. Once this is calibrated and understood, the argument can be inverted to give a rough independent estimate of the redshift of a distant radio source without knowing its optical spectrum. The idea is to measure the diffuse optical emission in the radio lobes of radio galaxies which is produced by inverse Compton scattering of synchrotron electrons on the black body background. Recently, P. Crane, J. A. Tyson, and I have found some possible evidence for diffuse optical emission in the radio lobes of 3C390.3 and 3C285. Its intensity is between about 2 per cent and 10 per cent of the night sky background. We do not know yet whether this optical emission is the visible extension of the radio synchrotron or if it is inverse Compton scattering of the microwave background. Either can be made consistent with the present data, but it should be possible to distinguish between them in future. By measuring the polarization and spectrum, one can estimate the fraction of optical radiation which could be caused by inverse Compton scattering. The ratio of optical inverse Compton to radio synchrotron emission from a small volume depends only on the electron spectrum, the magnetic field, and the redshift of the source (through the z dependence of the energy of the microwave background). One can estimate the electron spectrum if one is willing to make the usual extrapolation from the spectral index of the observed radio synchrotron emission. The magnetic field can be found independently from the angular size if the synchrotron emission is self-absorbed. Thus the redshift can be estimated roughly since the ratio of inverse Compton to synchrotron radiation varies as $(1 + z)^{3+\alpha}$ where α is the spectral index of the synchrotron radiation - typically $.5 \leq \alpha \leq 1$. Although we would not expect this to give very accurate redshift information, it might eventually distinguish between redshifts of, say, 0.1 and 1. However, until we understand the physics of radio sources better, it is more reasonable to invert the argument for cases where we know the redshift of the optical galaxy connected with the source. Then this procedure gives an estimate of the magnetic field in the source and constrains models further. Incidentally, the optical synchrotron part of the emission sets very short lifetimes on the optically radiating electrons. This requires them to be generated close to the region where they radiate. The main observational development we need for this is a high resolution device to measure polarization and colors (if not spectra) at less than 1 per cent of the night sky background.

So far this discussion has mainly been concerned with short wavelengths, i.e. $\lambda \leq 21$ cm. Observations at longer wavelengths could also provide interesting cosmological information. One such observation has been suggested by Sunyaev and Zeldovich (1974). This is to search for the redshifted 21 cm. radiation produced by protogalaxies. If galaxies formed according to fairly conventional ideas, most of their mass was once in the form of neutral hydrogen. If the velocity dispersion in a protocluster of this gas was $\sim 10^3$ km s^{-1}, it would lead to a 21 cm. line of width $\Delta \nu / \nu \lesssim 10^{-3}$. This would be redshifted to a wavelength of a couple meters if galaxies formed at redshifts around 5-10. The flux density in such a line would be between about one and a few millijanskys, depending on whether q_0 was of order unity or of order 0.1 . The detection of this radiation depends critically on it being in a very narrow line so it can be distinguished from the background. This in turn depends on the hydrogen in protoclusters existing in a neutral state only during a very narrow range of redshift. If there are many galaxies along the line of sight radiating over a large range of red-shift, the line will be correspondingly smeared out and difficult to find. Thus this type of observation would give information on the manner in which galaxies and clusters of galaxies formed, and on the relevant cosmological boundary conditions.

Similarly, the time may come when high resolution spectra of distortions of the microwave background reveal as much about the history of matter in the early Universe as the spectra of ionized gases revealed about the inside of the atom. There has been a large amount of theoretical work on this problem during the last decade, and it is discussed in IAU Symposium No.63 (Longair, ed. 1975) as well as in this symposium. So I won't describe it further here.

A quite different type of observation which may eventually become practical is to measure the dispersion in a pulse caused by ionized intergalactic gas (Haddock & Sciama, 1965). If the pulse comes from a distant source, the time delay in its reception at low and high frequen-cies is a function of q_0, H_0, z, and the electron density between the source and the observer. If z is measured independently, then observa-tions of many sources can provide a best fit to q_0, H_0, and n_{e0}. Here one must also determine the part of the dispersion caused by electrons in clouds surrounding the source. The ideal source would be a distant pulsar. But this would require a great increase in our present sensitiv-ity to pulsating signals. It is also conceivable that pulses are produced by the mechanisms which generate radio emission in galactic nuclei and extended radiogalaxies.

III. OBSERVATIONS FOR THE FAR FUTURE - IF EVER.

Taken at face value, one of the simplest ways to find q_0 is to measure it directly. One can readily show that the cosmological expansion changes the redshift of a distant source as seen by a local observer at the rate

$$\frac{dz}{dt_0} = H_0 (1 + z) \left[1 - (1 + 2q_0 z)^{\frac{1}{2}} \right]$$

for any redshift, z, in a matter dominated Friedman universe. This agrees with our expectation that if expansion does not accelerate, i.e. if $q_0 = 0$, the relative velocity of the source remains constant. For small z, the fractional change in redshift becomes d $\ln z/dt_0 = -H_0 q_0$. Thus by measuring z to one part in about 10^8 at intervals of about 100 years, one could hope to determine q_0 directly. At present radio redshift measurements are one to two orders of magnitude more accurate than optical redshifts. Davis (1975) has suggested applying this method to the 21 cm. absorption line of 3C286 ($z \approx 0.8$) which has a half-width of 25 KHz. The absorption redshift can be determined to one part in 10^6. With this accuracy we would have to wait about ten thousand years for second epoch measurements, and even astronomers are not that patient. However a factor of about 10 improvement in accuracy is possible now with narrower lines, a further improvement of order unity in waiting time is possible if higher redshift ($z \gtrsim 3$) lines are discovered, and if another factor of about 10 is granted for unexpected developments, then it is conceivable that this method could give a result over the next few decades. It might even become possible to apply it optically with new types of detectors.

To use this method practically, it would be necessary to apply it to a large number of sources. The reason is that the cosmological deceleration is only one cause of change in the source's redshift. There will also be changes in the internal structure of the source, and changes due to the local acceleration of the source by nearby sources. All these effects will have to be averaged out to find the residual cosmological deceleration.

There is another type of experiment which, if it can be done, would not only occur in future, but would tell us about the future of the Universe. This involves the question of whether it is possible to distinguish between Maxwell electrodynamics and Wheeler-Feynman absorber theory. They are equivalent if the Universe absorbs radiation completely along the future light cone, but not if the Universe is an incomplete absorber. Thus an experiment performed now would depend upon the advanced radiation produced along our future light cone. This would contradict our usual notions of causality. The first experiment to attempt to distinguish between the two theories was Partridge's (1973). He measured the power input to a microwave source as it alternatively radiated into free space (where very little energy was absorbed locally) and then into an efficient local absorber. The power input to the source remained constant to one part in 10^8 as the output was switched between these two regions. However it has been pointed out (Pegg, 1973; Heron and Pegg, 1974; Davies, 1975; Pegg, 1975) that not all forms of absorber theory can be distinguished by this experiment, and the null result may have been independent of the cosmological boundary conditions. One might also be able to account for any observed power difference using Maxwell's theory in terms of ordinary retarded radiation from background sources. At present, this subject is in a rather controversial state and people are attempting to devise more powerful experiments.

Finally, some of my colleagues have suggested that the easiest way to find out about the Universe would be to wait until we are discovered by a more advanced civilization. Then we could ask them. But even if this turns out to be true, surely it is much more fun to try to roll the Universe toward these questions ourselves.

Acknowledgments

Part of this review was written at the Aspen Center for Physics during the summer of 1976.

References

Davies, P.C.W.: 1975, J. Phys. A. Math. Nucl. Gen. 8, 272.

Davis, M.M.: 1975, Bulletin AAS 7, 236.

Fall, S.M.: 1976, Montly Notices Roy. Astron. Soc. 176, 181.

Fisher, J.R. and Tully, R.B.: 1975, Astron. Astrophys. 44, 151.

Haddock, F.T. and Sciama, D.W.: 1965, Phys. Rev. Letters 14, 1007.

Heron, M.C. and Pegg, D.T.: 1974, J. Phys. A. Math. Nucl. Gen. 7, 1965.

Lang, K.R.: 1976, Astrophys. J. Letters 206, L91.

Longair, M.S.: 1975, IAU Symp. 63.

Noerdlinger, P.D.: 1973, Phys. Rev. Letters 30, 761.

Partridge, R.B.: 1973, Nature 244, 263.

Peebles, P.J.E.: 1976, Astrophys. J. Letters 205, L109.

Pegg, D.T.: 1973, Nature Phys. Sci. 246, 40.

Pegg, D.T.: 1975, J. Phys. A. Math. Nucl. Gen. 8, L60.

Rees, M.J.: 1968, Astrophys. J. Letters 153, L1.

Sargent, W.L.W. and Turner, E.: 1976, Astrophys. J. submitted.

Sunyaev, R.A. and Zeldovich, Ya.B.: 1974, Monthly Notices Roy. Astron. Soc. 171, 375.

Wolfe, A.M., Brown, R.L. and Roberts, M.S.: 1976, Phys. Rev. Letters in press.

DISCUSSION

E.M. Burbidge: You have said nothing about all of the things we would
hope to find, in the intermediate future, with the "Large Space Telescope".

Setti: I have made computations of the expected optical fluxes from
radio components due to inverse Compton of the relativistic electrons
against the 2.7 K background photons. For instance, in the case of
equipartition between particle and field energies (protons/electrons = 1)
and for q_o = 1 and a spectral index α = -1, one finds

$$m_B \simeq 31.8 - 1.25 \log \left[\frac{S_{178} \; (Jy) \; (Hz)^4}{(\Delta\phi^{11}) \; z \; (Ho/50)} \right] \left(\frac{mag}{(arc \; sec)^2} \right)$$

where $\Delta\phi^{11}$ is the diameter of the radio component in arc sec and z is the
redshift. So that in some cases one may expect $m_B \approx 29$ mag/(arc sec)^2.
Bright and closeby sources, like Cygnus A, would of course provide a
stronger case. It is perhaps unfortunate that the location of Cygnus A
in the sky may make it difficult to test the existence of any weak optical
emission.

McCrea: The work of W. Baum should not be overlooked in any discussion
of possible changes in the constants of physics.

PROGRESS, PROBLEMS and PRIORITIES: A PERSONAL VIEW

H. van der LAAN
STERREWACHT LEIDEN

INTRODUCTION

Half a century after Hubble expansion was recognized and a quarter century after the identification of Cygnus A we must acknowledge that progress in observational cosmology is slow, the efforts are laborious, corrections are subtle and correlations elusive. Part of progress is the realization that there are no short-cuts, no easy methods or quick successes. The result of this awareness has been increasingly careful attention to the completeness of samples, to selection criteria and the hazards of intercomparisons from survey to survey. This week in Cambridge has shown both modest progress and progress in modesty.

There has been little or no discussion of cosmology in the restricted sense of space-time structure. Nor was there a lot of astrophysical analysis in this symposium. Mostly it was a symposium in the *phenomenology* of extragalactic radio sources; and quite properly so. This population, a violent constituent of the universe, is extremely diverse, spanning wide ranges of luminosities and sizes, *a fortiori* of emissivities; it shows a great variety of morphologies and of optical to radio coupling. Several properties, such as luminosity, size and duration, are plausibly suspected to depend on cosmological epoch in an evolving universe. It is necessary then to acquire very large source samples and very detailed multi-spectral data sets in order to unravel the suspected epoch-dependent relations and the inevitable selection effects. Slowly the dynamic range of source count fluxes and size count angular diameters becomes comparable to the characteristic width of the luminosity and linear size distributions.

Nearly two decades ago, in a brilliant analysis summarized in I.A.U. Symposium no. 9, Martin Ryle stated some essential conclusions based on early radio source counts, on the Copernican principle and the then available sky brightness limits. There were only nineteen extragalactic radio source identifications then and they were not essential to Ryle's arguments. He stated: "It ... seems very probable

D. L. Jauncey (ed.), Radio Astronomy and Cosmology, 389-395. All Rights Reserved.
Copyright © 1977 by the IAU.

that most of the radio sources belong to a class for which $P > 3 \times 10^{25}$ watts $(c/s)^{-1}$ $ster^{-1}$, ..." and: "If further observations confirm the present conclusions, the contribution of radio observations to cosmology will become of great importance." Ryle went on to illustrate this importance by showing how awkward even then the counts were for the steady state cosmology. In the discussion his opponents had to invoke an unknown galactic halo population to avoid these conclusions.

Martin Ryle in 1958 advocated more surveys, more measurements of angular diameters, further analysis of optical observations and 21 cm absorption-line measurements. It is clearly appropriate to discuss the subject here, in the new Cavendish Laboratory, to see how far we came and where we ought to go.

SURVEYS; ANISOTROPIES

That there is much progress in quantity, quality, spectral range and flux density-reach of surveys is evident from the first half dozen papers in this volume. Some nagging problems remain: bright source surveys are incomplete for large (say $\theta > 100''$ arc) angular size sources; flux density scales do not match perfectly; there are declination-dependent calibration errors and resolution corrections to worry about; earth-rotation synthesis telescopes have resolution and flux-corrections which are still partially discordant. These problems need further attention from several teams. A careful reviewpaper of all surveys containing, say, more than 100 sources, describing their scope, content, quality, selection bias, etc. would be very helpful at this stage and would serve their use, diminish their abuse, for statistical purposes.

For several reasons, among them the spectral-index-dependent V/Vmax tests, it is of great importance to perform a substantial survey with a very sensitive large antenna at a high frequency, say $\sim$ 15 GHz. So far we have been in either the transparent domain ('steep spectrum sources') or the partially opaque domain ('flat spectrum sources', a substantial fraction of the 5 GHz survey sources). At 15 GHz we may expect to nearly complete the transition from transparent to opaque and to find a substantial number of ultracompact ('inverted spectrum') sources. These are also of great interest for further X-ray and optical work. The superb 100 m Bonn/Effelsberg telescope seems ideal for this high priority task.

The search for anisotropies or non-Poissonian characteristics in the distribution of survey samples or well-defined subsets continues. There are several teasing effects on the 2 to 2.5 σ level, but so far none are convincing of a universal anomaly. I do not advocate ignoring the anomalies found, although in the past equally intriguing effects have gone away when the samples were enlarged. (An example is the very flat integral source counts for part I of the Parkes 2700 MHz survey.) The anomalous subsample can be isolated, remeasured at other frequencies, identified optically. Thus can the nature of e.g. source number excess be understood. The tremendous range of luminosities results in a great interval of z being contained in a single sample of small flux density interval. Since it is obvious that particular cosmological non-uniformities, if there be any, are likely restricted to small redshift intervals, it is clear that count and spectral index anomalies are best evaluated by adding optical information to the set of data.

For the same reason, the great width of the luminosity distribution, radio source counts do not lend themselves for the determination of stringent isotropy limits. Nevertheless, as is summarized in Webster's paper and is corroborated by limits on microwave background anisotropies, we have no emperical evidence now that the universe does not conform to the cosmological principle. Appealing to the criterion of simplicity it is justified therefore to work with homogeneous isotropic world models for interpretive purposes, for the time being.

Apart from the high frequency survey advocated above, and the value of a low frequency survey sensitive to low surface brightness objects of large angular size such as reported here by Baldwin, one may question the need for further surveys. The discussion is not urgent in practice, because the earth rotation aperture synthesis radio telescopes survey inevitably. In the course of the next several years for example the Westerbork telescope alone will map hundreds of $\sim 1^{\circ}$ diameter fields all over the Northern sky, with flux limits at the 1 mJy level. These samples, fringe benefits of other programs, provide a data reservoir that lends itself to diverse statistical tests and contains source material for deep optical identification and redshift efforts.

OPTICAL IDENTIFICATIONS

In this area of our symposium's concern, progress has been most impressive. This is due to (a) the $\lesssim 1''$ arc positional accuracies provided by the several interferometers, (b) the accurate tie-in of optical and radio astrometric frames and (c) the very deep plates provided by both image tubes and, especially on the 4 m Mayall telescope at Kitt Peak, by direct photography.

The problem of misidentification due to random coincidences in large radio-position error boxes is a thing of the past. (In the Southern Hemisphere the new Fleurs synthesis radio telescope can now provide $< 1''$ arc positions for Parkes and Molonglo survey sources where the identifications may be in doubt; a systematic effort to establish the statistical reliability of the thousands of identifications might well be undertaken there for the sake of economy of optical telescope time). A new problem arises to take the place of the old identification confusion. Combining very sensitive radio surveys with very deep plates leads to inter-galaxy angular distances of the same order as or less than the radio sources' angular sizes. Especially for very faint objects, where large z is suspected and many hours of spectrophotometry are called for, this is a problem. It was illustrated by the spontaneous

discussion on the final day of the symposium concerning 3C 303. The
way to overcome the problem is not to rely upon a radio centroid
position, but to use a synthesis telescope that combines good angular
resolution (< 3" arc) and dynamic range (better than 20 db) with the
best point source sensitivity. Detecting the central source in doubles,
coincident with the optical object, can remove all identification
doubts. The 5 km telescope, the Westerbork SRT and the VLA, each with
a different mix of these qualities, will solve most such problems
north of -30°.

An important condition for further progress is effective collaboration between optical and radio observers. Some systematic exchange
of information on deep fields and plates of outstanding quality,
perhaps in an IAU bulletin, seems overdue.

RADIO SOURCE COUNTS; SPECTRAL INDEX DISTRIBUTIONS

The contribution by Willis et al. and the review by Wall demonstrate
the depth and the spectral range, source counts have now attained. Such
counts played a very important rôle when nearly two decades ago the
prime cosmological question was: 'does the Universe conform to the
'Perfect Cosmological Principle', i.e. is it represented by the Steady
State model, yes or no?' As mentioned in our introduction, the early
counts pointed to a decided no, although they left many sceptics till
the microwave background discovery in 1965. The next simple question:
'if the Universe be represented by a Friedman cosmology, then is it
open or closed?' is not amenable to simple tests by the use of radio
source counts. Epochal population evolution overwhelms geometric
distinctions and the source counts now play a rather different rôle:
given a specified cosmic evolution of radiation and matter density,
what can the source counts teach us about radio source formation and
evolution, and thus indirectly about the formation of galaxies and
their active nuclei.

Questions of the type just raised can only be pursued by combining
a lot of radio and optical data. Of particular importance is the
completeness of 3 CR and of its identifications. From it comes the local
luminosity distribution. Together with the source counts it can be used
to specify various z-dependent source population evolutions. These then
generate predicted distributions of redshift and apparent magnitude
as a function of flux interval, predictions that can in principle be
tested.

The various spectral index distributions shown at this symposium
are most remarkable for their lack of flux density dependence over
considerable ranges of S, ranges large enough for the z- and P-
distributions to change markedly. It is striking, even disturbing,
given the presumed universality of the microwave photon flux, that
there is no evidence of radiative loss spectral steepening for even the
deepest samples! In the different $g_\aleph(\alpha)$ distributions shown there are

some S-dependent effects (see e.g. Jauncey et al., 1972, Astron. J. and Pauliny-Toth, Maslowski,this volume) which require further investigations. Similar remarks as were made above for possible anisotropies apply to these situations. Clearly the different source count results and the $\alpha(\nu_i,\nu_j)$ distributions can be tested for compatibility, including radio K-corrections, using programs of the generality described by Wall.

LUMINOSITY FUNCTIONS

Impressive progress was reported in this area, witness especially the paper by Perola and Fanti. Great statistical improvements, careful complete sample selection, an extension of the radio luminosity distribution to much lower powers and redshift data for ellipticals over a large range of absolute luminosities constitute these advances. The fact that the RLF has a very similar shape for different bins of ΔM_V, yet scales strongly in amplitude with M_V, appears of fundamental importance. That cluster and non-cluster populations are not distinguish-able in this respect provides additional clues concerning the physics which couples stellar populations to activity in stellar system nuclei. So does the relation of the radio core power to absolute magnitude. Clearly the extension of this work to more comprehensive samples and the concentration of theoretical investigations upon the relations found thus far deserve high priority. It is unfortunate that the program could not provide more time for these modern perspectives this work provides. The reorientation of the subject is ahead of the symposium program!

The quasar luminosity distribution overlaps with that of radio galaxies only at the high power end where the space density is so low that a current epoch RLF is not defined. There are two items concerning quasars as a population upon which I wish to comment. One is the m-z relation presented by Bahcall and Turner. We are used to see this relation for the brightest ellipticals in clusters and for strong radio galaxies. We easily forget that these relations are in fact envelope curves for all other ellipticals, not shown, at lower z for a given m-value, i.e. rather like Figure 1 in Bahcall's paper but with better statistics. The Hubble diagram for quasars provides a well defined relation, one which is expected in the so-called conventional redshift interpretation. The 'non conventionalists' have yet to accommodate this relation.

Another noteworthy item is Schmidt's V/V_{max} test for flat spectrum quasars from the NRAO 5 GHz survey. Contrary to steep radio spectrum quasars and to optically selected ones whose $<V/V_{max}> \sim 0.65$, indicative of a strongly epoch dependent LF, the flat spectrum quasars give a value ~ 0.5 which in naive parlance is referred to as the Euclidian value. My view is that the rather lower value of V/V_{max} for this sample poses an interesting problem of epoch-dependent astrophysics. Steep spectrum radio sources radiate from large volumes of low optical depths at radio frequencies. Optically selected and flat spectrum quasars radiate from ultra compact regions with moderate to high opacity at radio frequencies. Schmidt's result may imply that

optical photons and relativistic particles can get out of the quasar
over a large z range, but radio waves encounter greater (possibly
free-free) opacity in the compact region at earlier epochs. The flat
spectrum quasars are then relatively suppressed. We need to survey at
higher frequencies, that is lower opacities, to recover them.

To further understand the radio to optical coupling the determination of the function Ψ (R) (see Perola and Fanti) is important. For
this purpose more apparently-bright quasars need to be found, one of
several reasons why an unbiased optical survey such as the one going on
at Caltech is so valuable.

ANGULAR DIAMETER TESTS

Radio source number counts have gone very nearly as far as they
can. As mentioned above they can provide limited additional insight
for physical cosmology and then only in conjunction with difficult
optical work on very faint samples. An alternative, in fact complemen-
tary pursuit is the determination of angular size distributions of
complete samples as a function of limiting flux and/or as a function
of redshift. Especially Swarup and Kapahi at the Ooty Observatory
have in recent years explored and fruitfully posed the method of
angular size counts. Their work and related efforts at Jodrell Bank
and Cambridge initiated what promises to be a broad multi-observatory
programme using aperture synthesis and intermediate baseline interferometers as well as lunar occultation radio data to be matched by optical
identifications and redshift determinations. The observational bottle-
neck will be optical rather than radio telescope time. In the course
of the next decade the structure of the N(S, Θ, z) distribution function
will slowly emerge. When it does, will it be interpretable?

There are two snags: one is the astrophysics of the hot spots in
the lobes of the strong symmetric double sources. Hewish suspects them
to have a tolerably low intrinsic linear diameter dispersion. The
challenge is to learn the formation and dissipation processes of these
bright but radio synchrotron-transparent regions. Particularly, is
that evolution independent of epoch-dependent environmental conditions?
If not, is it predictably dependent or will physical reality again
hide geometric distinctions of alternative cosmologies? Here and
elsewhere Roeder has emphatically drawn our attention to the other
barrier facing angular size counts interpretation in terms of geometric
cosmology. It may well be that such a laboriously acquired set of data
in the end tells us something about density contrasts distribution
while the elusive q_o slips through our fingers once more. But that too
is cosmology.

OPTICAL SPECTROPHOTOMETRY

Of all progress evident at this symposium, results reported by
Boksenberg and by several other workers, especially from Lick Observatory
stand out. Brightest galaxies now can have their redshifts measured

even if $m_V > 21$; Minkowski's old z-value record for 3C 295 has been broken in half a dozen cases and we may expect the z = 1 mark to be passed soon. Devices of high quantum-efficiency and good stability enabling accurate sky-subtraction are slowly finding their way to the domes of large telescopes. Since the number of telescopes with apertures exceeding 3 meters is increasing from three to at least nine in this decade, and since each of them will presumably be equipped with one or more of these near-optimum devices, we are entering a new era of observational cosmology. It is clear that there is every opportunity for radio and optical astronomers to work as fruitfully together in the near future as in the recent past. As I said earlier, optical telescope time will continue to be the scarcest item: even with coarse spectral resolution just sufficient for a z-determination, photons per pixel from a twentysecond magnitude galaxy are few and far between. It is all the more important then that when the best dozen optical telescopes are engaged for radio source studies, the most accurate and reliable radio data sets from well defined complete samples be used as source material. Such samples, from 3CR levels to eight magnitude deeper are now becoming plentiful.

At very great spectral resolution the new spectrophotometers show the wealthy detail of emission and absorption line spectra of quasars and of active nuclei in galaxies. Data are needed for many strong radio sources, for Seyfert and Markarian galaxies. The energy trans-formations there pose perhaps the most challenging problem of modern astrophysics. Radio VLBI observations and simultaneous multifrequency variable source monitoring as well as X-ray data hoped for from HEAO-B will supplement the optical spectrophotometry.

Opaque radio components in galactic nuclei are the most telling indicators of *current* activity. Radio detections of central components in already identified double sources as well as in Markarian galaxies being discovered in the ongoing objective prism surveys should be systematically attempted. Such programs will help to select the most promising objects deserving costly spectroscopic investigation. Exchange of observing plans and early results, between teams working in different regions of the spectrum, will enhance the chances for progress.

IN CONCLUSION

In one of the final papers of I.A.U. symposium nr. 9 McVittie said: "In my view of cosmology, observation enables us to reject certain classes of models but it does not permit us to pinpoint some particular model among those that remain". At the end of this symposium that view is reaffirmed. Even narrowing the options is difficult but remains an exciting enterprise. Exciting not only because cosmology has held a fascination worthy of hard labour since the dawn of scholarship. But rewarding also because its pursuit ever leads to surprising new objects and to processes we did not anticipate in astrophysical reality.

The complete sample of 166 3CR sources selected as satisfying the following conditions:

a) $S_{178} \geqslant 10$ Jy from Kellermann et al.
(Astrophys. J., **157**, 1, (1969)).

b) $\delta \geqslant 10^{o}$

c) $|b| \geqslant 10^{o}$

3C	S_{178}	Ref	3C	S_{178}	Ref	3C	S_{178}	Ref
6.1	17.3	PH	133	22.3	JPR	239	13.2	PH
9	17.8	PH	138	22.2	ER	241	11.6	JPR
13	12.0	JPR	147	60.5	ER	244.1	20.3	JPR
14	10.4	JPR	153	15.3	PH	245	14.4	JPR
16	11.2	JPR	171	19.5	PH	247	16.8	JPR
19	12.1	JPR	172	15.1	JS	249.1	12.5	PH
20	42.9	JPR	173.1	15.4	JPR	250	11.9	JPR
22	12.1	JPR	175	17.6	JPR	252	11.0	JPR
28	16.3	RP	175.1	11.4	JPR	254	19.9	PH
31	16.8	JPR	181	14.5	PH	263	15.2	PH
33	54.4	HM	184	13.2	JPR	263.1	18.2	JPR
33.1	13.0	JPR	184.1	13.0	RP	264	26.0	N
34	11.9	JPR	186	14.1	RP	265	19.5	JPR
35	10.5	JPR	190	15.0	JPR	266	11.1	JPR
41	10.6	L	191	13.0	JPR	267	14.6	JPR
42	12.0	JPR	192	21.0	JPR	268.1	21.4	JPR
43	11.6	JPR	196	68.2	PH	268.3	10.7	PH
46	10.2	JPR	200	14.0	JPR	268.4	10.3	PH
47	26.4	PH	204	10.5	PH	270.1	13.6	JPR
48	55.0	ER	205	12.6	PH	272	10.3	JPR
49	10.3	JPR	207	13.6	PH	272.1	19.5	JPR
55	22.7	JPR	208	18.5	JPR	274	1050.0	T1
61.1	31.2	HM	212	15.1	JPR	274.1	16.5	JPR
65	15.2	L	215	11.4	PH	275.1	18.3	JPR
66B	30.0	N	216	20.2	JPR	277.2	12.0	JPR
67	10.0	PH	217	11.3	JPR	277.3	12.4	PH
68.1	12.8	JPR	219	41.2	T2	280	23.7	PH
68.2	13.0	PH	220.1	15.8	JPR	280.1	11.9	JPR
76.1	12.2	JPR	220.3	15.7	JPR	284	11.4	RP
79	30.5	RP	223	14.7	RP	285	11.4	JPR
83.1B	26.0	RP	225B	20.0	JPR	286	24.0	ER
84	62.6	ER	226	15.0	JPR	287	16.0	ER
98	47.2	JPR	228	21.8	JPR	288	18.9	PH
109	21.6	RP	231	14.6	H	289	12.0	JPR
123	189.0	PH	234	31.4	RP	293	12.7	JPR
132	13.7	JPR	236	11.3	PH	294	10.3	JPR

D. L. Jauncey (ed.), Radio Astronomy and Cosmology, 397-398. All Rights Reserved.
Copyright © 1977 by the IAU.

3C	S_{178}	Ref	3C	S_{178}	Ref	3C	S_{178}	Ref
295	83.5	PH	336	11.5	PH	401	20.9	JPR
299	11.8	RP	337	11.8	JPR	427.1	26.6	PH
300	17.9	RP	338	46.9	JPR	432	11.0	JPR
303	11.2	PH	340	10.1	JPR	433	56.2	PH
303.1	12.4	JPR	341	10.8	JPR	436	17.8	RP
305	15.7	PH	343	12.4	ER	437	14.6	L
305.1	12.4	JPR	343.1	11.5	ER	438	44.7	JPR
309.1	22.7	ER	345	10.8	ER	441	12.6	L
310	56.0	JPR	346	10.9	PH	442A	17.0	JPR
314.1	10.6	JPR	349	13.3	JPR	449	11.5	JPR
315	18.9	N	351	13.7	RP	452	54.4	RP
318	12.3	ER	352	11.3	JPR	454	11.6	JPR
319	15.3	JPR	356	11.3	JPR	454.3	13.0	ER
321	13.5	JPR	368	13.8	JPR	455	12.8	JPR
322	10.1	JPR	380	59.4	JPR	460	10.3	PH
324	15.8	JPR	381	16.6	RP	465	37.8	RB
325	15.6	JPR	382	19.9	RB	469.1	12.4	L
326	12.1	–	386	23.9	JPR	470	10.1	RP
330	27.8	JPR	388	24.6	PH			
334	10.9	JPR	390.3	47.5	HM			

REFERENCES TO 5-KM PAPERS ON THE 166 SAMPLE

ER   Elsmore, B. & Ryle, M., 1976. Mon. Not. R. Astr. Soc., 174, 411

H   Hargrave, P.J., 1974. Mon. Not. R. Astr. Soc., 168, 491.

HM   Hargrave, P.J. & McEllin, M., 1975. Mon. Not. R. Astr. Soc.,
      173, 37.

JPR   Jenkins, C.J., Pooley, G.G. & Riley, J.M., 1977. Mem. R. Astr.
      Soc., in press.

JS   Jenkins, C.J. & Scheuer, P.A.G., 1976. Mon. Not. R. Astr. Soc.,
      174, 327.

L   Longair, M.S., 1975. Mon. Not. R. Astr. Soc., 173, 309.

N   Northover, K.J.E., 1973. Mon. Not. R. Astr. Soc., 165, 369.

PH   Pooley, G.G. & Henbest, S.N., 1974. Mon. Not. R. Astr. Soc.,
      169, 477.

RB   Riley, J.M. & Branson, N.J.B.A., 1973. Mon. Not. R. Astr. Soc.,
      164, 271.

RP   Riley, J.M. & Pooley, G.G., 1976. Mem. R. Astr. Soc., 80, 105.

T1   Turland, B.D., 1975. Mon. Not. R. Astr. Soc., 170, 281.

T2   Turland, B.D., 1975. Mon. Not. R. Astr. Soc., 172, 181.